FOOD ENGINEERING

Emerging Issues, Modeling, and Applications

Innovations in Agricultural and Biological Engineering

FOOD ENGINEERING
Emerging Issues, Modeling, and Applications

Edited by

Murlidhar Meghwal, PhD
Megh R. Goyal, PhD, PE

Apple Academic Press Inc. | Apple Academic Press Inc.
3333 Mistwell Crescent | 9 Spinnaker Way
Oakville, ON L6L 0A2 | Waretown, NJ 08758
Canada | USA

© 2017 by Apple Academic Press, Inc.

First issued in paperback 2021

No claim to original U.S. Government works

ISBN-13: 978-1-77463-620-6 (pbk)
ISBN-13: 978-1-77188-368-9 (hbk)

Library and Archives Canada Cataloguing in Publication

Food engineering : emerging issues, modeling, and applications / edited by Murlidhar Meghwal, PhD, Megh R. Goyal, PhD, PE.

(Innovations in agricultural and biological engineering)
Includes bibliographical references and index.
Issued in print and electronic formats.
ISBN 978-1-77188-368-9 (hardcover).--ISBN 978-1-77188-369-6 (pdf)
1. Food industry and trade. I. Meghwal, Murlidhar, author, editor II. Goyal, Megh Raj, editor III. Series: Innovations in agricultural and biological engineering

TP370.F65 2016   664   C2016-905358-X   C2016-905359-8

CIP data on file with US Library of Congress

Apple Academic Press also publishes its books in a variety of electronic formats. Some content that appears in print may not be available in electronic format. For information about Apple Academic Press products, visit our website at **www.appleacademicpress.com** and the CRC Press website at **www.crcpress.com**

CONTENTS

LIST OF CONTRIBUTORS

Abdennour Abbas, PhD
Assistant Professor, Biosensors and BioNanotechnology Laboratory, Department of Bioproducts and Biosystems Engineering, University of Minnesota, Twin Cities, 2004 Folwell Ave, St. Paul, MN, 55108, USA, Tel.: +1 612 624 4292, E-mail: aabbas@umn.edu; Website: http://www.abbaslab.com.

Snober Ahmed, PhD
Research Fellow, Biosensors and BioNanotechnology Laboratory, Department of Bioproducts and Biosystems Engineering, University of Minnesota, 2004 Folwell Hall, Falcon Heights, MN 55108, USA. Tel.: +1 612 626 7501, E-mail: ahmed580@umn.edu.

Asaad R. S. Al-Hilphy, PhD
Assistant Professor, Food Science Department, Agriculture College, Basrah University, Basrah, Iraq. Tel.: +96 4772696458, aalhilphy@yahoo.co.uk, asaad197013@gmail.com.

Alaa A. Al-Seraih, PhD
Research Fellow, Food Science Dept., Agriculture College, Basrah University, Basrah, Iraq. Tel.: +9647712555097, E-mail: alseraihalaa@yahoo.com.

Ammar B. R. Al-temimi, PhD
Research Fellow, Food Science Dept., Agriculture College, Basrah University, Basrah, Iraq. Tel.: +96 47705693829, E-mail: ammaragr@siu.edu.

Dawn C. P. Ambrose, PhD
Principal Scientist, ICAR-Central Institute of Agricultural Engg-Regional Center, Coimbatore 641 003, Tamil Nadu, India, E-mail: dawncp@ yahoo.com.

Satarupa Banerjee, PhD
Institute Research Scholar, School of Medical Science and Technology, Indian Institute of Technology, Kharagpur 721302, India, Tel.: +91-9474005265, E-mail: satarupabando@gmail.com.

John Brockgreitens, PhD
2004 Folwell Hall, Falcon Heights, MN 55108, USA, Tel.: +1 612 626 7501, E-mail: brock240@umn.edu.

Minh-Phuong Ngoc Bui, PhD
Research Fellow, Biosensors and BioNanotechnology Laboratory, Department of Bioproducts and Biosystems Engineering, University of Minnesota, 2004 Folwell Hall, Falcon Heights, MN 55108, USA, Tel.: +1 612 626 7501, E-mail: mbui@umn.edu.

Runu Chakraborty, PhD
Professor, Department of Food Technology and Biochemical Engineering, Jadavpur University, Kolkata 700032, India. Tel.: +91 (033) 2414 6822; fax: +91 (033) 2414 6822, E-mail: rchakraborty@ftbe.jdvu.ac.in, crunu@hotmail.com.

Jyotirmoy Chatterjee, PhD
Associate Professor, School of Medical Science and Technology, Indian Institute of Technology, Kharagpur 721302, India. E-mail: jchatterjee@smst.iitkgp.ernet.in.

Arpita Das, PhD
Research Fellow, Faculty of Chemistry and Chemical Engineering, Babes-Bolyai University, 400028, Cluj-Napoca, Romania. Tel.: +40 758653629, E-mail: arpita_84das@yahoo.co.in.

Harita R. Desai, PhD

Research Scientist, Department of Pharmaceutical Sciences and Technology, Institute of Chemical Technology, Nathalal Parekh Marg, Matunga, 400 019 Mumbai, India, Tel.: +91 22 3361 2211/+91 9969805272, fax: +91 2233611020, E-mail: hdesai27@gmail.com, Skype: harita.desai274@skype.com.

Devendra M. Ghodki, PhD

Professor, Department of Mechanical Engineering, N.I.T., Rourkela 769008, Odisha, India. E-mail: devendra.ch2@gmail.com.

Bhupendra M. Ghodki, PhD

Agricultural and Food Engineering Department, IIT Kharagpur, Khargpur 721302, India. E-mail: bhupendramghodki@gmail.com.

Tridib Kumar Goswami, PhD

Professor, Agricultural and Food engineering Department, Indian Institute of Technology, Kharagpur 721302, West Bengal, India. E-mail: tkg@agfe.iitkgp.ernet.in.

Megh R. Goyal, PhD

Retired Professor in Agricultural and Biomedical Engineering from General Engineering Department, University of Puerto Rico—Mayaguez Campus; Senior Technical Editor-in-Chief in Agriculture Sciences and Biomedical Engineering, Apple Academic Press Inc., USA. E-mail: goyalmegh@gmail.com

C. V. Kavitha Abirami, PhD

Associate Professor and Head, Academics and Human Resource Development, Indian Institute of Crop Processing Technology, Ministry of Food Processing Industries, Government of India, Pudukkottai Road, Thanjavur 613 005, Tamil Nadu, India. Tel.: +91 4362 228155, fax: +91 4362 227971, E-mail: director@iicpt.edu.in.

Vivek Kumar, PhD

Research Fellow, Agricultural and Food Engineering Department, IIT, Kharagpur 721302, India. Tel.: +91 9734982094; E-mail: vivek.btag@gmail.com.

Murlidhar Meghwal, PhD

Assistant Professor, Food Science and Technology Division, Center for Emerging Technologies, Jain University, Jain Global Campus, Kanakapura Main Road, Ramanagara Dist., Jakkasandra 562112, Karnataka, India, Tel.: +91 9739204027, E-mail: murli.murthi@gmail.com.

Gayatri Mishra, PhD

Research Scholar, Agric. & Food Eng. Department, Indian Institute of Technology, Kharagpur 721302, West Bengal, India, Tel.: +91 9434507295/+91 8895058298, E-mail: gayatri.mishra21@gmail.com.

Brajesh Kumar Panda, PhD

Professor, Agricultural and Food Engineering Department, IIT Kharagpur, Khargpur 721302, India. Tel.: +91 9474003623, E-mail: brajeshkumarpnd2@gmail.com.

Ashok R. Patel, PhD

Professor, Vandemoortele Center for Lipid Science & Technology, Lab. of Food Tech. & Eng., Faculty of Bioscience Eng., Ghent University, Coupure Links 653, 9000 Gent, Belgium, E-mail: Patel.Ashok@Ugent.be.

A. Santhakumaran, PhD

Professor and Director, Academics and Human Resource Development, Indian Institute of Crop Processing Technology, Ministry of Food Processing Industries, Government of India, Pudukkottai Road, Thanjavur 613005, Tamil Nadu, India. Tel.: +91 4362 228155, fax: +91 4362 227971, E-mail: director@iicpt.edu.in.

Satya Vir Singh, PhD

Associate Professor, Department of Chemical Engineering & Technology, Indian Institute of Technology (Banaras Hindu University), Varanasi 221005, Uttar Pradesh, India, E-mail: svsingh.che@itbhu.ac.in.

Prem Prakash Srivastav, PhD

Associate Professor, Department of Agricultural and Food Engineering, Indian Institute of Technology, Kharagpur 721302, West Bengal, India. Tel.: +91 3222283134, E-mail: pps@agfe.iitkgp.ernet.in.

Ashok Kumar Verma, PhD

Professor, Department of Chemical Engineering & Technology, Indian Institute of Technology (Banaras Hindu University), Varanasi 221005, India, E-mail: akverma.che@itbhu.ac.in.

Deepak Kumar Verma, PhD

Research Scholar, Department of Agric. and Food Engineering, Indian Institute of Technology, Kharagpur 721302, West Bengal, India. Tel.: +91 7407170260/+91 9335993005/+91 3222 281673, fax: +91 3222 282224, E-mail: rajadkv@rediffmail.com, deepak.verma@agfe.iitkgp.ernet.in.

LIST OF ABBREVIATIONS

ALA	α-Linolenic acid
AM	additive manufacturing
BAL	bronchoalveolar lavage
BBI	Bowmin–Birk inhibitor
BSHs	bile salt hydrolases
CAD	computer-aided design
CLIP	continuous liquid interface production
CLTV	controlled low-temperature vacuum dehydration
COX-2	cyclooxygenase 2
ECGC	epigallocatechin-3-gallate
ELISA	enzyme-linked immunosorbent assays
EMC	equilibrium moisture content
EMT	epithelial mesenchymal transition
ESL	extended shelf-life
FDM	fused deposition modeling
FFA	free fatty acids
FOS	fructo-oligosaccharides
GC	gas chromatography
GCBs	green coffee beans
GMO	genetically modified organism
GOS	galacto-oligosaccharides
HHPP	high hydrostatic pressure processing
HIPL	high-intensity pulsed light
HPLC	high-performance liquid chromatography
HPP	hydrostatic pressure processing
IR	infrared
LOM	laminated object manufacturing
LSPR	localized surface plasmon resonance
MAP	modified atmosphere packaging
MLS	microbial luminescence system
NK	natural killer
PE	potential energy
PE	polyethylene
PEF	pulsed electric fields

PGPR	polyglycerol polyricinoleate
PL	pulsed light
PPO	polyphenol oxidase
PSI	pounds/square Inch
PUFA	polyunsaturated fatty acids
RFID	radiofrequency identification
RMSE	root mean square error
SC	sodium caseinate
SCFAs	short-chain fatty acids
SDE	simultaneous distillation extraction
SLM	selective laser melting
SLS	selective laser sintering
SMRC	Systems and Materials Research Corporation
SPME	solid-phase microextraction
TAG	triacylglycerol
TNF	tumor necrosis factor
TTIs	time–temperature indicators
UV	ultraviolet
VEGF	vascular endothelial growth factor
VL	visible
WHO	World Health Organization
XOS	xylooligosaccharide

PREFACE 1

Food engineering has been gaining significant role in the university curricula as well as in the food industries. In food process engineering, products from plant and animal origin are processed under several operations where products gain values additions and quality enhancement. In such process, there may be material gain or weight loss and transfer of heat, mass, and energy with surroundings.

The objective of this book is to offer academia/engineers/technologists/users from different disciplines information to gain recent and emerging knowledge on the breadth and depth of this multifaceted field. The field of food engineering is interdisciplinary, as it requires knowledge of physics, engineering, economics, agronomy, crop science, food science, biotechnology, nanotechnology, sociology, and manufacturing technology. There is an urgent need to explore and investigate the current shortcomings and challenges of the current innovations and challenges. The mathematical model equations and models are used for selected food processing operations.

On recent advances and emerging trends in food process engineering, there was not found a suitable book which can be useful for researchers, scholars, students, professors, industry professionals for product development and also for developing machineries for food processing and production. This book volume, in short, explores and conveys the key concepts on food engineering that are presented in four parts.

Part I: Modeling in Food Engineering has unique value and addresses novel food processing technologies that are of immense interest in relation to food safety and quality. With rapid adaptation, modification, and infusion of new processes and instrumentation, consumers can have access to safe, nutritious, high-quality products through governing principles of mathematical modeling, modeling of water absorption in chickpeas during soaking, and modeling in foods. The fundamental principles and associated numerical approaches are some of the key elements addressed in this volume.

Part II: Review of Research Advances in Food Engineering includes the latest development made in food engineering such as the role of encapsulation in food and nutrition and innovative and intelligent food packaging technologies.

Part III: Role of Food Engineering in Human Health mainly deals with the effect of processed food products on health and engineering ways to produce such products such as three-dimensional printing of food; structuring edible oil using food-grade oleogelators; extraction technology for rice volatile aroma compounds; nonthermal process: pulsed electric fields, pulsed light, high hydrostatic pressure, and ionizing radiation; biosensors in food engineering; and milk pasteurization by microwave.

Part IV: Emerging Issues and Applications in Food Engineering highlights the application of the most recent trends and emerging technologies in food processing sector such as the application of probiotic and prebiotic for human health; effect of functional food, nutraceuticals on human health; phytochemicals, functional food, and nutraceuticals for oral cancer chemoprevention; uses and application of aromatic and medicinal plants in food engineering.

I would like to thank all contributing authors for their sincere contribution of time and effort. It has been our pleasure to put together all of their efforts in this book volume. Many thanks again to all contributors.

—Murlidhar Meghwal, PhD
December 31, 2015

PREFACE 2

Food engineering (FE) is a multidisciplinary field of applied physical sciences which combines science, microbiology, and engineering education for food and related industries. FE includes, but is not limited to, the application of agricultural engineering, mechanical engineering, and chemical engineering principles to food materials. Food engineers provide the technological knowledge transfer essential to the cost-effective production and commercialization of food products and services. Physics, chemistry, and mathematics are fundamental to understanding and engineering products and operations in the food industry. FE encompasses a wide range of activities. Food engineers are employed in food processing, food machinery, packaging, ingredient manufacturing, instrumentation, and control. Firms that design and build food processing plants, consulting firms, government agencies, pharmaceutical companies, and health-care firms also employ food engineers. Specific FE activities include: research and development of new foods, biological, and pharmaceutical products; development and operation of manufacturing, packaging, and distributing systems for drug/food products; design and installation of food/biological/pharmaceutical production processes; design and operation of environmentally responsible waste treatment systems; marketing and technical support for manufacturing plants; etc.

According to the ***Institute of Food Technologists (IFT),*** "Food science (FS) draws from many disciplines such as biology, chemical engineering, and biochemistry in an attempt to better understand food processes and ultimately improve food products for the general public. As the stewards of the field, food scientists study the physical, microbiological, and chemical makeup of food. By applying their findings, they are responsible for developing the safe, nutritious foods and innovative packaging that line supermarket shelves everywhere." "Food Technology (FT): The food you consume on a daily basis is the result of extensive food research, a systematic investigation into a variety of foods' properties and compositions. After the initial stages of research and development comes the mass production of food products using principles of food technology. All of these interrelated fields contribute to the food industry—the largest manufacturing industry in the United States."

<http://www.ift.org/knowledge-center/core-sciences/food-engineering. aspx> has listed the following focus areas in FE/FS/FT, but not limited to

1. Current food production challenges.
2. Drying, evaporation, freezing, cooking.
3. Education and professional development: education, extension, and outreach; industry, government, and academia: project management, communication skills, teaching strategies; quality assurance projects and image analysis; new findings, new innovations, new breakthroughs, and new conversations.
4. Food additives: overall quality, safety, nutritive value, appeal, convenience, and economy of foods.
5. Food health and nutrition: development of foods to maintain and improve health; dietary guidelines; food and nutrition labeling; food products and technologies; foods for the prevention and management of diseases and cancer; functional foods; health benefits and processing of lipid based nutritionals; hospital food service and initiatives; market trends; medical foods; microbiome—diet and health; nutraceuticals; nutrigenomics; prebiotics; sodium reduction; weight management.
6. Food processing and packaging: crops, meat, dairy, marine.
7. Food safety and defense.
8. Food sustainability.
9. FutureFood 2050.
10. Heat and mass transfer.
11. Kinetic models for microbial survival during processing.
12. Measurement, modeling, and control of food processing systems.
13. Moving boundaries in food engineering.
14. Product development and ingredient innovations.
15. Public policy and regulations.

Therefore, I conclude that scope of FE is wide enough, and focus areas may overlap one another. More information on these focus areas can be explored on google.com or can be obtained from: Institute of Food Technologists (IFT), 525 W. Van Buren, Ste 1000 Chicago, IL 60607, USA, Tel.: +1 312 782 8424; fax: +1 312 782 8348, E-mail: info@ift.org. The mission of this book volume is to introduce the profession of food engineering. We cannot guarantee the information in this book series will be enough for all situations.

At the 49th annual meeting of the Indian Society of Agricultural Engineers at Punjab Agricultural University (PAU) during February 22–25 of 2015, a group of ABEs and FEs convinced me that there is a dire need to publish book volumes on focus areas of agricultural and biological engineering (ABE). This is how the idea was born on new book series titled, "Innovations in Agricultural & Biological Engineering." This book on *Food Engineering: Modeling, Emerging issues and Applications* is the second volume under this book series, and it contributes to the ocean of knowledge on food engineering.

The contributions by all the cooperating authors to this book volume has been most valuable in the compilation. Their names are mentioned in each chapter and in the list of contributors. This book would not have been written without the valuable cooperation of these investigators, many of them are renowned scientists who have worked in the field of food engineering throughout their professional careers. I am glad to introduce Dr Murlidhar Meghwal, who is an Assistant Professor in the Food Science and Technology Division, of Center for Emerging Technologies at Jain University - Jain Global Campus in District Karnataka, India. With several awards and recognitions including from President of India, Dr Meghwal brings his expertise and innovative ideas in this book series. Without his support, leadership qualities as Lead Editor of the book volume, and extraordinary work on food engineering applications, readers will not have this quality publication.

I will like to thank editorial staff, Sandy Jones Sickels, Vice President and Ashish Kumar, Publisher and President at Apple Academic Press, Inc., for making every effort to publish the book when the diminishing food resources are a major issue worldwide. Special thanks are due to the AAP Production Staff as well.

I request that readers offer your constructive suggestions that may help to improve the next edition.

I express my deep admiration to my family for understanding and collaboration during the preparation of this book volume. One of my college mates (Dr. R. Paul Singh) at PAU can be distinguished as among top five food engineers in USA. At present, Dr. Singh is a Distinguished Emeritus Professor of the Food Engineering Department of Biological and Agricultural Engineering, Department of Food Science and Technology at the University of California, Davis, CA 95616, USA. I invite readers to consult him at rpsingh@ucdavis.edu or visit his website at www.rpaulsingh.com, whenever they need.

Can anyone live without food?

As an educator, there is a piece of advice to one and all in the world: "Permit that our almighty God, our Creator, provider of all and excellent Teacher, feed our life with His Grace…; and Get married to your profession…"

—Megh R. Goyal, PhD, PE,
Senior Editor-in-Chief
December 31, 2015

FOREWORD

Food engineering is a multidisciplinary field of applied physical sciences that combines science, microbiology, and engineering for the food and related industries. Food engineering includes the application of agricultural engineering, food science, food mechanical engineering, and chemical engineering principles to food materials. Food engineers provide the technological knowledge essential for the cost-effective production of food products and services. Physics, chemistry, and mathematics are fundamental to understanding and engineers the products and operations in the food industry.

In the development of food engineering, one of the many challenges is to employ modern tools, technology, and knowledge to develop new products and processes while simultaneously, improving the quality, safety, and security issues. New packaging materials and techniques are being developed to provide more protection to foods, and novel preservation technology is emerging. Additionally, process control and automation regularly appear among the top priorities identified in food engineering. Advanced monitoring and control systems are needed to facilitate automation and flexible food manufacturing.

Since many years, food engineers have attempted to describe physical phenomena such as heat and mass transfer that occur in food during unit operations by means of mathematical models. Foods are hierarchically structured and have features that extend from the molecular scale to the food plant scale. In order to reduce computational complexity, food features at the fine scale are usually not modeled explicitly, but incorporated through averaging procedures into empirical models. As a consequence, detailed insight into the processes at the microscale is lost.

In this book, volume titled Food Engineering: Emerging Issues, Modeling, and Applications, by Murlidhar Meghwal and Megh R. Goyal, the editors have covered many of the above-discussed vital issues. The basic topics like governing principles of mathematical modeling, modeling of unit operations, overall concepts of mathematical modeling in foods, and biosensors in food engineering make this book unique among many other books on this topic. Applied research topics like encapsulation in food and nutrition, emerging trends in packaging and packaging technologies, application of probiotic and prebiotic, functional food, nutraceuticals for oral cancer chemoprevention, and food engineering for aromatic and medicinal plants are highly useful on updating the knowledge of the researchers, teachers, and students. Novel concepts and ideas to be implemented in the food sector like three-dimensional printing of food, structuring edible oil using food grade oleogelators, extraction of rice volatile aroma compounds, and various nonthermal processes adds to the value of this book as an important publication on food.

I congratulate the authors and the publisher for taking up this important task of writing a book on *food engineering* and their efforts have yielded this publication, which will be a very useful reference book for students, researchers, and academicians working in food sector.

R. T. Patil, PhD
Chief Technical Adviser—Khyati Foods Pvt. Ltd., Bhopal
Chairman & ED, Benevole Welfare Society for Post-harvest
Technology, Bhopal
Former Group Director, Technocrats Institute of
Technology-MBA, Bhopal (MP)
Former Director, Central Institute of Post-harvest Engineering
& Technology (CIPHET), Ludhiana
Former Vice President, Association of Food Scientists
& Technologists (I), Mysore
Former Vice President, Indian Society of Agricultural Engineers,
New Delhi

R. T. Patil, PhD
Bhopal, India
December 31, 2015

WARNING/DISCLAIMER
READ CAREFULLY

The goal of this volume on Food Engineering: Modeling, Emerging Issues and Applications is to guide the world community on how to manage efficiently the technology available for different processes in food engineering. The reader must be aware that dedication, commitment, honesty, and sincerity are most important factors in a dynamic manner for complete success.

The editors, the contributing authors, the publisher, and the printer have made every effort to make this book as complete and as accurate as possible. However, there still may be grammatical errors or mistakes in the content or typography. Therefore, the contents in this book should be considered as a general guide and not a complete solution to address any specific situation in food engineering. For example, one type of food technology does not fit all case studies in food engineering.

The editors, the contributing authors, the publisher, and the printer shall have neither liability nor responsibility to any person, any organization, or entity with respect to any loss or damage caused, or alleged to have caused, directly or indirectly, by information or advice contained in this book. Therefore, the purchaser/reader must assume full responsibility for the use of the book or the information therein.

The mention of commercial brands and trade names is only for technical purposes. It does not mean that a particular product is endorsed over another product or equipment not mentioned.

All weblinks that are mentioned in this book are active (last accessed December 31, 2015). The editors, the contributing authors, the publisher, and the printing company shall have neither liability nor responsibility, if any of the weblinks is inactive at the time of reading of this book.

ABOUT LEAD EDITOR

 Murlidhar Meghwal, PhD, is a distinguished researcher, engineer, teacher, and professor at the Food Science and Technology Division, Centre for Emerging Technology, Jain Global Campus, Jain University, Bangalore, India. He is the lead editor for this book volume. He received his B. Tech. degree (Agricultural Engineering) in 2008 from College of Agricultural Engineering Bapatla, Acharya N. G. RANGA Agricultural University, Hyderabad, India; his M. Tech. degree (Dairy and Food Engineering) in 2010 and PhD degree (Food Process Engineering) in 2014 from Indian Institute of Technology Kharagpur, West Bengal, India.

He worked for one year as a research associate at INDUS Kolkata for development of a quicker and industrial-level parboiling system for paddy and rice milling. In his PhD research, he worked on ambient and cryogenic grinding of fenugreek and black pepper by using different grinders to select a suitable grinder.

Currently, Dr. Meghwal is working on developing inexpensive, disposable, and biodegradable food containers using agricultural wastes; quality improvement, quality attribute optimization and storage study of kokum (*Garcinia indica Choisy*); and freeze drying of milk. At present, he is actively involved in research and course coordinator for the MTech (Food Tech) degree program and also teaching at the Food Science and Technology Division, Jain University Bangalore, India. He has written one book and many research publications in food engineering. He has attended many national and international seminars and conferences. He is a reviewer and member of editorial boards of reputed journals.

He is recipient of the Bharat Scout Award from the President of India as well as the Bharat Scouts Award from Governors. He received meritorious the "Foundation for Academic Excellence and Access (FAEA-New Delhi) Scholarship" for his full undergraduate studies from 2004 to 2008. He also received a Senior Research Fellowship awarded by Ministry of Human Resources Development (MHRD), Government of India, during 2011–14;

and Scholarship of Ministry of Human Resources Development (MHRD), Government of India research during 2008–2010.

He is a good sport person, mentor, social activist, critical reviewer, thinker, fluent writer and well-wishing friend to all. Readers may contact him at: murli.murthi@gmail.com

ABOUT SENIOR EDITOR-IN-CHIEF

Megh R. Goyal, PhD, PE, is a Retired Professor in Agricultural and Biomedical Engineering from the General Engineering Department in the College of Engineering at University of Puerto Rico–Mayaguez Campus; and Senior Acquisitions Editor and Senior Technical Editor-in-Chief in Agriculture and Biomedical Engineering for Apple Academic Press Inc.

He has worked as a Soil Conservation Inspector and as a Research Assistant at Haryana Agricultural University and Ohio State University. He was first agricultural engineer to receive the professional license in Agricultural Engineering in 1986 from College of Engineers and Surveyors of Puerto Rico. On September 16, 2005, he was proclaimed as "Father of Irrigation Engineering in Puerto Rico for the twentieth century" by the ASABE, Puerto Rico Section, for his pioneer work on micro irrigation, evapotranspiration, agroclimatology, and soil and water engineering. During his professional career of 45 years, he has received many prestigious awards. A prolific author and editor, he has written more than 200 journal articles and textbooks and has edited over 35 books. He received his BSc degree in engineering from Punjab Agricultural University, Ludhiana, India; his MSc and PhD degrees from Ohio State University, Columbus; and his Master of Divinity degree from Puerto Rico Evangelical Seminary, Hato Rey, Puerto Rico, USA.

ENDORSEMENTS FOR THIS BOOK VOLUME

This book provides a comprehensive coverage of the various aspects of food engineering. Topics including modeling, food preservation, and recent research on food engineering and health-related aspects will be very useful to students and professionals in food engineering. Increasing awareness on food processing and preservation and the growing processed food market make this book an excellent source for reference in these areas.

—Narendra Reddy, PhD
Professor and Ramalingaswami Fellow
Centre for Emerging Technologies, Jain University,
Jain Global Campus, Jakkasandra Post, Bangalore, India

OTHER BOOKS ON AGRICULTURAL & BIOLOGICAL ENGINEERING BY APPLE ACADEMIC PRESS, INC.

Management of Drip/Trickle or Micro Irrigation
Megh R. Goyal, PhD, PE, Senior Editor-in-Chief

Evapotranspiration: Principles and Applications for Water Management
Megh R. Goyal, PhD, P.E., and Eric W. Harmsen, Editors

Book Series: Research Advances in Sustainable Micro Irrigation
Senior Editor-in-Chief: Megh R. Goyal, PhD, P.E.

Volume 1: Sustainable Micro Irrigation: Principles and Practices
Volume 2: Sustainable Practices in Surface and Subsurface Micro Irrigation
Volume 3: Sustainable Micro Irrigation Management for Trees and Vines
Volume 4: Management, Performance, and Applications of Micro Irrigation Systems
Volume 5: Applications of Furrow and Micro Irrigation in Arid and Semi-arid Regions
Volume 6: Best Management Practices for Drip Irrigated Crops
Volume 7: Closed Circuit Micro Irrigation Design: Theory and Applications
Volume 8: Wastewater Management for Irrigation: Principles and Practices
Volume 9: Water and Fertigation Management in Micro Irrigation
Volume 10: Innovation in Micro Irrigation Technology

Book Series: Innovations and Challenges in Micro Irrigation
Senior Editor-in-Chief: Megh R. Goyal, PhD, P.E.

Volume 1: Principles and Management of Clogging in Micro Irrigation
Volume 2: Sustainable Micro Irrigation Design Systems for Agricultural Crops: Methods and Practices

Volume 3: Performance Evaluation of Micro Irrigation Management: Principles and Practices

Volume 4: Potential of Solar Energy and Emerging Technologies in Sustainable Micro Irrigation

Volume 5: Micro Irrigation Management: Technological Advances and Their Applications

Volume 6: Micro Irrigation Engineering for Horticultural Crops: Policy Options, Scheduling, and Design

Volume 7: Micro Irrigation Scheduling and Practices

Book Series: Innovations in Agricultural & Biological Engineering
Senior Editor-in-Chief: Megh R. Goyal, PhD, P.E.

- Dairy Engineering: Advanced Technologies and Their Applications
- Developing Technologies in Food Science: Status, Applications, and Challenges
- Engineering Interventions in Agricultural Processing
- Engineering Practices for Agricultural Production and Water Conservation: An Interdisciplinary Approach
- Flood Assessment: Modeling and Parameterization
- Food Engineering: Emerging Issues, Modeling, and Applications
- Food Process Engineering: Emerging Trends in Research and Their Applications
- Food Technology: Applied Research and Production Techniques
- Processing Technologies for Milk and Milk Products: Methods, Applications, and Energy Usage
- Soil and Water Engineering: Principles and Applications of Modeling
- Soil Salinity Management in Agriculture: Technological Advances and Applications
- Modeling Methods and Practices in Soil and Water Engineering
- Emerging Technologies in Agricultural Engineering
- Technological Interventions in Management of Irrigated Agriculture
- Technological Interventions in the Processing of Fruits and Vegetables

EDITORIAL

Apple Academic Press Inc., (AAP) will be publishing various book volumes on the focus areas under the book series titled *Innovations in Agricultural and Biological Engineering.* Over a span of 8–10 years, Apple Academic Press Inc. will publish subsequent volumes in the specialty areas defined by *American Society of Agricultural and Biological Engineers* (<asabe.org>).

The mission of this series is to provide knowledge and techniques for agricultural and biological engineers (ABEs). The series aims to offer high-quality reference and academic content in **Agricultural and Biological Engineering** (ABE) that is accessible to academicians, researchers, scientists, university faculty, and university-level students and professionals around the world. The following material has been edited/ modified and reproduced below [From: "Goyal, Megh R., 2006. Agricultural and biomedical engineering: Scope and opportunities. Paper Edu_47 Presentation at the Fourth LACCEI International Latin American and Caribbean Conference for Engineering and Technology (LACCEI' 2006): Breaking Frontiers and Barriers in Engineering: Education and Research by LACCEI University of Puerto Rico—Mayaguez Campus, Mayaguez, Puerto Rico, June 21–23"]:

WHAT IS AGRICULTURAL AND BIOLOGICAL ENGINEERING (ABE)?

"Agricultural Engineering (AE) involves application of engineering to production, processing, preservation and handling of food, fiber, and shelter. It also includes transfer of technology for the development and welfare of rural communities," according to <isae.in>. "ABE is the discipline of engineering that applies engineering principles and the fundamental concepts of biology to agricultural and biological systems and tools, for the safe, efficient and environmentally sensitive production, processing, and management of agricultural, biological, food, and natural resources systems," according to <asabe.org>. "AE is the branch of engineering involved with the design of farm machinery, with soil management, land development, and mechanization and automation of livestock farming, and with the efficient planting, harvesting, storage, and processing of farm commodities," definition by: <http://dictionary.reference.com/browse/agricultural+engineering>.

"AE incorporates many science disciplines and technology practices to the efficient production and processing of food, feed, fiber and fuels. It involves disciplines like mechanical engineering (agricultural machinery and automated machine systems), soil science (crop nutrient and fertilization, etc.), environmental sciences (drainage and irrigation), plant biology (seeding and plant growth management), animal science (farm animals and housing) etc.," by: <http://www.ABE.ncsu.edu/academic/agricultural-engineering.php>.

"According to https://en.wikipedia.org/wiki/Biological_engineering: "BE (Biological engineering) is a science-based discipline that applies concepts and methods of biology to solve real-world problems related to the life sciences or the application thereof. In this context, while traditional engineering applies physical and mathematical sciences to analyze, design and manufacture inanimate tools, structures and processes, biological engineering uses biology to study and advance applications of living systems."

SPECIALTY AREAS OF ABE

Agricultural and Biological Engineers (ABEs) ensure that the world has the necessities of life including safe and plentiful food, clean air and water, renewable fuel and energy, safe working conditions, and a healthy environment by employing knowledge and expertise of sciences, both pure and applied, and engineering principles. Biological engineering applies engineering practices to problems and opportunities presented by living things and the natural environment in agriculture. BA engineers understand that the interrelationships between technology and living systems have available a wide variety of employment options. The <asabe.org> indicates that *"ABE embraces a variety of following specialty areas."* As new technology and information emerge, specialty areas are created, and many overlap with one or more other areas.

1. **Aquacultural Engineering**: ABEs help design farm systems for raising fish and shellfish, as well as ornamental and bait fish. They specialize in water quality, biotechnology, machinery, natural resources, feeding and ventilation systems, and sanitation. They seek ways to reduce pollution from aquacultural discharges, to reduce excess water use, and to improve farm systems. They also work with aquatic animal harvesting, sorting, and processing.

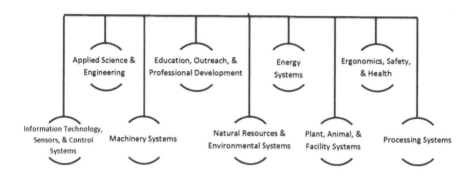

2. **Biological Engineering** applies engineering practices to problems and opportunities presented by living things and the natural environment.

3. **Energy:** ABEs identify and develop viable energy sources—biomass, methane, and vegetable oil, to name a few—and to make these and other systems cleaner and more efficient. These specialists also develop energy conservation strategies to reduce costs and protect the environment, and they design traditional and alternative energy systems to meet the needs of agricultural operations.

4. **Farm Machinery and Power Engineering**: ABEs in this specialty focus on designing advanced equipment, making it more efficient and less demanding of our natural resources. They develop equipment for food processing, highly precise crop spraying, agricultural commodity and waste transport, and turf and landscape maintenance, as well as equipment for such specialized tasks as removing seaweed from beaches. This is in addition to the tractors, tillage equipment, irrigation equipment, and harvest equipment that have done so much to reduce the drudgery of farming.

5. **Food and Process Engineering:** Food and process engineers combine design expertise with manufacturing methods to develop economical and responsible processing solutions for industry. Also food and process engineers look for ways to reduce waste by devising alternatives for treatment, disposal, and utilization.

6. **Forest Engineering**: ABEs apply engineering to solve natural resource and environment problems in forest production systems and related manufacturing industries. Engineering skills and expertise are needed to address problems related to equipment design and manufacturing, forest access systems design and construction; machine–soil interaction and erosion control; forest operations analysis and improvement; decision modeling; and wood product design and manufacturing.

7. **Information & Electrical Technologies engineering** is one of the most versatile areas of the ABE specialty areas, because it is applied to virtually all the others, from machinery design to soil testing to food quality and safety control. Geographic information systems, global positioning systems, machine instrumentation and controls, electromagnetics, bioinformatics, biorobotics, machine vision, sensors, spectroscopy: These are some of the exciting information and electrical technologies being used today and being developed for the future.

8. **Natural Resources:** ABEs with environmental expertise work to better understand the complex mechanics of these resources, so that they can be used efficiently and without degradation. ABEs determine crop water requirements and design irrigation systems. They are experts in agricultural hydrology principles, such as controlling drainage, and they implement ways to control soil erosion and study the environmental effects of sediment on stream quality. Natural resources engineers design, build, operate, and maintain water control

structures for reservoirs, floodways, and channels. They also work on water treatment systems, wetlands protection, and other water issues.

9. **Nursery and Greenhouse Engineering**: In many ways, nursery and greenhouse operations are microcosms of large-scale production agriculture, with many similar needs—irrigation, mechanization, disease and pest control, and nutrient application. However, other engineering needs also present themselves in nursery and greenhouse operations: equipment for transplantation; control systems for temperature, humidity, and ventilation; and plant biology issues, such as hydroponics, tissue culture, and seedling propagation methods. And sometimes the challenges are extraterrestrial: ABEs at NASA are designing greenhouse systems to support a manned expedition to Mars!

10. **Safety and Health:** ABEs analyze health and injury data, the use and possible misuse of machines, and equipment compliance with standards and regulation. They constantly look for ways in which the safety of equipment, materials, and agricultural practices can be improved and for ways in which safety and health issues can be communicated to the public.

11. **Structures and Environment:** ABEs with expertise in structures and environment design animal housing, storage structures, and greenhouses, with ventilation systems, temperature and humidity controls, and structural strength appropriate for their climate and purpose. They also devise better practices and systems for storing, recovering, reusing, and transporting waste products.

CAREER IN AGRICULTURAL AND BIOLOGICAL ENGINEERING

One will find that university ABE programs have many names, such as biological systems engineering, bioresource engineering, environmental engineering, forest engineering, or food and process engineering. Whatever the title, the typical curriculum begins with courses in writing, social sciences, and economics, along with mathematics (calculus and statistics), chemistry, physics, and biology. Students gain a fundamental knowledge of the life sciences and how biological systems interact with their environment. One also takes engineering courses, such as thermodynamics, mechanics, instrumentation and controls, electronics and electrical circuits, and engineering design. Then students add courses related to particular interests, perhaps including mechanization, soil and water resource management, food and process engineering, industrial microbiology, biological engineering or pest management. As seniors, engineering students work in a team to design, build, and test new processes or products.

For more information on this series, readers may contact:

Ashish Kumar, Publisher and President	Megh R. Goyal, PhD, PE
	Book Series Senior
Sandy Sickels, Vice President	Editor-in-Chief
Apple Academic Press, Inc.,	*Innovations in Agricultural and*
Fax: 866-222-9549; E-mail:	*Biological Engineering*
ashish@appleacademicpress.com	E-mail: goyalmegh@gmail.com
http://www.appleacademicpress.com/	
publishwithus.php	

PART I
Modeling in Food Engineering

CHAPTER 1

GOVERNING PRINCIPLES OF MATHEMATICAL MODELING

A. SANTHAKUMARAN

Academics and Human Resource Development, Indian Institute of Crop Processing Technology, Ministry of Food Processing Industries, Government of India, Pudukkottai Road, Thanjavur 613005, Tamil Nadu, India. E-mail: director@iicpt.edu.in

CONTENTS

ABSTRACT

In this chapter, general principles have been described for the development of a mathematical model. An example of a free falling object in vacuum at sea level on the surface of the earth was used to compare the mathematical model to the empirical model and simulation model.

1.1 INTRODUCTION

The purpose of the chapter is different from the traditional course in food process engineering. This chapter provides basic concepts emphasis the fundamentals of mathematical modeling. The objective is to induce self-thinking and knowledge to implement the creativity in the working discipline among the readers who possess minimum activation capacity to read and power of understanding. The governing principles of creating mathematical models are

- idealization,
- formation,
- manipulation,
- reformation if any,
- evaluation,
- justification, and
- validation

1.2 MATHEMATICAL MODELS

Mathematical models are the golden chance for finding the fact from outcomes of experimental data. The nonmathematical nature of the outcomes of the physical experiments is the birth of mathematical modeling. Mathematical models are a logical description of a system how it performs.[2] Mathematical models are also a symbolic representation of the nonmathematical form of the real life problems, which explains the behavior and helps to understand the features of a system. Mathematical model can be classified into deterministic and predictive models. Mathematical models or deterministic models are based on assumptions. Predictive models are obtained from the outcome of the experimental data. Predictive models are also known as empirical models. Simulation study can be used to generate the outcome of the experimental data and to test the analytical solutions.

Simulation is the imitation of the operation of a real world process or system over time for obtaining the fact by generating numerical outcome of the physical experiments without complicated integrations and differential equations with the help of computer software. Modeling on the physical experimental data reduces time and is less expensive, compared to nonmathematical form of the physical experimental data. Modeling is a value addition and adding scope to the nonmathematical form of the physical observations. Mathematical models have

a. analytical solution,
b. graphical solution, and
c. numerical solution:
    i. finite difference method,
    ii. finite element method,
    iii. finite volume method, and
    iv. simulation method or Bootstrap method.

Analytic solution gives how a mathematical model behaves under all circumstances or conditions. It is also known as closed form. It helps to standardize or optimize the outcomes of the nonmathematical form of the physical experimental data.

1.3 BUILDING MATHEMATICAL MODEL: AN EXAMPLE

Building mathematical model depends upon the objectives for studying a particular problem. Then based on these objectives, researcher prepares a list of causes and their effects of the problem for building the mathematical model. For example, we are interested to study the effect of velocity on a free falling object from a moderate height in vacuum at sea level on the surface of the earth. On the earth, the causes are: (1) distance traveled, (2) time of travel, (3) gravitational force, (4) air resistance, (5) air density, (6) air flow, (7) size of the object, (8) shape of the object, (9) mass of the objects if any, and (10) initial velocity at $t = 0$. The velocity of the falling object depends on these causes. The mathematical model is built based on these causes.

1.4 GOVERNING PRINCIPLES TO BUILD A MATHEMATICAL MODEL

1.4.1 IDEALIZATION

For considering a mathematical model, a researcher makes an idea to proceed the causes of the motion of a free falling object. All causes are not important for constructing a model. If some of the causes are not significant, then these are not included in the model. The size, shape, and mass of the object have no effect on the motion of a free falling object. The object is falling in a vacuum so that the velocity of a free falling object is not influenced by the causes: size, shape, mass, air resistance, air flow, and air density. The remaining causes (distance traveled, time of travel, and gravitational force) solely affect the velocity of a free falling object. Idealization is the process of removing the unimportant or less significance causes and thereby the significance causes alone are considered for constructing the mathematical model.

> **Assumption:** A fact is accepted as true or as certain to happen without proof.
>
> **Axiom:** A statement or proposition, which is regarded as being established, is accepted or self-evidently true.
>
> **Hypothesis:** A supposition or proposed explanation made on the basis of limited evidence as a starting point for further investigation.

1.4.2 FORMULATION

Build a mathematical model, one wish to describe the assumptions or statements or principles or hypotheses or axioms. We want to build a model for the motion of a free falling object from moderate height in vacuum at sea level on the surface of the earth. We assume that rate of change of distance (x) of an object at any time (t) is proportional to the distance already fallen, that is

$$\frac{dx}{dt} \propto x \text{ or } \frac{dx}{dt} = kx \tag{1.1}$$

where k is a constant of proportionality of the falling object.

1.4.3 MANIPULATION

Based on the formulation, one uses the mathematical knowledge to manipulate for arriving the solution of the mathematical model using analytical or graphical or numerical methods. The analytical solution of the motion of free falling object is obtained by variable and separable method of integration with respect to the variables x and t.

$$\int \frac{[dx]}{x} = k \int dt + \operatorname{Log} c \text{ or}$$

$$\log x = k(t) + \log c \text{ or}$$

$$\log(x/c) = k(t) \text{ or}$$

$$x = c[e]^{kt} \tag{1.2}$$

Using the initial boundary condition in Equation (1.2):

For $t = 0$, then $x = 0$; we get, $c = 0$.

Thus $x = 0$ for all $t \geq 0$ $\tag{1.3}$

This implies that the object remains in the same place for all the time, that is, the object does not move. Here, the manipulation is perfect, but the equation, $x = 0$ for all $t \geq 0$, is not meaning full. Thus, the assumption is wrong. Now one needs to reformulate the assumption for the motion of a free falling object.

1.4.4 REFORMULATION

The reformulate the assumption that rate of change of distance (x) of s free falling object at any time (t) is proportional to the time (t). The object has been falling, that is

$$\frac{dx}{dt} \propto x \text{ or } \frac{dx}{dt} = kx,$$

where k is a constant of proportionality of the falling object.

The constant k is same for all objects. It does not matter, what the objects is. Weight of the body ($W = mg$ (N)) is the only force acting on the object, when the object is falling. Newton's Second Law of motion is defined as follows:

$$F = ma$$

where F is the force in N—weight of an object in our case (W), 1 kgf $= 9.81$ N, m = mass of an object in kg, and a = acceleration of an body in m/s². Therefore, we have

$$a = F/m, \text{ or}$$

$$a = W/m = mg/m = g = 9.81 \text{ m/s}^2 \tag{1.4}$$

Equation (1.4) is independent of mass of an object. Therefore, the mass of an object has no effect on the motion of the free falling object. Thus, the proportionality constant, $k = g$

Now the differential Equation (1.1) is rewritten as

$$dx = ktdt + c \text{ or}$$

$$dx = gtdt + c \tag{1.5}$$

After the integrating Equation (1.5), we get $\int [dx]/x = gt \int dt + \text{Log } c$ or

$$x = g(t^2/2) + c \tag{1.6}$$

Using the initial condition, we get, $c = 0$
Equation (1.6) reduces to

$$x = g(t^2\}/2), \text{ in m} \tag{1.7}$$

Velocity is obtained by differentiating Equation (1.7) with respect to t.

$$v_t = \text{distance traveled in time interval of } dt = dx/dt = gt, \text{ in m/s} \tag{1.8}$$

If additional drag force acts on the motion of a free falling object, then velocity is

$$v_t = gt + v_0(t), \text{ m/s} \tag{1.9}$$

where $v_0(t)$ is the initial velocity of an object and g = acceleration due to gravity on the planet earth = 9.81 m/s².

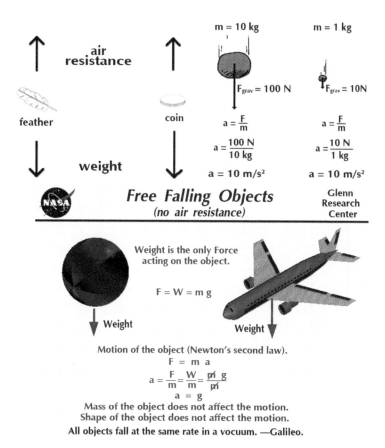

FIGURE 1.1   A falling object: (Left—top) free body diagram showing all forces acting on an object; (right—top and bottom) showing mass of a body does not affect an acceleration of a falling object.[1,3]

1.4.4.1   MATHEMATICAL RESULTS

A free falling object is shown in Figure 1.1. Table 1.1 shows the mathematical results based on the analytical solution: velocity $v_t = gt$, where $g = 9.81$ m/s², for given values of time (t in s) and distance (in m), $x = 1/2(g)t^2$. Note that equation for velocity is a linear equation that represents a straight line passing through an origin and with a line slope of 9.81 m/s².

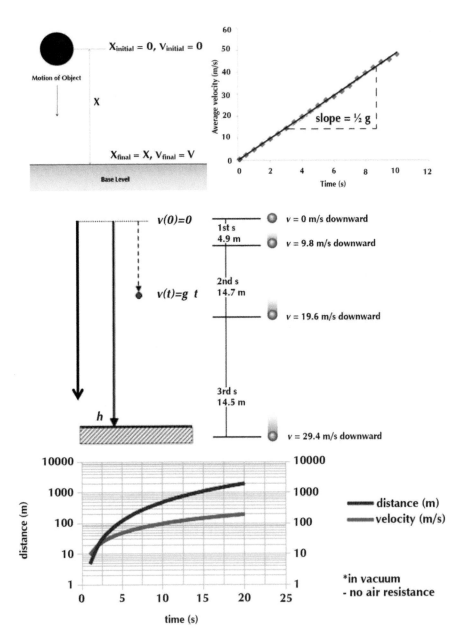

FIGURE 1.2 Velocity and distance traveled by a free falling object as a function of elapsed time, for $g = 9.81$ m/s^2 and using data from Table 1.1. <http://en.wikipedia.org/wiki/Equations_for_a_falling_body>.

The equation for the distance traveled is a polynomial of degree two passing through the origin. Slope of this quadratic equation is 9.81 m/s² at any instant of time. The values in Table 1.1 are plotted in Figure 1.2.

TABLE 1.1   Deterministic Data for a Free Falling Object.

Variable	At elapsed time from $t = 0 - 10$ s										
	0	**1**	**2**	**3**	**4**	**5**	**6**	**7**	**8**	**9**	**10**
a (m/s²)	9.8	9.8	9.8	9.8	9.8	9.8	9.8	9.8	9.8	9.8	9.8
x (m)	0	4.9	19.6	44.4	78.4	122.5	176.4	240.01	313.6	397.31	490.5
v (m/s)	0	9.81	19.6	29.4	39.24	49.05	58.86	68.67	78.40	88.29	98.1

The equation for distance traveled shows that, after one second, an object will have fallen a distance of $1/2 \times 9.8 \times 1^2 = 4.9$ m. After 2 s, it will have fallen $1/2 \times 9.8 \times 2^2 = 19.6$ m; and so on. The equation becomes grossly inaccurate at great distances. If an object fell 10,000 m to Earth, then the results will differ by only 0.08%. However, if it fell from geosynchronous orbit (which is 42,164 km), then the difference is almost 64%.

The remarkable observation that all free falling objects fall at the same rate was first proposed by **Galileo,** nearly 400 years ago. Galileo conducted experiments using a ball on an inclined plane to determine the relationship between the time and distance traveled. He found that the distance depended on the square of the time and that the velocity increased as the ball moved down the incline. The relationship was the same regardless of the mass of the ball used in the experiment. (However, if the experiment had been attempted from the Leaning Tower of Pisa, he would have observed that one ball would have touched before the other!) Falling cannon balls are not actually free falling: they are subject to air resistance and would fall at different terminal velocities, according to <http://www.grc.nasa.gov/WWW/K-12/airplane/mofall.html>.

1.4.4.2   PHYSICAL EXPERIMENT

Initially stationary object, a ball is allowed to fall freely under gravity. It drops a distance, which is proportional to the square of the elapsed time. This image of the ball, spanning half a second, was captured with a flash camera at 20 flashes per second. During the first 1/20th of a second the ball

Food Engineering: Emerging Issues, Modeling, and Applications

drops one unit of distance (a unit is about 12 cm); for 2/20th, it has dropped at total of 4 units; for 3/20th, 9 units; and so on. To take the picture, the ball—about the size of a tennis ball—was suspended by a short length of black thread and was released as the shutter was opened and the flash triggered. The shutter remained open for the whole of the half-second period, during which time the flash was fired multiple times to capture the ball at 1/20 s intervals. Table 1.2 illustrates the experimental data. The visual representation of the motion of a ball falling from moderate height is shown (Fig. 1.3).

TABLE 1.2 Experimental Data for a Free Falling Ball.

Position (t)	Snap time	Distance traveled (unit)	Distance traveled (cm)	vt cm/ (t/20)th s	vt (cm/s)	vt (two successive positions) (cm/s)
0	0	0	0	0	0	0
1	1/20	1	12	12	12	12
2	2/20	4	48	24	48	36
3	3/20	9	108	36	108	60
4	4/20	16	192	48	192	84
5	5/20	25	300	60	300	108
6	6/20	36	432	72	432	132
7	7/20	49	588	84	588	156
8	8/20	64	768	96	768	180
9	9/20	81	972	108	972	204
10	10/20	100	1200	120	1200	228

Using the SPSS, one can fit the linear regression line [$v(t)$ versus t: $v_t = 9.751(t) = dx/dt$, m/s]. Use $t = 1, 2, 3, ..., 10$. The velocity equation can be integrated for initial condition (For $t = 0$, we have $c = 0$) to give us the equation for distance traveled (m): $x = 4.875t^2$, where $t = 1, 2, 3, ..., 10$. Linear regression analysis between velocity and elapsed time is shown in Figure 1.4.

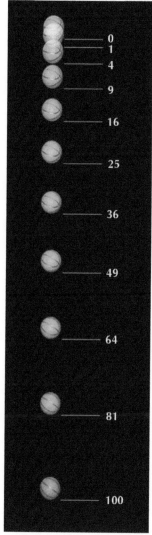

FIGURE 1.3   The visual representation of the motion of a falling ball.

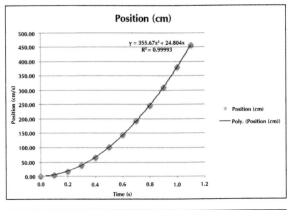

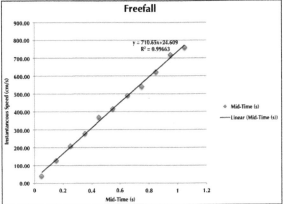

FIGURE 1.4   Linear regression analysis between velocity and elapsed time.

1.4.5   EVALUATION

Any shortcoming or reality in the model is evaluated by proportion of variability (coefficient of determination, R^2), which lies in the desirable interval 80–100%. The desirable index interval is considered on the basis of the human body water level. The normal human body consists of water approximately close to 80%. If the proportion of variability does not lie in the interval, then one may guess that the model has been affected by some kind of shortcomings. Further adjusted coefficient of determination is defined as

$$R^2 = R^2 - (1 - R^2)\{p/(n - p - 1)\} \tag{1.10}$$

where n is the number of trials, p is the number of independent variables (explanatory).

Adjusted R^2 can be negative value, but R^2 cannot be a negative value. When adjusted R^2 increases, then there is an indication that a new additional explanatory is included in the future selection stage of model building.

TABLE 1.3   Observed and Expected Values.

Variable	t: At elapsed time from $t = 0$–10 s										
	0	1	2	3	4	5	6	7	8	9	10
Observed, v_t (m/s)	0	9.81	19.6	29.4	39.24	49.05	58.86	68.67	78.40	88.29	98.1
Estimated, v_t (m/s)	0	9.75	19.5	29.25	39.00	48.75	58.50	68.25	78.01	87.75	97.51

1.4.5.1   COEFFICIENT OF DETERMINATION

Coefficient of determination is the correlation between the observed values and the expected values.[4,5] It is denoted by R^2. From Table 1.3,

$$R^2 = \left[\frac{\sum_{n}^{i=1}(E_i - \bar{O})^2}{\sum_{n}^{i=1}(O_i - \bar{O})^2} \right]$$

(1.11)

where $\bar{O} = \sum_{n}^{i=1}(O_i)/n$, O_i is the ith observed value, E_i is the ith estimated value, N = number of trials.

Therefore, coefficient of determination (R^2) was estimated as 0.95. It lies in the desirable interval 80–100%. The effect of velocity is explained about 95% on the cause of the time. The left out unexplained factor is only 5%. It is negligible because of least value near to zero. Therefore, one can conclude that the velocity (m/s) of a falling object is $v_t = 9.751(t)$, for all $t \geq 0$. This equation is evaluated as the best fit to the observed data in Table 1.3.

1.4.6   JUSTIFICATION

Justification of the predictive model is justified by root mean square error (RMSE). It means that the observed and estimated values are close to one another. It can be easily visualized by drawing the curves of the observed and the estimated data. When the curves resemble to one another and are of same nature, then the model is justified as the best one.

On the other hand for graphical method, the RMSE of the model with less error is appropriate, that is, RMSE lies in the desirable interval 0–1, rather than the best model is justified by RMSE tends to zero or the coefficient of variation of RMSE = [RMSE/average of the observed values] lies in the desirable interval 0–0.05. Since person to person only the difference is 5% apart from 23 pair of chromosomes. This concludes that the model is an apt to the physical experimental observations. From Table 1.3[4,5]:

$$\text{RMSE} = \left\{ \frac{\left[\sum_{n}^{i=1} (E_i - \bar{O})^2 \right] 0.5}{n} \right\} = 0.3594$$

This value of RMSE lies in the desirable interval 0–1. Thus, the predictive model, $v(t) = 9.751(t)$, is justified more appropriate to the physical experimental observations. Coefficient of variation of RMSE = (RMSE)/(average of observed values) = (0.3594)/(53.95) = 0.0067, which also lies in the desirable interval 0–0.05.

1.4.7 VALIDATION

For a same set of physical observations, there is more than one model that can be suitable for a physical experimental data. In such a case, Chi-square statistic is used to test the goodness of fit of the model at 5% level of error or other valid method for selecting the best one, which is identified by the physical experimental data.

1.4.7.1 CHI-SQUARE TEST[4,5]

Validation is tested by Chi-square (χ^2) test of goodness of fit. The Chi-square test statistic is defined below:

$$\chi^2 = \frac{[\sum_{n}^{i=1} (O_i - E_i)^2]}{[E_i]} \; \chi^2, \tag{1.12}$$

The χ^2 is distributed with $(n - k - 1)$ degrees of freedom, where n is the number of classes and k is the number of parameters estimated. Here the calculated χ^2 value is equal to 0.019. The critical χ^2 value for the χ^2 statistic is 15.50 with 8 degrees of freedom at 5% level of error. The acceptance region is 0–15.50 and rejection region is more than 15.50. The calculated Chi-square value falls in the acceptance region of 0–15.50. This shows that the velocity, $v_t = 9.751t$, from the experimental data is validated at 5% level of error.

TABLE 1.4 Simulated and Experimental Observations.

Velocity, v_t, two successive positions (cm/s)	Probability density $f(t)$	Probability density $f(t)$ (%)	Cumulative probability (%)	Generated random interval velocity	Generated velocity, v_t, two successive positions (cm/s)
12	0.1	10	10	00–09	180 (70)
36	0.1	10	20	10–19	108 (46)
60	0.1	10	30	20–29	108 (48)
81	0.1	10	40	30–39	132 (57)
108	0.1	10	50	40–49	60 (21)
132	0.1	10	60	50–59	132 (51)
156	0.1	10	70	60–69	180 (71)
180	0.1	10	80	70–79	132 (55)
204	0.1	10	90	80–89	204 (86)
228	0.1	10	100	90–99	60 (26)
Total = 1200	–	–	–	–	Total = 1244

1.5 SIMULATION RESULTS[4,5]

The distance of an object between two successive positions is equally likely and uniformly distributed when capture the images in each snap for the motion of free falling object. This is expected when it drops from a moderate height in vacuum at sea level on the surface of the earth. This gives the conceptual idea that the probability density function of the falling positions of the object is

$$f(t) = 1/10, \text{ for } t = 1, 2, 3, \ldots, 10 \tag{1.13}$$

$$= 0, \text{ otherwise}$$

Using the probability density of the positions, Table 1.4 is constructed. Table 1.5 shows a spaceman random number: <http://www.stattrek.com/statistics/random-number-generator.aspx>.

TABLE 1.5 Spaceman Random Number.

13962	70992	65172	28053	–
43905	46911	72300	11641	–
00504	48658	38051	59408	–
61274	57238	47267	35303	–
43753	21159	16239	50595	–
835305	51662	21636	68192	–
36807	71420	35804	44862	–
19110	55680	18792	41487	–
82615	86980	93290	87971	–
05621	26584	63013	68181	–
–	–	–	–	–

<http://www.stattrek.com/statistics/random-number-generator.aspx>

Choose a column or row arbitrary or diagonal from the random number (Table 1.5) and combine two-digit numbers. Here, the second column in Table 1.5 is chosen. Now jointly consider 10 two-digit numbers successively: 70, 46, 48, 57, 21, 51, 71, 55, 86, and 26. The number 70 falls in the interval 70–79 in Table 1.4. The value against the interval gives the velocity, which is 180 cm/s. It is shown in Table 1.4. Similarly, the velocities are simulated for the rest of the trails. Thus with the simulation method, the velocity of the free falling object in vacuum at sea level is $v_t = 12.44$ m/s.

1.6 CONCLUSIONS

The motion of the free falling objects in vacuum from moderate height at sea level on the surface of the earth is summarized in Table 1.6.

Different methods of obtaining the velocity and distance for a free falling object yielded values that are very close to each other. Therefore, in this case, the investigator can use the mathematical model for the motion of a free falling object in vacuum at sea level on the surface of the earth:

$$v(t) = gt \text{ (m/s)}$$

$$x = [gt^2]/2 \text{ (m) for } g = 9.81 \text{ m/s}^2.$$

TABLE 1.6 Comparison of Results.

Mathematical method		Experimental method	Simulation method
		Equations	
Velocity (m/s)	$v_t = 9.81(t)$	$v_t = 9.751(t)$	–
Distance (m)	$x = 4.9(t^2)$	$x = 4.875(t^2)$	–
		For $t = 1.26$ s	
$x = 7.70$ m		$x = 7.74$ m	$x = 7.74$ m
$v_t = 12.36$ m/s		$v_t = 12.29$ m/s	$v_t = 12.44$ m/s

Differentiating twice the distance equation, we get the acceleration: $a = g$ (m/s^2). These equations are independent of the mass of the body, for a free falling object in vacuum at sea level on the surface of the earth. This also confirms statement by Galileo: "All objects fall at the same rate in a vacuum."

KEYWORDS

- acceleration
- air resistance
- analytical model
- distance
- drag force
- empirical models
- evaluation
- experimental models
- free falling object
- Galileo

- **mass**
- **mathematical model**
- **mathematical modeling**
- **NASA**
- **Newton's Second Law**
- **acceleration due to gravity**
- **simulation**
- **vacuum**
- **validation**
- **velocity**
- **weight**

REFERENCES

1. http://en.wikipedia.org/wiki/Equations_for_a_falling_body.
2. Meyer, J. W. *Concepts of Mathematical Modeling*. McGraw Hill Book Co., 2004.
3. NASA. 2016, https://www.grc.nasa.gov/www/k-12/airplane/ffall.html.
4. Santhakumaran, A. *Fundamentals of Testing Statistical Hypotheses*. Atlantic Publishers: New Delhi, India, 2001.
5. Santhakumaran, A. *Probability Models and Their Parametric Estimation*. K. P. Jam Publication: Chennai, 2004. https://www.scribd.com/santhakumarana.

CHAPTER 2

MODELING OF WATER ABSORPTION IN CHICKPEAS DURING SOAKING

C. V. KAVITHA ABIRAMI and A. SANTHAKUMARAN[*]

Academics and Human Resource Development, Indian Institute of Crop Processing Technology, Ministry of Food Processing Industries, Government of India, Pudukkottai Road, Thanjavur 613 005, Tamil Nadu, India. E-mail: director@iicpt.edu.in

[*]*Corresponding author.*

CONTENTS

ABSTRACT

In this chapter, water absorption during the soaking of chickpeas has been studied using three methods: mathematical, physical experiment, and simulation. The diffusion coefficient and radii of chickpeas before soaking and after soaking were determined and compared for these three methods.

2.1 INTRODUCTION

Grains are biological solid materials, which are composed of organic and inorganic substances such as carbohydrates, proteins, vitamins, fats, ash, water, minerals, salts, enzymes, etc. Paddy and wheat are rich in carbohydrates, whereas legumes are rich in proteins. Legumes have short chain carbohydrates and are difficult to digest in the human body. It leads to flatulence in stomach of the human. Legumes are dried in sunlight and kept in gunny bags and stored in warehouses for preserving to get longer shelf life. Dried legumes have loss of physical and chemical properties. Dried legumes are processed themselves for easier preparation, greater processing efficiency, high quality yield and high water holding capacity. Before cooking, dried legumes are soaked in water for 6–7 h. Also soaking in water is helpful in the removal of antinutritional factors present in the dried legumes. Soaking may not be alone enough for reaching the equilibrium moisture of the legumes leading to reduction of the soaking time and thereby minimizing the cooking time.

Water activity is derived from the thermodynamics equilibrium principles. They are mass, momentum, and energy. Mass is continuum of elements in a substance. Continuum means a sequence in which adjacent elements are same, but the extremes are quite distinct. Momentum is that an object has a tendency of a moving substance to continue moving. For an object moving in a line, the momentum is the mass × velocity. Energy is a capacity to do work. Work is transfer of energy. Equilibrium is the requirements for describing the water activity. The requirements of water activity are pure water. Water activity ($a_w = 1$) is the standard state as the system in equilibrium with constant temperature. Temperature is a measure of warmth or coldness in a material. Water activity is a measure of energy status of water in substances. When dealing with water, water activity (a_w) is the ratio of escaping tendency of water in a material to escaping tendency of water with no curvature at same temperature. For practical purposes, under general conditions, water activity is closely approximated to vapor pressure. Equilibrium is obtained in a substance when energy per molecule of substance is same everywhere in the substance. Equilibrium between the liquid and the vapor phases implies that energy per molecule of liquid and vapor phases are the same. It addresses the fact that allows the measurement of the vapor phases to determine the water activity of the sample, that is, a_w is the ratio of the vapor pressure of water in a substance to the vapor pressure of pure water at the same temperature:

$$a_w = \frac{[\text{vapor pressure of water in a material}]}{[\text{vapor pressure of pure water}]} \qquad (2.1)$$

Relative humidity of air is the ratio of the vapor pressure of air (p) to its saturation vapor pressure (p_0) of the air. When vapor is equal to temperature, the water activity a_w of the sample is equal to the relative humidity of the air surrounding the sample in a sealed measurement chamber:

$$a_w = \left[\frac{p}{p_0}\right] \times 100 = \text{Equilibrium relative humidity} \qquad (2.2)$$

Thus, water activity of a material is dependent on temperature. Some materials increase water activity with increasing temperature, others decrease it with increasing temperature, but most high-moisture substances have negligible change with temperature. Heat is a form of energy associated with the motion of atoms or molecules. Therefore, heat is associated with temperature. Heat energy is the amount of energy needed to raise the temperature of 1 g of water by 1°C.

In a system, the total energy is composed of the internal, potential, and kinetic energy. Internal energy is the energy associated with random motion of molecules. It refers to invisible microbic energy of the atomic or molecular scale. Internal energy is the sum of its potential and kinetic energies. Stored energy is called potential energy. Potential energy can be transferred from potential energy to kinetic energy and between objects (relative to another objects). Potential energy can be viewed as motion waiting to happen. Kinetic energy is the energy possessed by an object during its motion. Molecules moving in space have kinetic energy. Kinetic means motion and when two magnets are apart, they have more potential energy than they are close together. When a person stands on the surface of an earth, the potential energy is zero. Otherwise potential energy (PE: J, N m):

$$PE = mgh \qquad (2.3)$$

where m is the mass of a body in kg, g is the acceleration due to gravity $=$ 9.81 m/s^2 on the surface of earth, and h is the height in m. If we take ground level as datum, then potential energy is zero.

2.2   DEVELOPMENT OF MATHEMATICAL MODEL OF WATER ABSORPTION FOR CHICKPEAS DURING SOAKING PROCESS

Random motion is named after the botanist Robert Brown as Brownian motion. In the year 1827, he looked through a microscope at particles found in pollen grains in water. As an energy measurement, internal energy is a driving force for water movement from regions of high water activity to regions of lower water activity. Water activity of absorption is that moisture migration when legumes are soaked in water. In the year 1905, Albert Einstein confirmed the existence of atoms and molecules in pollen grains in water. There are two parts to Albert Einstein's Theory. One is the formulation of a diffusion equation for Brownian molecules in which the diffusion coefficient is related to the mean squared displacement of Brownian particle. The second part consists in relating the diffusion coefficient to measurable physical quantities.

When various transport mechanisms occur, it is difficult to separate individual mechanisms. The rate of moisture movement is described by an effective diffusivity (D_0) irrespective of which mechanism is really involved in moisture movement. Fick's Second Law of diffusion is very practical and convenient to describe moisture transport in Brownian motion. Fick's laws of diffusion describe diffusion and were derived by Adolf Fick in 1855. They can be used to solve for the diffusion coefficient. Fick's First Law can be used to derive his second law which in turn is identical to the diffusion equation. The diffusion equation is a partial differential equation which describes density dynamics in a material undergoing diffusion. It is also used to describe processes exhibiting diffusive-like behavior, for instance the "diffusion" of alleles in a population in population genetics. In random motion of multidimensional molecules, the following assumptions are considered:

- Diffusivity of mass transfer is only in distance with respect to time.
- Moisture is initially distributed uniformly throughout the mass of a sample.
- Mass transfer is symmetric with respect to center of a sample.
- Surface moisture content of the sample reaches equilibrium with the condition of surrounding air.

- Resistance of the mass transfer at the surface is negligible compared to internal resistance of the sample.
- Mass transfer is by diffusion only.
- Diffusion coefficient is constant throughout the mass of the sample.
- Shrinkage in the product during is negligible.

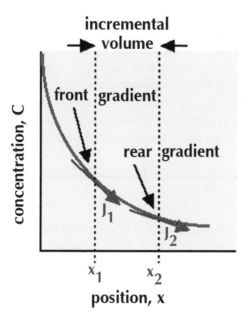

Based on Fick's Second Law of diffusion, the rate of change of concentration with respect to the time is directly proportional to the rate of change of gradient:

$$\frac{\partial C(x,t)}{\partial t} = D_0 \frac{\partial^2 C(x,t)}{\partial x^2} \tag{2.4}$$

where $C(x,t)$ is concentration of mass in a system (concentration gradient), partial differential on left-hand side is a change in concentration with respect to time (t), D_0 is a constant of proportionality called diffusivity in m²/s, and rate of change of gradient.

Assuming that all the particles start from the origin at the initial time, $t = 0$, the diffusion can be integrated to obtain:

$$C(x,t) = \left\{ \frac{[C_0]}{[2q_i\pi D_0 t]^{0.5}} \right\} \left\{ e^{-\{x[2/(2q_i D_0 t)]\}} \right\}, \qquad \text{for} \quad t \geq 0 \tag{2.5}$$

where q_i is the diffusivity dimension, $i = 1, 2, 3,..., \infty$, and C_0 is the concentration at $t = 0$. For example, if $q_1 = 2$, mass transfer is in one dimension (length), if $q_2 = 4$, mass transfer is in two dimensions (area), and if $q_3 = 6$, mass transfer is in three dimensions (volume).

The system is considered as infinite solid material, that is, lower limit of the mass transfer is $-\infty$ to upper limit of the mass transfer is $+\infty$, that is, $-\infty < x < +\infty$. Equation 2.5 allows calculating the potential and kinetic energy of legumes' internal mass transfer through the moments mean and variance directly. The first moment, mean is to vanish, which implies Brownian particle is equally likely to move to the left as it is to move the right, that is, Mean $= 0$ (location measure, potential energy $= 0$). The second moment, variance (kinetic energy) is given by $\sigma^2 = q_i D_0 t$. This expresses the mean squared displacement in terms of the time elapsed and the diffusivity. In case of infinite slab, D_0 is given by squared deviation from mean, $\sigma^2 = q_i D_0 t$, or:

$$D_0 \left(\text{m}^2/\text{s}\right) = \frac{\left[r(t) - r(0)\right]^2}{\left[q_1 t\right]} \tag{2.6}$$

In this case, $q_1 = 2$ and $r(t)$ is the thickness of the slab for reaching equilibrium water absorption in time t, while soaking in water and $r(0)$ is the thickness of the slab before soaking at time $t = 0$. In case of infinite sphere or cylinder:

$$D_0 \ (\text{m}^2/\text{s}) = \frac{\left[r(t) - r(0)\right]^2}{\left[q_3 t\right]} \tag{2.7}$$

where during soaking, $r(t)$ is the radius of the sphere or cylinder for reaching equilibrium water absorption at time t; $r(0)$ is the radius of the sphere or cylinder before soaking at time $t = 0$ and in $q_3 = 6$.

Therefore, D_0 (m) is the average distance moved in a given time during the motion of molecules in m²/s. In case of finite solid material of slab, the solution to Fick's equation is expressed as

$$MR = \left[8/\pi^2\right] \sum_{n=0}^{\infty} \left[\frac{1}{(2n+1)^2}\right] \exp\left[-(2n+1)^2 (D_0 \pi^2 t)/L^2\right] \tag{2.8}$$

where L is the thickness of the slab in m. In the case of cylinder or sphere:

$$MR = \left[6/\pi^2\right] \sum_{n=0}^{\infty} \left[\frac{1}{(n)^2}\right] \exp\left[-n^2 \left(D_0 \pi^2 t\right)/r(0)^2\right] \qquad (2.9)$$

where $r(0)$ is the radius of the sphere or cylinder before soaking in water at $t = 0$.

Dimensionless moisture ratio (MR) is defined as

$$MR = \frac{\left[M_t - M_e\right]}{\left[M_0 - M_e\right]} \qquad (2.10)$$

where M_0 is the initial moisture content in percent, M_e is the equilibrium moisture content in percent = $[kg_{water}]/[kg_{dry}]$, M_t is the moisture content at time t in seconds = $[W_t - W_0]/[W_0]$, W_t is the weight (kg_{water}) of material at time t, and W_0 is the weight (kg_{dry}) of the material at $t = 0$.

Chickpea is a legume of the family *Fabaceouse*. It is cultivated since ancient times and is found in the Middle East. Chickpeas production is growing rapidly across the developing countries especially in the West Asia during the last three decades. India is the largest producer and importer of chickpeas for human diet in the world. One hundred grams of mature boiled chickpeas contain 164 cal, 2.6 g of fat (of which only 0.27 g is saturated fat), and 7.6 g of protein. Research has shown that chickpeas consumption may lower blood cholesterol. There is a great deal of opportunity to expand value addition in chickpeas. If chickpeas are used for human diet as well as value addition of the chickpeas productions, their physical and nutritional quality must be improved from the rehydrated chickpeas.

2.3 METHODS AND MATERIALS: PHYSICAL PROPERTIES OF CHICKPEAS

Chickpeas were soaked in water at uniform size, constant initial temperature, and constant initial moisture content. Before soaking the chickpeas, initial moisture content (dry basis, %) was found using an oven with conditions of surrounding air. Then the moisture content was determined as follows:

$$M_0 (\%) = 100 \times \frac{\left[\text{Weight of the sample at 130°C in oven for 2.5h}\right]}{\left[\text{Weight of the sample at ambient temperature}\right]} \qquad (2.11)$$

Physical properties of the chickpeas (length in cm, breadth in cm, and height in cm) were determined using a Vernier caliper and average values of five observations were recorded. In this study, the chickpeas were considered as spherical biological solid materials. In order to take into account the roundness index of chickpeas, the radius of an equivalent sphere was used in this study. Chickpeas are almost spherical in shape and to make the readings more accurate, the average radius of the spherical chickpeas was determined using the averages of geometric mean (G_m), arithmetic mean (A_m) and squared mean (S_m) as follows:

$$r = \left[\frac{1}{2}\right]\left[\frac{(G_m + A_m + S_m)}{3}\right] \qquad (2.12)$$

where r = radius of chickpea, $G_m = (L \times B \times H)^{1/3}$, $A_m = (L + B + H)/3$, and $S_m = [(LB + BH + HL)/3]^{1/3}$, diameter of chickpea = $2 \times r$.

The roundness index of the chickpeas was calculated before and after soaking the chickpeas in water. The roundness index of the chickpeas was determined as follows:

$$\text{Roundness index} = [4\pi A]/[P^2] \qquad (2.13)$$

where A = area of aggregate and P = perimeter of aggregate. Equation (2.13) is analogous to hydraulic diameter of noncircular pipes in fluid mechanics.

2.3.1 SOAKING PROCEDURE OF CHICKPEAS

Chickpeas were purchased in the open market. The sample was weighed and the measurements for length in cm, breadth in cm and height in cm were made before soaking chickpeas in water. Table 2.1 summarizes the physical properties of chickpeas before soaking in water.

The average diameter of the chickpeas before soaking is 9.88/13 = 0.76 cm. The radius of the chickpeas before soaking is $r_0 = 0.76/2 = 0.38$ cm. A sample of chickpeas with a fresh weight of 34 g was soaked in 100 ml of water in a beaker at ambient temperature. Thus, 20 samples in 20 beakers in series were used. Every half an hour, physical measurements of length, breadth, and height were taken for each beaker. The values are listed in Table 2.1. The soaking time at the time of measurements was recorded. Table 2.2 shows the soaking time, weight, moisture content, and moisture ratio. The average diameter of chickpeas after soaking was 17.4/12 = 1.45 cm. The radius was $r_t = 1.45/2 = 0.73$ cm.

TABLE 2.1   Physical Characteristics of Chickpeas.

S. No.	Soaking time	Length	Breadth	Height	Arithmetic mean	Geometric mean	Squared mean	Diameter
		L	B	H	A_m	G_m	S_m	D
	min				cm			
				Before soaking in water				
1	n/a	0.95	0.69	0.65	0.76	0.75	0.83	0.78
2	n/a	0.73	0.65	0.63	0.73	0.72	0.81	0.75
3	n/a	0.99	0.69	0.64	0.77	0.76	0.84	0.79
4	n/a	0.91	0.68	0.63	0.74	0.74	0.82	0.76
5	n/a	0.91	0.69	0.64	0.75	0.74	0.82	0.77
6	n/a	0.94	0.66	0.64	0.75	0.74	0.82	0.77
7	n/a	0.94	0.68	0.65	0.76	0.75	0.83	0.78
8	n/a	0.94	0.64	0.66	0.75	0.74	0.82	0.77
9	n/a	0.94	0.63	0.64	0.74	0.72	0.81	0.76
10	n/a	0.91	0.72	0.63	0.72	0.71	0.80	0.74
11	n/a	0.92	0.61	0.60	0.71	0.70	0.79	0.73
12	n/a	0.93	0.60	0.61	0.71	0.70	0.79	0.73
13	n/a	0.93	0.64	0.63	0.73	0.72	0.81	0.75
Avg.	**n/a**	–	–	–	–	–	–	**0.76**
				After soaking in water				
1	30	1.30	1.10	1.08	1.16	1.13	1.10	1.13
2	60	1.65	1.30	1.18	1.38	1.36	1.34	1.36
3	90	1.79	1.35	1.28	1.47	1.46	1.29	1.41
4	120	1.82	1.36	1.30	1.49	1.48	1.30	1.42
5	150	1.84	1.42	1.35	1.54	1.52	1.33	1.46
6	180	1.85	1.45	1.36	1.55	1.54	1.34	1.48
7	210	1.86	1.47	1.37	1.57	1.55	1.35	1.49
8	240	1.86	1.50	1.40	1.59	1.58	1.36	1.51
9	270	1.86	1.52	1.42	1.60	1.59	1.37	1.50
10	300	1.87	1.54	1.47	1.62	1.63	1.38	1.54
11	330	1.87	1.55	1.50	1.64	1.63	1.39	1.55
12	360	1.87	1.55	1.50	1.63	1.61	1.38	1.55
Avg.	–	–	–	–	–	–	–	**1.45**

TABLE 2.2 Experiment Results for Chickpeas after Soaking in Water.

S. No.	Soaking time (t)	Weight (W_t) (d.b.)	Moisture content (Y)	Moisture ratio (MR)
	min	g	%	–
1	0	34.0	7.01	1.00
2	30	45.2	32.94	0.41
3	60	49.1	44.41	0.34
4	90	52.3	53.82	0.30
5	120	55.4	62.94	0.24
6	150	58.0	70.59	0.20
7	180	59.8	75.88	0.18
8	210	61.7	81.49	0.16
9	240	62.5	83.82	0.13
10	270	63.9	87.94	0.11
11	300	67.0	97.06	0.04
12	330	68.3	108.88	0.04
13	360	70.0	108.88	0.02
Avg.			67.23	

Moisture ratio was used to obtain the preexponent constant D_0 (diffusivity coefficient). It is further useful to determine the minimum activation energy of chickpeas. The average water absorption was estimated as 67.23%, using moisture content in percent. The relationship between moisture absorption and soaking time of chickpeas follows a square root response. The coefficient of determination for the predictive model (R^2) was 0.992 and RMSE was 0.7167. Figure 2.1 shows the visual presentation of moisture content against the soaking time.

$$y = 7.3198 = 4.445(t)^{1/2} + 0.0461(t), \quad \text{for } t = 0, 30, 60, \ldots, 330$$
$$= 0, \qquad\qquad\qquad\qquad\qquad \text{otherwise} \tag{2.14}$$

2.4 MATHEMATICAL METHOD

The radii of the chickpeas before and after soaking in water were $r(t) = 0.73$ cm and $r(0) = 0.38$ cm. With Albert Einstein infinite volume control method, the diffusivity coefficient (D_0) is given as follows:

$$D_0 = \frac{(0.73 - 0.38)^2}{6 \times 330 \times 60} = 1.03114 \times 10^{-6} \text{ (cm}^2\text{/s)} = 1.0311 \times 10^{-10} \text{ (m}^2\text{/s)}$$

where 330×60 s are required to reach the equilibrium.

2.5 EXPERIMENTAL METHOD

Using *finite volume control method*, the diffusion coefficients can be predicted taking into account only the first term of the infinite Fourier series expansion of the solution (Eq. (2.5)) of the partial differential Equation (2.4), within the boundary of solid material region $[0 \leq x \leq d(t)$, for $t \geq 0$, where $d(t)$ is the average diameter of the chickpeas at time t during soaking in water.

Equations (2.9) and (2.10) are defined previously to estimate moisture ratio (MR). Using $n = 1$ in Equation (2.9), we get following Equation (2.15):

$$MR = [6/\pi^2]\exp[-\{(D_0\pi^2 t)/r(0)^2\}] \qquad (2.15)$$

where $r(0)$ is the average radius of the chickpeas before soaking $= 0.38$ cm $= 0.0038$ m, $D_0 =$ preexponent diffusivity coefficient $= 1.8547 \times 10^{-6}$ cm^2/s $= 1.8547 \times 10^{-10}$ m^2/s, and $MR =$ moisture ratio.

Using values of D_0 and $r(0)$ in Equation (2.15), we get

$$MR = [6/\pi^2]\exp[-\{(1.8547 \times 10^{-6})(\pi^2 t)/(0.38)^2\}] \qquad (2.16)$$

MATLAB software was used to solve the nonlinear regression Equation (2.16). The comments from MATLAB are tabulated below:

t	0	1800	3600	5400	7200	9000	10,800	12,600	14,400	27,060	18,000	19,800	21,600
MR	1.0	0.41	0.34	0.30	0.24	0.20	0.20	0.18	0.16	0.13	0.11	0.04	0.04

2.6 SIMULATION METHOD

Simulation is the imitation of the operation of a system over time. Simulation methods first require a model be developed. This model represents the key characteristics or behavior functions of the selected physical system. The model represents the system itself, whereas the simulation represents the operation of the system over time: Water absorption of chickpeas while soaking is uniformly distributed. This is true because of the assumption in Fick's Second Law of diffusion. The developed model for simulating the diffusivity of the absorption of water of chickpeas is that the mass concentration of chickpeas has equally likely for the 13 uniform water absorption time intervals. It is represented as

$$f(t) = 1/13, \text{ where } t = 0, 30, 60, 90, 120, 150, 180, 210,$$

$$240, 270, 300, 330, 360 = 0, \text{ otherwise} \tag{2.17}$$

Based on the probability model, Equation (2.17) simulates the random diameters of the chickpeas before soaking in water. The random diameters are independent in time and are generated by constructing Table 2.3. First three-digit random numbers are chosen from the first column of the spacemen random numbers in Table 2.5. In Table 2.3, the last column indicates the three-digit numbers from column one in Table 2.5 and the corresponding diameters of the chickpeas against the intervals in which the random numbers are fallen, and they are given in brackets. For example, 139 is the random number, it falls in the second row of the last column of Table 2.3. The diameter against the random number 139 is 0.75 and the process is repeated until to complete the last row of Table 2.3. Moisture content (%, Y-axis) versus soaking time (min, X-axis) curve is shown in Figure 2.1.

TABLE 2.3   Simulated Diameters before Soaking in Water.

S. No.	Observed diameter (cm)	Chance of happening	Chance of happening in 1000	Cumulating chance in 1000	Generated random interval	Generated diameter
1	0.78	0.077	77	077	000–076	139 (0.75)
2	0.75	0.077	77	154	075–153	439 (0.77)
3	0.79	0.077	77	231	154–230	005 (0.78)
4	0.76	0.077	77	308	231–307	612 (0.77)
5	0.77	0.077	77	385	308–384	437 (0.77)
6	0.77	0.077	77	462	385–461	835 (0.73)
7	0.78	0.077	77	539	462–538	368 (0.77)
8	0.77	0.077	77	616	539–615	191 (0.79)
9	0.76	0.077	77	693	616–692	826 (0.73)
10	0.74	0.077	77	770	693–769	056 (0.78)
11	0.73	0.077	77	847	770–846	069 (0.78)
12	0.73	0.077	77	923	847–922	849 (0.73)
13	0.73	0.076	76	1000	923–999	663 (0.76)

The total value of the generated diameter before soaking is: $0.75 + 0.77 + 0.78 + 0.77 + 0.77 + 0.73 + 0.77 + 0.79 + 0.73 + 0.78 + 0.78 + 0.73 + 0.76 = 9.91$ cm. The average diameter of the chickpeas is $d(0) = 9.91/13 = 0.76$ cm, and the average radius $r(0) = d(0)/2 = 0.76/2 = 0.38$ cm.

During soaking of chickpeas in water, the simulated diameters are generated in Table 2.4. The uniform concentration distributed over the uniform time intervals is

$$f(t) = (1/12)t, \text{ where } t = 30, 60, 90, 120, 150, 180, 210, 240,$$

$$270, 300, 330, 360 = 0, \text{ otherwise} \tag{2.18}$$

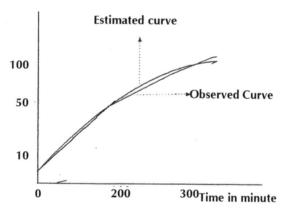

FIGURE 2.1 Moisture content (%, Y-axis) versus soaking time (min, X-axis).

In Table 2.4, the last column is obtained as in Table 2.3 using the second column of the three-digit random numbers in Table 2.5.

TABLE 2.4 Simulated Diameters while Soaking in Water.

S. No.	Soaking time, t (min)	Observed diameter (cm)	Chance of happening	Chance of happening in 1000	Cumulating chance in 1000	Generated random interval	Generated diameter
1	30	1.13	0.083	83	083	000–083	709 (1.50)
2	60	1.36	0.083	83	166	083–165	469 (1.48)
3	90	1.41	0.083	83	249	166–248	486 (1.48)
4	120	1.42	0.083	83	332	249–331	572 (1.49)
5	150	1.46	0.083	83	415	332–414	211 (1.41)
6	180	1.48	0.083	83	498	415–497	516 (1.49)

TABLE 2.4 *(Continued)*

S. No.	Soaking time, t (min)	Observed diameter (cm)	Chance of happening	Chance of happening in 1000	Cumulating chance in 1000	Generated random interval	Generated diameter
7	210	1.49	0.083	83	581	498–580	714 (1.50)
8	240	1.51	0.083	83	664	581–663	336 (1.49)
9	270	1.50	0.083	83	747	664–746	869 (1.55)
10	300	1.54	0.083	83	830	747–829	265 (1.42)
11	330	1.55	0.083	83	913	830–912	372 (1.46)
12	360	1.55	0.083	83	996	913–995	604 (1.51)

The total value of the generated diameters after soaking is 1.50 + 1.48 + 1.48 +1.49 + 1.41 + 1.49+ 1.50 + 1.49+ 1.55 + 1.42 + 1.46 + 1.51 = 17.78 cm. The average generated diameter is $d(t) = 17.78/12 = 1.48$ cm. The average radius of the chickpeas is $r(t) = 1.48/2 = 0.74$ cm. Table 2.5 shows the spacemen random numbers.

TABLE 2.5   Spacemen Random Numbers.

13962	70992	–
43905	46941	–
00504	48658	–
61274	57238	–
43753	21159	–
83503	51662	–
36807	71420	–
19110	55680	–
82615	86984	–
05621	26584	–
06936	37293	–
84981	60458	–
66354	88441	–
–	–	–
–	–	–

These random numbers are already available or can be generated with the help of software. From the simulation method, the generated random radii of the chickpeas are $r(t) = 0.74$ cm and $r(0) = 0.38$ cm. Based on these values, the simulated diffusivity coefficient is

$$D_0 = [0.74 - 0.38]^2/[6 \times 330 \times 60] = 1.0909 \times 10^{-6} \text{ cm}^2/\text{s}$$
$$= 1.0909 \times 10^{-10} \text{ m}^2/\text{s}$$

Here, 6 represents the diffusivity dimension for a sphere and 330×60 s are required to reach the equilibrium water absorption level.

The water absorption of chickpeas with mathematical, experimental, and simulation method are shown Table 2.6.

TABLE 2.6   Results of Chickpeas Water Absorption.

Method		
Mathematical	**Experimental**	**Simulation**
$D_0 = 1.0312 \times 10^{-10}$ m^2/s	$D_0 = 1.8547 \times 10^{-10}$ m^2/s	$D_0 = 1.0909 \times 10^{-10}$ m^2/s
$r(t) = 0.73$ cm	Initial moisture = 7.01%	$r(t) = 0.74$ cm
$r(0) = 0.38$ cm	$r(0) = 0.38$ cm	$r(0) = 0.38$ cm

Diffusivity coefficient in experiment method is slightly larger as compared to the value based on other two methods, because the calculation of diffusivity coefficient in Equation (2.15) is considered only first term in the expansion of the Fourier series expansion.

2.7   CONCLUSIONS

For water absorption of chickpeas during the soaking process in water, the preexponent constants D_0 were estimated using three methods: mathematical, physical experiment, and simulation. In all the three methods, the internal diffusivity coefficient is close to each other. The diffusivity coefficient of the chickpeas can be easily obtained by the Albert Einstein method. It does not involve any complication procedure. But for the small biological spherical materials, the limitation is getting the exact value of the squared displacement distance in a given interval of time during random motion, which is associated with before and after soaking in water, radii of the spherical materials. Scanning processes can be used for the measurements of the radii values and the preexponent diffusivity coefficient can be achieved as close to the exact results. Water activities of the biological solid materials concern over the determination of

- Physical aspects (mass, momentum, and energy)
- Chemical changes (organic and inorganic)
- Biological changes (plant genetics, plant biology, and plant breeding), and
- Quality characteristics (sensory studies)

Thus food processing needs inspiration and innovation of emerging technology, which involves cooperation multidisciplinary areas of engineering.

KEYWORDS

- arithmetic mean
- Brownian motion
- chickpeas
- coefficient of determination
- cylinder
- diffusion coefficient
- energy
- equilibrium water absorption
- experiment method
- Fick's First Law
- Fick's laws
- Fick's Second Law
- food processing
- kinetic energy
- legumes
- mass
- mathematical method
- moisture content
- moisture ratio

- **momentum**
- **nonlinear regression method**
- **plant biology**
- **plant breeding**
- **plant genetics**
- **potential energy**
- **RSME**
- **sensory studies**
- **simulation method**
- **sphere**
- **water absorption**

REFERENCES

1. Ayranci, E.; Ayranci, G.; Dogantan, Z. Moisture Sorption Isotherms of Dried Apricot, Figs and Raisins at 20°C and 30°C. *J. Food Sci.* **1990,** *55*(6), 1591–1593.
2. Barrer, R. M. *Diffusion in and Through Solids*. The MacMillan Company: London, UK, 1941.
3. Chen, C. S.; Clayton, J. T. The Effect of Temperature on Sorption Isotherms of Biological Materials. *Trans. ASAE* **1967,** *14*(5), 309–319.
4. Chuang, L.; Toledo, R. T. Predicting the Water Activity of Mulitcomponent Systems from Water Sorption Isotherms of Individual Components. *J. Food Sci.* **1976,** *41*(4), 922–927.
5. Crank, J. *The Mathematics of Diffusion*. Oxford University Press: Oxford, UK, 1975.
6. Fick, A. *Phil. Mag.* **1855,** *10*, 30.
7. Fontan, C. F.; Chirife, J.; Boquet, R. Water Activity in Multicomponent Non-electrolyte Solution. *J. Food Technol.* **1981,** *18*(5), 553–559.
8. Henderson, S. M. A Basic Concept of Equilibrium Moisture. *Agric. Eng.* **1952,** *33*(1), 29–32.
9. Maria, S. *The Diffusion Coefficient of Water, a Neutron Scattering Study Using Molecular Dynamics Simulations*. Dissertation of the Degree of Master in Physics, University of Surrey: Surrey, UK, 1952.
10. Ross, K. D. Estimation of Water Activity in Intermediate Moisture Foods. *Food Technol.* **1975,** *29*(3), 26–34.
11. Zogzas, N. P.; Maroulis, Z. B.; Marinos-Kouris, D. Moisture Diffusivity Methods of Experimental Determination: A Review. *Dry. Technol.* **1994,** *12*(3), 483–515.

CHAPTER 3

MATHEMATICAL MODELING IN FOODS: REVIEW

SATYA VIR SINGH and ASHOK KUMAR VERMA[*]

Department of Chemical Engineering & Technology, Indian Institute of Technology (Banaras Hindu University), Varanasi 221005, Uttar Pradesh, India. E-mail: svsingh.che@itbhu.ac.in, akverma.che@ itbhu.ac.in

[*]Corresponding author.

CONTENTS

ABSTRACT

The food is necessity of every person. Unlike conventional process streams in a chemical industry the food materials do not have constant transport properties. The thermophysical properties of food materials change with time. This change may be due physiochemical changes during processes such as cooking, baking, etc. Though, the laws of conservation of momentum, mass and energy is same for all fluids, the modeling require a specific treatment due to time-dependent behavior of food materials, viscoelastic nature of food and enzymatic, chemical, or microbial reactions during food processing.

The momentum balance approach to solve problems of flow of food in liquid form provides the velocity profiles and pressure drop. The former is needed to solve equation of continuity for mass-transfer operations and equation of change of energy for heat-transfer operations. The rheology of the fluid affects both. Rheological properties of the fluid were presented.

Heat-transfer operation is used for heating or cooling the food material. These may be heat sensitive or may go through changes in their thermal properties. The model equations based on equation of change of energy for heating of the food due to conduction and convection are discussed. Freezing and thawing is used for reduction of enzymatic processes, microbial growth, condensation of flavors, preservation of foods, etc. Heat-transfer by radiation prevails over conduction and convection. By incorporating generation terms in the equation of change of thermal energy microwave heating can be modeled.

Equation of change of mass is useful in modeling dispersion of various ingredients in the food. The processes are modeled as unsteady state diffusion equations. A number of food processing use packed beds. The flow through packed beds may be described by Darcy's law or equations such as Karman–Kozeny equation. A brief account of adsorption isotherms and models to describe adsorption isotherms were presented. One-dimensional model for adsorption equipment was presented.

Several models require equation of change for mass and energy to be solved simultaneously. Models for drying of food depend upon the type of drying. The models for freeze drying and spray drying were discussed briefly. The cooking or boiling of rice involves consideration of different parts of rice as different phase. These models involve moving boundary problem. The model equations are one-dimensional diffusion and convection equations. Earlier models were shrinking core and diffusion controlled models. The baking process involves heating, movement of moisture toward the surface of the bread where it is lost due to evaporation. The recent models are known as evaporation–condensation–diffusion model in which the moisture movement is due to evaporation and condensation of moisture at the bubble surface and by diffusion in dough. Models for deep frying consider convection due to bubbles generated during phase change of water onto vapor form. It is absent at initial and final stages when only conduction and convection of oil are to be considered.

A large number of enzymatic reactions are described by Michaelis–Menten kinetics. Microbial reactions use of live cells and biomass growth is described by Monod's kinetics.

In general, all the models are unsteady state. One-dimensional models are also applicable in many cases. The 3D models are solved using CFD models. The temperature dependence of properties of food should also be considered.

3 1   INTRODUCTION

The food is necessity of every person. It contains nutrients, proteins, oils, carbohydrates, vitamins, and several other ingredients required for growth and maintenance of our body. The food can be obtained from various parts of the plants, for example, leaves, roots, fruits, etc., from animal sources in form of meat, milk, eggs, and sea food. It may be in form of liquid, solid, and mixture of the two. The presence of gaseous phase provides a particular texture and a feel when the food is consumed. Due to wide variation of the properties of food material and its variation during processing the Modeling in food requires special attention. A detailed description of fluid material is available elsewhere.[25]

Mathematical modeling to describe food processing is never simple. In general, following trends are observed.

a. The average properties of food are not of much significance since presence of even a small amount of under- or over-processed portion in food will make it to fail in meeting the standards. The whole batch will be rejected. It indicates that distributed parameter models are to be preferred over lumped parameter model.

b. Unlike conventional process streams in a chemical industry, the food materials do not have constant transport properties. The thermophysical properties of food materials change with time. It may be due to physicochemical changes during processes such as cooking, baking, etc. Due to time-varying properties of food unsteady-state models are used more frequently than the steady-state models.

c. The processing time is also important. After a definite time the food is over-processed and its acceptance reduces. The boundary condition $t \to \infty$ for the differential equations should be used with care.

d. Since entire food is to be processed, the spatial boundary conditions are finite. For example, bread should be treated as finite block or slab. The importance of processing time is shown in Figure 3.1.

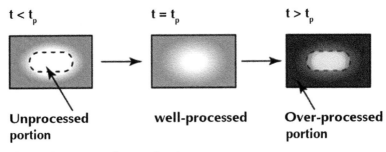

FIGURE 3.1   Importance of processing time.

Thus, in general, models for food are unsteady-state, distributed parameter models with finite spatial coordinates and time.

Application of laws of conservation of momentum, mass, and heat transfer in the form of differential equations are applicable within a single phase. During several food-processing operations the phase and morphology of the food changes so much that it is convenient to treat various parts of the food items as separate items. Accordingly, different coordinate systems, boundary conditions, and set of conservations equations are written for each phase. For example, during cooking of rice, initially there are two phases, water and the rice at the start of the cooking. During cooking the cooked portion behaves as gel and is treated as third phase.[8]

In this chapter, recent trends in modeling of various aspects of food have been presented.

3.2   FLOW AND MOMENTUM BALANCE EQUATIONS

The food processing involves several unit operations. The problems related to flow of the liquids, syrups, and suspensions may be modeled by applying law of conservation of momentum. The rheological properties of the food determine the pressure drop and velocity profile in an equipment or pipe/tubes.

Momentum balance equations are based on the fact that the momentum of a fluid in a phase is conserved. Since the food is not present in gas phase, the incompressible form of the momentum balance equations is applicable within the food items. Balance equations for compressible flow should be applied to describe the flow of air during processes such as drying or chilling of food items. However, in such situations CFD Modeling may be required.

The momentum balance approach to solve problems of flow of food in liquid form, for example, juices, pulp, emulsions, provides the velocity profiles and pressure drop. While the latter is helpful in estimation of power requirement, the former is needed to solve equation of continuity for mass-transfer and equation of change of energy for mass- and heat-transfer operations. The rheology of the fluid affects both.

3.2.1 RHEOLOGY OF JUICES, PULP, AND EMULSIONS

The food items as clear liquids are less frequently observed. Most of the liquid food items are suspensions and emulsions. For example, tomato puree is a suspension and milk is an emulsion. Treating their behavior as liquids is justifiable due to small size of the solids. It also reduces the complexity of the problem.

When shear strain is developed due to applied shear stress, often the stress is related as a function of shear strain. If the shear strain varies linearly with shear stress, the fluid is known as Newtonian fluids.

$$\tau_{yx} = -\mu \frac{dv_x}{dy} \tag{3.1}$$

Here, τ_{yx} is the x-component of the shear stress at y plane and μ is known as viscosity of the fluid. The velocity gradient is known as strain. All other fluids which do not follow Equation (3.1) are known as non-Newtonian fluids. The stress/shear strain ratio is known as apparent viscosity in all cases. It is not a constant and depends upon stress applied. The momentum balance equations require the definition of shear rate. It is to be expressed in terms of the variables already present in the momentum balance equations. These variables are spatial coordinates and time. Velocities are rate of change of spatial position. The definitions of shear rate in terms of spatial coordinates and time are called as rheological models. Some of the frequently used models are described here. Power Law or Ostwald–de Wale Model fluids are given by

$$\tau_{yx} = -m \left| \frac{dv_x}{dy} \right|^{n-1} \frac{dv_x}{dy} \tag{3.2}$$

When the value of $n < 1$, the fluid is also known as pseudoplastic. For $n > 1$, it is called dilatants. Power law fluid reduces to Newtonian fluid for $n = 1$.

Several fluids such as tomato puree require a threshold to flow. If after applying threshold, stress–strain relationship exhibit linear relationship, the fluids are known as Bingham plastic and its stress–strain behavior are described by

$$\tau_{yx} = -\mu \frac{dv_x}{dy} + \tau_0 \ \text{if} \left| \tau_{yx} \right| > \tau_0 \tag{3.3}$$

$$\frac{dv_x}{dy} = 0 \ \text{if} \left| \tau_{yx} \right| < \tau_0$$

A rheological model combining threshold and power law relationship was proposed by Herschel and Berlkley.[12] Few fluids behave as Newtonian fluid at very low and at very high stresses. If the viscosities at the extreme values are denoted by μ_0 and μ_∞, respectively, the rheology of the fluid can be modeled as following.

$$-\frac{dv_x}{dy} = \left(\frac{1}{\mu_\infty + (\mu_0 - \mu_\infty)/1 + \left(\tau_{yx} / \tau_s \right)^2} \right) \tau_{yx} \tag{3.4}$$

These and few more rheological models have been reported in literature.[6,12,30]

3.2.2 MODELS BASED ON MOMENTUM BALANCE EQUATION

The law of conservation of momentum is based on the assumption that the momentum of the fluid remains constant. It, however, may be generated and converted to other forms, for example, the momentum is converted to thermal energy due to viscous dissipation. Momentum balance equations can be written in Cartesian, cylindrical, and spherical coordinates and are presented in books on transport phenomena.[6] It consists of three equations one for each velocity component. For example, x-component of the momentum balance equation in Cartesian coordinates can be written as

$$\rho\left(\frac{\partial v_x}{\partial t}+v_x\frac{\partial v_x}{\partial x}+v_y\frac{\partial v_x}{\partial y}+v_z\frac{\partial v_x}{\partial z}\right)=-\frac{\partial p}{\partial x}+\left(\frac{\partial \tau_{xx}}{\partial x}+\frac{\partial \tau_{yx}}{\partial y}+\frac{\partial \tau_{zx}}{\partial z}\right)+\rho g_x \quad (3.5)$$

Here, v_x is fluid velocity in x-direction, ρ is the density of the fluid, p is pressure and g_x is x-component of acceleration due to gravity. Similar equations for y and z components of momentum are also written. These equations are solved taking into consideration the equation of continuity which in Cartesian coordinates is as follows:

$$\frac{\partial p}{\partial t}+\frac{\partial}{\partial x}\left(\rho v_x\right)+\frac{\partial}{\partial y}\left(\rho v_y\right)+\frac{\partial}{\partial z}\left(\rho v_z\right)=0 \quad (3.6)$$

Equation (3.5) requires the definition of shear stress, which may be obtained from rheological models as described above. Analytical solutions for several flow problems for non-Newtonian fluids may be seen in literature.[6,12,30] Its use in Modeling of few several food processes is illustrated later in this chapter. Equation (3.5) does not include generation terms.

3.2.3 MODELS FOR TIME-DEPENDENT BEHAVIOR OF LIQUID FOOD MATERIALS

Several fluids exhibit variation of viscosity with time. These fluids are called time-dependent fluids. Thixotropic fluids are those for which the shear rate decreases with time and are also known as shear-thinning liquids. Rarely, food items are shear thickening (rheopectic).[26] These changes generally takes place due to change in the structure of the food item. The viscosity of time-dependent food material is described by a function similar to power law fluid modified by a time-dependent term. One such relationship given by Mason et al. is reproduced as[11]

$$\tau_{yx}=-m\left|\frac{dv_x}{dy}\right|^{n-1}\frac{dv_x}{dy}\left[1+\left(b_0t\frac{dv_x}{dy}-1\right)\frac{\sum b_i\exp(-t/\lambda_i)}{\sum b_i}\right] \quad (3.7)$$

Here, t is time. Other quantities are constants. Waltman correlated the shear stress as a function of time by using the following equation.[15]

$$\tau = A - B\log t \qquad (3.8)$$

It should be kept in mind that using a more complicated rheology model increases complexity of the momentum balance equation and hence the computational effort.

3.2.4 PASTES AND DOUGH: VISCOELASTIC BEHAVIOR

The rheological models discussed above involve shear stress only. A number of food items such as gel and egg-fluid exhibit presence of normal stresses. These fluids are called as viscoelastic fluids. Thick pastes and dough are some of the examples of viscoelastic fluids. The effect of stress changes with time due to viscoelastic behavior. The strain, shear stress, and normal stress are combined in one equation. Maxwell's equation expresses strain-rate as a combination of two terms.[15]

$$\left(\frac{d}{dt}\frac{\partial v_x}{y}\right) = \frac{1}{\eta}\tau + \frac{1}{G}\frac{\partial \tau}{\partial t} \qquad (3.9)$$

Kelvin–Voigt model for viscoelastic model combines shear stress and strain in the following manner[11].

$$\tau = \frac{1}{\eta}\left(\frac{d}{dt}\frac{\partial v_x}{\partial y}\right) + G\frac{\partial v_x}{\partial y} \qquad (3.10)$$

A detailed account of rheological models for food items may be obtained from several books.[6,11,12,15,26,30]

3.3 EQUATION OF CHANGE OF ENERGY

Heat-transfer operation is used for heating or cooling the food material. The food items may be heat sensitive or may go through changes in their thermal properties. The thermal conductivity in general is not isotropic. Heat transfer takes place through three mechanisms: conduction, convection, and radiation.

The equation of change of energy for heat transfer in Cartesian coordinates applicable within a continuous food material may be written as

$$\rho C_p \left(\frac{\partial T}{\partial t} + v_x \frac{\partial T}{\partial x} + v_y \frac{\partial T}{\partial y} + v_z \frac{\partial T}{\partial z} \right) = k_h \left(\frac{\partial^2 T}{\partial^2 x} + \frac{\partial^2 T}{\partial^2 y} + \frac{\partial^2 T}{\partial^2 z} \right) \quad (3.11)$$

where T is the temperature, C_p is specific heat of the material, and k_h is thermal conductivity of the material.

The right-hand side of Equation (3.11) represents heat transfer due to conduction which generally dominates in case of solids. The left-hand side of the equation contains the unsteady term and convective heat-transfer terms. The convective heat transfer dominates in liquids having large velocities. Heat generation due to chemical reaction or any other phenomena such as microwave heating may be added to right-hand side of the equation. Equation (3.11) is applicable for fluids for which thermal conductivity is isotropic and is temperature independent. Heat transfer due to radiation is not included in this equation because radiative heat transfer takes place from one phase to another and Equation (3.11) is applicable within a single phase. Hence, radiative heat transfer appears in the boundary conditions. Three types of boundary conditions are encountered while solving Equation (3.11).

a. *Specifying the temperature*: Most of the boundary conditions of this type assume constant surface temperature. For steam-heated equipment, the heat-transfer surface such as jacket or heating coils is at constant temperature. However, the temperature may be assumed to vary with time and location at the heat-transfer surface. It is also known as boundary condition of first type or Dirichlet condition.

b. *Specifying heat transfer flux*: The flux is specified as constant or described by Fourier's law. It is observed in case of radiative heat transfer (solar heater) or electrically heated equipment. This boundary condition is known as a boundary condition of the second kind or Neumann condition.

c. *Specifying heat-transfer coefficient*: Several times the food item is surrounded by a fluid medium which is flowing. Hydrodynamic conditions determine the heat-transfer coefficient. The boundary condition then specifies the heat-transfer flux (expressed in terms of temperature gradient) as a function of heat-transfer coefficient. This kind of boundary condition is also called as of third kind.

Frequently, heat transfer from surrounding to food material takes place by convection as well as by radiation. The flux is then expressed considering both mechanisms to act in parallel.

The equation of change of energy is applicable in a single phase. However, no food material is homogenous and consists of single phase. For modeling purpose due to fine structure of the food material it is assumed as a continuous phase. The properties such as thermal conductivity, density, specific heat, etc. are measured and reported considering the food material as continuous phase. This treatment simplifies the situation and the equation of change becomes applicable within the entire food material. As an example while drying peas, each pea is single phase. The equation of change is applicable only within each pea. The boundary conditions specify the processes taking place at the boundary of the pea. It helps in understanding the conditions of peas exposed to different surroundings.

Equation (3.11) with appropriate boundary conditions is solved to obtain the temperature profile in the food item. This knowledge may be used to estimate many other parameters required to know the progress of processing of food.

3.3.1 THERMAL CONDUCTIVITY AND CHANGING FOOD STRUCTURE DURING THERMAL PROCESSING

The equation of change of energy requires the knowledge of thermal conductivity of food material. All foods are made of many constituents. Water is one of the major constituents. Since the thermal conductivity of water and ice are quite different, the thermal conductivity of frozen food and unfrozen food differs. The thermal conductivity depends upon temperature also and expressed as second-order polynomial.[16] To estimate the thermal conductivity of food material using that of its constituents the assumption of food behaving as layered material parallel to the direction of heat flow has been found to be more accurate than other methods. The two other models assume the layers of food constituent perpendicular to the direction of flow and case of dispersed constituents.[16]

If the food is fibrous, then the thermal conductivity in the direction of heat transfer is different than that in the direction perpendicular to the direction of fiber arrangement. However, this difference is up to 10% only.[16]

3.3.2 HEAT TRANSFER DURING THE PHASE CHANGE: FREEZING AND THAWING

Cooling or refrigeration processes are used for reduction of enzymatic processes, microbial growth, condensation of flavors, preservation of foods, etc. Though, these are also heat-transfer operations, but due to thermophysical properties being different than that of heated material, it requires a separate treatment.

Food material is frozen to increase the shelf life so that it can be transported at larger distances and be consumed in another season. The entire process involves two steps. Freezing involves decreasing the temperature of food item below freezing point and thawing involves increasing temperature of the frozen food above freezing point. Both steps involve phase change of some of the constituents, mainly water, in food. The heat transfer takes place by using air or sometimes water as a cooling/heating medium. Use of impinging air stream is used to increase heat-transfer coefficient. Heat-transfer and mass-transfer coefficients during freezing and thawing were reviewed recently.[18]

It is quite common to believe that radiation is important at high temperature only. But it is also important during freezing of the food material because the heat transfer by conduction or convection is too small which makes radiation comparable. At the surface, evaporation of water is responsible for loss or gain of heat in form of latent heat. Under these, an effective heat-transfer coefficient can be defined in the following manner.[18]

$$h_{\text{eff}} = h \frac{\left(T_{\text{air}} - T_s\right)}{\left(T_{\text{max}} - T_s\right)} + F \epsilon \sigma \frac{\left(T_{\text{rad}}^4 - T_s^4\right)}{\left(T_{\text{max}} - T_s\right)} + k \Delta H^{\text{vap}} \frac{\left(P_{T_d} - a_{ws} P_{T_s}\right)}{\left(T_{\text{max}} - T_s\right)} \qquad (3.12)$$

The three terms on right-hand side of Equation (3.12) are due to convection, radiation, and evaporation, respectively (Fig. 3.2). The other symbols are h and h_{eff} representing convective and effective heat-transfer coefficients respectively, F is view factor, ϵ is emissivity of the solid surface, σ is Stefan–Boltzmann constant, k is mass-transfer coefficient, ΔH^{vap} is latent heat of vaporization and P_{T_d} and P_{T_s} are vapor pressure estimated at dew point and surface temperature, respectively.

The equation of change of energy remains same. The heat-transfer coefficients are used in the boundary conditions. Suitable correlations for convective heat transfer and mass-transfer coefficients can be used to estimate the convective and evaporative terms. View factor depends upon the geometry and arrangement of food item when kept for drying.

Various mechanisms for heat transfer during freezing of food material are shown in Figure 3.2. At the surface of the food item convection, evaporation, and radiation takes place. Near the surface, the food is frozen and only conduction is important since all the pores are filled with ice and other solidified constituents. The porous food which is in the core of the food item cannot accessed by flowing air. The boundary between the frozen and unfrozen parts moves from surface to the core, as the time progresses. During thawing, the frozen part is in the core and the porous unfrozen part is adjacent to the surface. Under such circumstances, it is possible to increase convective transport by using impingement of air.

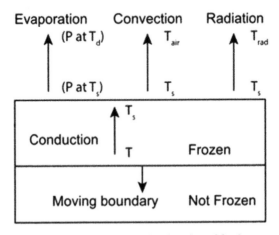

FIGURE 3.2   Mechanisms of heat transfer during freezing of food.

Partial thawing can be modeled by considering heat transfer only.[22] In general, heat transfer within the solid food takes place by conduction only.[11]

$$\frac{\partial T}{\partial t} = \frac{k_h}{\rho C_p}\left(\frac{\partial^2 T}{\partial^2 x} + \frac{\partial^2 T}{\partial^2 y} + \frac{\partial^2 T}{\partial^2 z}\right) + \frac{Q}{C_p} \tag{3.13}$$

where Q is heat generated by any phenomena, for example heat of respiration.

Equation (3.13) is obtained while neglecting the effect of ice-nucleation and mass transfer due to moisture diffusion. Separate equations are to be used in unfrozen and frozen portions of the food using thermal conductivity k_{uf} and k_{ff}, respectively. The problems related to moving boundaries have been dealt by Farid.[14] The boundary between the two portions is moving and the distance of the boundary front Y is given by.[11,14]

$$k_{uf}\left(\frac{\partial T}{\partial y}\right)\Big|_{Y+} - k_{ff}\left(\frac{\partial T}{\partial y}\right)\Big|_{Y-} = \rho\lambda\frac{dY}{dt} y = Y \tag{3.14}$$

Equation (3.13) is generally solved numerically. Analytical solutions are available only for simple geometries and for conditions that no evaporation and heat generation takes place. The solutions are available in terms of Biot number which is the ratio of convective heat-transfer rate outside the food item to heat-transfer rate due to conduction inside it.[11]

Farid[14] provided the following equation for all the three coordinate systems. It incorporates the moving boundary conditions and is suitable to convert into difference equation.

$$\frac{dT}{dt} = \frac{y}{Y}\frac{dY}{dt}\frac{\partial T}{\partial y} + \frac{\alpha}{y^n}\left[\frac{\partial}{\partial y}\left\{y^n\left(\frac{\partial T}{\partial y}\right)\right\}\right] \tag{3.15}$$

By substituting the value of $n = 0$, 1, and 2, equations are obtained in Cartesian, cylindrical, and spherical coordinates, respectively. Here, $y = r$ is used in coordinates other than Cartesian coordinates.

When thawing is done using impingent technology, the moisture evaporates and mass transfer for moisture should be taken into account. Anderson et al.[2] used heat transfer and mass transfer of moisture in cylindrical coordinates. The mass transfer considers unsteady two-dimensional diffusion (radial and axial) of moisture. Both equations were solved simultaneously using finite difference technique.

3.3.3 HEAT TRANSFER WITH HEAT GENERATION: MICROWAVE HEATING

Several food processing methods use radiofrequency, microwave to generate heat in the food material. Microwave assisted drying, freeze drying are used to increase the removal of moisture from the surface of the food material. The governing equation in absence of convection is given by Equation (3.13). However, in case of microwave heating the generation term depends upon the distance of the point from the surface where the adsorption of microwave is maximum. At any other point inside the food material, Lambert's law recommended[5,15]:

$$Q = Q_0\exp(-x/\delta). \tag{3.16}$$

Here δ is the penetration depth, which is property of the food material. Due to complex shapes of food material and nonlinear nature of thew generation term most of the model equations are solved numerically. A good summary of the investigations are available in literature.[11] In presence of evaporation due to loss of temperature, the property of the food material is the penetration depth and is also a function of distance from the surface. Consequently, the generation term can be written as following.[5]

$$Q = \frac{Q_0}{\delta(x)} \exp\left(-\int_0^x \frac{dx}{\delta(x)}\right)$$

(3.17)

Model for continuous microwave heating includes the convective term also. For microwave heating in a cylindrical tube 1D model equation is a steady-state energy balance given as following.[2]

$$\frac{\partial T}{\partial t} = \alpha\left[\frac{1}{r}\frac{\partial}{\partial r}\left(\frac{\partial T}{\partial r}\right)\right] + Q_0 \exp\left(-\frac{(R-r)}{\delta}\right)$$

(3.18)

3.4 EQUATION OF CHANGE OF MASS

The equation of change of mass in Cartesian coordinates is obtained by applying law of conservation of mass for each chemical species present in the system.

$$\rho\left(\frac{\partial C_i}{\partial t} + v_x\frac{\partial C_i}{\partial x} + v_y\frac{\partial C_i}{\partial y} + v_z\frac{\partial C_i}{\partial z}\right) = D_i\left(\frac{\partial^2 C_i}{\partial^2 x} + \frac{\partial^2 C_i}{\partial^2 y} + \frac{\partial^2 C_i}{\partial^2 z}\right)$$

(3.19)

Here, C_i is the concentration of ith species and D_i is the diffusivity of ith species in the medium. The right-hand side represents mass transfer due to diffusion and left-hand side denotes convective mass transfer except the first term which represents accumulation term. The equation is useful in studying the diffusion of flavors and transfer of moisture during drying, thawing, etc. The convective term is useful in flow conditions.

3.4.1 MODELS FOR DIFFUSION IN FOOD: DISPERSION

Many of food items are solid or gel in which the convection is absent. The mechanism of dispersion of additives such as salt, color, etc. in the food items is through the diffusion only. Since the food material may have the shape resembling sphere, cylinder or sheet, the model equation for unsteady state diffusion in various coordinate system can be written as

$$\rho\frac{\partial C_i}{\partial t} = D_i \left(\frac{\partial^2 C_i}{\partial^2 x} + \frac{\partial^2 C_i}{\partial^2 y} + \frac{\partial^2 C_i}{\partial^2 z} \right) \tag{3.20}$$

$$\rho\frac{\partial C_i}{\partial t} = D_i \left[\left(\frac{1}{r}\frac{\partial}{\partial r}\left(r\frac{\partial C_i}{\partial r} \right) + \frac{1}{r^2}\frac{\partial^2 C_i}{\partial^2 \theta} + \frac{\partial^2 C_i}{\partial^2 z} \right) \right] \tag{3.21}$$

$$\rho\frac{\partial C_i}{\partial t} = D_i \left[\left(\frac{1}{r^2}\frac{\partial}{\partial r}\left(r^2\frac{\partial C_i}{\partial r} \right) + \frac{1}{r^2\mathrm{Sin}\theta\partial\theta}\frac{\partial}{}\left(\mathrm{Sin}\theta\frac{\partial C_i}{\partial \theta} \right) + \frac{1}{r^2\mathrm{Sin}^2\theta}\frac{\partial^2 C_i}{\partial^2 \varphi} \right) \right] \tag{3.22}$$

Equations (3.20), (3.21), and (3.22) are in Cartesian, cylindrical, and spherical coordinates, respectively. Analytical solutions to these equations for simple boundary conditions are available in several books.[6,12,30] Spread of various food additive and flavors can be modeled using diffusion equation.

3.4.2 MOLECULAR TRANSPORT IN POROUS MATERIALS

A number of unit operations used for processing of food involve flow through porous media. Filtration, flow through packed bed, membrane separation, etc. are some of these. Though, a vast literature on these processes is available, few simple laws require a mention here. The flow through a packed bed may be modeled by Darcy's law. The flow rate, Q, is proportional to the pressure drop, ΔP.[5]

$$Q = \frac{A\Delta P}{\mu R} \tag{3.23}$$

Here, μ is viscosity of the fluid and R is the resistance of the porous media which is proportional to its thickness.

The resistance of the porous media is experimentally determined. If the porous media is made of particulate matter of known size then other equations can also be used. Karman–Kozeny equation generally used for filtration is given as[5]

$$Q = \frac{A}{K\mu S^2} \frac{\varepsilon^3}{(1-\epsilon)^2} \frac{\Delta P}{L} \tag{3.24}$$

where ϵ is void fraction, K is a constant, S is surface area per unit volume of porous medium, and L is thickness of the medium.

By comparing Equations (3.23) and (3.24), an expression for the resistance of the filter medium, R, can be obtained. To determine the values of S and ϵ, size, shape, and method of packing is required.

3.4.3 MODELS FOR ADSORPTION PROCESSING

One of the methods for separation of useful compounds from food material is adsorption. To predict the performance of adsorbers, it is necessary to known the equilibrium data. Polyamides, polyvinyl pyrollidone, nylon polymers, β-cyclodextrin polymers, α-cyclodextrin polymers, cellulose acetate, cellulose acetate derivatives like Cellulose acetate butyrate, cellulose triacetate, cellulose ester in gel form, polystyrene divinyl benzene resins, etc. can be used as adsorbants. The equilibrium data are modeled by equations known as isotherms. A few useful isotherms, frequently described in literature,[3,4,31] are given in Table 3.1.

TABLE 3.1   Adsorption Isotherms.[3,4,31]

Isotherm	Equation	
Langmuir	$q_e = \dfrac{aC_e}{1+bC_e}$	(3.25)
Freundlich	$q_e = K_f C_e^{1/n}$	(3.26)
Redlich–Peterson	$q_e = \dfrac{K_R C_e}{1+a_R C_e^{\beta}}$	(3.27)
Dubinin–Radushkevich	$q_e = q_D \exp\left(-B_D \varepsilon_d^2\right)$ where $\varepsilon_d = RT \ln\left(1+\dfrac{1}{C_e}\right)$	(3.28)
Temkin	$q_e = \dfrac{RT}{b} \ln\left(K_T C_e\right)$	(3.29)
Toth	$q_e = \dfrac{AC_e}{\left(B+C_e^D\right)^{1/D}}$	(3.30)

q_e = amount of material adsorbed at equilibrium, C_e = Concentration of adsorbate in solution at equilibrium, R = gas constant, T = temperature (K)

Langmuir isotherm is applicable for monolayer adsorption on adsorbent surface and also for all adsorption sites equivalent in terms of adsorption energy. It assumes that there is no interaction between adjacent adsorbed molecules. Freundlich isotherm incorporates role of substrate and interaction on the surface. However, it is an empirical expression. At low concentrations the Redlich–Peterson isotherm reduces to Henry's law and at high concentrations it reduces to Freundlich isotherm. The value of β lies between 0 and 1.

Dubinin–Radushkevich isotherm is based on adsorption mechanism to deduce the heterogeneity of the surface energies of adsorption and the characteristic porosity of the adsorbent. Mean free energy of sorption, E, is calculated from the following relationship.

$$E = \frac{1}{\sqrt{2B_D}} \tag{3.31}$$

For chemisorptions, $8 < E < 16$ kJ/mol and for physical adsorption, $E < 8$ kJ/mol. Temkin isotherm contains a factor that explicitly taking into the account of adsorbent–adsorbate interactions.

In a batch-adsorption apparatus, the concentration of the adsorbate changes with time. It is described by the kinetic models. A few kinetic models for adsorption behavior of amount of material adsorbed at time t, q_t, are given in Table 3.2.[3,29]

Pseudo first-order model is the most widely used rate equation based on solid capacity for calculating the adsorption rate of an adsorbate from a liquid phase and is known as the Lagergren rate equation. Banghams equation is useful when pore diffusion is the only rate controlling step. Here, C_b is initial concentration of adsorbate, m is the amount of adsorbent and V is volume of solution. Elovich's model is suitable for chemisorptions. Here, α is the initial adsorption rate and ω is the adsorption rate. Ribeiro's model assumes that adsorption is proportional to solute concentration in solution and to the fraction of unoccupied surface.

Singh's modified adsorption shell model considers that adsorption in macroporus resin occurs in the three distinct zones: (1) outer zone in which microspheres are almost completely saturated, (2) middle zone where adsorption is taking place at a given moment, and (3) the inner zone where adsorption is yet to take place. After some time (time delay) of the development of the three zones the outer saturated zone increases and inner zone shrinks and adsorption zone (middle zone) moves inward. The thickness of adsorption zone is finite.

TABLE 3.2   Kinetic Models for Adsorption Isotherms.[3,29]

Model	Equation	
Pseudo first-order model	$\dfrac{dq_t}{dt} = k_f\,(q_e - q_t)$	(3.32)
Pseudo second-order model	$\left(\dfrac{dq}{dt}\right) = k_s\,(q_e - q_t)^2$	(3.33)
Bangham model	$\log\left\{\log\left(\dfrac{C_b}{C_b - q_t m}\right)\right\} = \log\left(\dfrac{k_o m}{2.303V}\right) + \sigma\log(t)$	(3.34)
Elovich's kinetic model	$\dfrac{dq}{dt} = \propto \exp(-\alpha q)$	(3.35)
Ribeiro et al. model	$\dfrac{dC_t}{dt} = -k_1\left(1 - \dfrac{q}{q_{max}}\right)C_t$	(3.36)
Singh's Modified adsorption shell Model	$\left[\dfrac{1}{(1-u_t)^{2/3}} - 1\right] = \dfrac{2KD_p \in C_o}{5(1-\in)q_e R_p^2}\,t$	(3.37)
Intraparticle diffusion model	$q_t = k_d t^{1/2} + t$	(3.38)
Boyd's diffusivity model	$\ln\left[\dfrac{1}{1-u^2(t)}\right] = \dfrac{\pi^2 D_e t}{R_a^2}$	(3.39)

The model considers diffusion resistances in macropores and micro-pores. The equation given in Table 3.2 is only for very thin shell. Here, D_p is macropore diffusivities, R_p is the average radius of the adsorbent particles, r_c is radius of microspheres, ϵ is void fraction in adsorbent particle. The model fits the data over a wide range of experimental conditions.

Intraparticle diffusion model is widely applied for the analysis of adsorption kinetics. Here, k_d is the intraparticle diffusion rate constant and q_t is concentration of adsorbate in solid phase.

During the process of adsorption, the transportation of the adsorbate particles to the surface of the adsorbent surface takes place in several steps. The adsorption process may be controlled by film or external surface diffusion, pore diffusion, surface diffusion, and adsorption on the pore surface or a combination of one or more steps. In a rapidly stirred batch process, the diffusive mass transfer can be replaced by an apparent diffusion coefficient which will fit experimental adsorption rate data.

Boyd's diffusivity model is based on ion-exchange kinetics. It is based is based on diffusion through the boundary liquid film, considering adsorption kinetics as a chemical phenomenon. It is used to determine the rate-controlling step. Here, $u(t) = q_t/q_e$ and D_e = rate constant.

3.4.4 MODELING ADSORPTION PROCESS

To model adsorption equipment, it is required to use the model of the apparatus and equilibrium data. The equilibrium data are the various isotherms discussed in previous section. To illustrate this, let us consider adsorption in a fixed bed. Let us consider one-dimensional axial dispersion model, that is, a constant axial velocity, v_z, independent of radial position and dispersion in axial direction exists.[10]

$$\frac{\partial C}{\partial t} = D_L \frac{\partial^2 y}{\partial x^2} - v_z \frac{\partial C}{\partial z} - \left(\frac{1-\varepsilon}{\varepsilon}\right)\frac{3K_f}{R}(C - C_R) \tag{3.40}$$

Here, D_L is the dispersion coefficient, R is the radius of the particle and C_R is the concentration at the surface of the particle. The initial and boundary conditions are as following:

$$\text{at } t = 0; \; C(z,0) = 0$$

$$\text{at } z = 0; \; C = C_0 + \frac{D_L \varepsilon}{v_z}\frac{\partial C}{\partial z}$$

$$\text{at } z = H; \; \frac{\partial C}{\partial z} = 0 \tag{3.41}$$

The surface concentration was expressed by using Langmuir adsorption isotherm.[10]

$$\frac{\partial q}{\partial t} = k_1 C(q_{max} - q) - k_2 q \tag{3.42}$$

Here, k_1 and k_2 are intrinsic kinetic constants.

A number of other models for fixed-beds are reported elsewhere.[20] The other isotherms given in Table 3.1 can also be used. By combination of these, other models were developed and used.[20] For example, Thomas model do not consider axial dispersion and uses second-order kinetics.

3.5 PROCESSES INVOLVING SIMULTANEOUS HEAT AND MASS TRANSFER

Food processes such as drying, cooking, baking, etc. involve combination of mass and heat transfer taking place together. Thermal processing is often coupled with movement of moisture within the food material, various reactions taking place with water and other ingredients present in the food. From modeling point of view, the processes cannot be described by heat-transfer or mass-transfer phenomena alone. The heat- and mass-transfer equations need to be solved simultaneously. Modeling of frequently used food processing operations is discussed in this section.

3.5.1 DRYING

A vast literature on drying exists. Food material is considered as porous and hygroscopic. Foods are dried to control the moisture content and to increase its shelf life. The model for drying can be considered as combination of two sub-models. One for the drying equipment and another for the individual food item present inside the drying. Several drying methods such as freeze drying, microwave-assisted drying, spray drying, fluidized bed drying, and vacuum drying are also available. All of these drying processes involve certain common phenomena.

During air drying, food items are first washed and dried. The drying curve consists of three regimes. During heating of the food item, the temperature of the food item increases. Evaporation rate increases resulting in increased drying rate. The time at which the food item attains temperature of the surrounding air the drying rate becomes constant. During this period, the surface water is evaporated. This period can be reduced by wiping out extra water on the surface of the food item. After the entire water is evaporated the moisture content present within food item is removed. Several models have been proposed to understand the phenomena of drying of food. The models proposed are based on the following steps (Fig. 3.3).

   i. Evaporation of moisture takes place at the surface. It is accounted in form of boundary condition. During air drying, the temperature and humidity of the air is required to know the driving force for removal of moisture on the surface.

ii. Heat transfer from surface toward the core. It may takes place either by conduction or may account for evaporation of moisture within the food.

iii. Movement of moisture from core toward the surface. It is by diffusion only. The moisture may be in form of liquid or vapor or both.

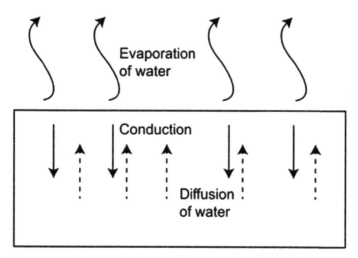

FIGURE 3.3   Mechanisms of heat transfer during conventional drying of food item.

3.5.1.1   GENERAL MODEL OF DRYING

A general approach to model drying is presented by Curcio.[13] Treating water vapor and liquid water as different species unsteady state diffusion with generation term for both species were written. The generation term is the rate of evaporation. The equations were written in vector form. Let us write the equation of change for concentration of liquid and vapor form of water, c_1 and c_v in Cartesian coordinates in the following equations:

$$\frac{\partial C_1}{\partial t} = \frac{\partial}{\partial x} D_i \frac{\partial C_1}{\partial x} + \frac{\partial}{\partial y} D_i \frac{\partial C_1}{\partial y} + \frac{\partial}{\partial z} D_i \frac{\partial C_1}{\partial z} - w \qquad (3.43)$$

$$\frac{\partial C_v}{\partial t} = \frac{\partial}{\partial x} D_i \frac{\partial C_v}{\partial x} + \frac{\partial}{\partial y} D_i \frac{\partial C_v}{\partial y} + \frac{\partial}{\partial z} D_i \frac{\partial C_v}{\partial z} + w \qquad (3.44)$$

The energy balance can be written as following:

$$\rho C_p \frac{\partial T}{\partial t} = \frac{\partial}{\partial x} k_{eff} \frac{\partial T}{\partial x} + \frac{\partial}{\partial y} k_{eff} \frac{\partial T}{\partial y} + \frac{\partial}{\partial z} k_{eff} \frac{\partial T}{\partial z} + \Delta H^{sbl} w \qquad (3.45)$$

These equations can be solved numerically. The equations allow us to use spatial variation of the properties of the food material.

3.5.1.2 MODEL FOR FREEZE DRYING

The steps involved in freeze drying are different than conventional drying. The steps are shown in Figure 3.4.

i. Evaporation of moisture is due to sublimation of ice.
ii. The heat required for sublimation is received by the food surface by radiation from a surface.
iii. To maintain a low vapor pressure of water pressure air a condenser is provided to condense the evaporated moisture.

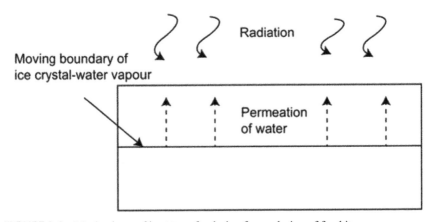

FIGURE 3.4 Mechanisms of heat transfer during freeze drying of food item.

Considering the food item as a slab consisting of dry portion and portion having ice crystals separated by a front, rate of heat transfer due to heat of sublimation and due to conductive transport of heat from the slab surface to sublimation front we get.[5]

$$q = A\rho_w\left(w_i - w_f\right)\Delta H^{\text{sbl}}\frac{dz}{dt} = \frac{kA\left(T_0 - T_i\right)}{z} \qquad (3.46)$$

Here, w_i and w_f are initial and final water content, respectively, k is the permeability of the dry food material and ΔH^{sbl} is latent heat of sublimation. This equation is integrated to get time of drying. Similarly equating rate of mass transfer due to sublimation to water vapor permeated through dry layer we get[5]:

$$A\rho_w\left(w_i - w_f\right)\frac{dz}{dt} = \frac{\Pi A\left(p_i - p_0\right)}{z} \qquad (3.47)$$

3.5.1.3 MODEL FOR SPRAY DRYING

Spray drying is generally used for drying of thick pastes, for example, as concentrated milk and tomato puree. The paste is atomized at the top so that they, in form of small spherical drops, fall downward at terminal velocity. Hot air flows co-currently. The water evaporates from the surface of the droplets. For the droplet, the increase in temperature is due to convective heat transfer to the surface and heat loss from it in form of latent heat.[24]

$$mC_p\frac{dT}{dt} = hA\left(T_{\text{air}} - T\right) + \frac{dm}{dt}\Delta H^{\text{vap}} \qquad (3.48)$$

Any suitable correlation to estimate heat transfer coefficient can be used. The particle velocity can be determined by usual method by balancing gravity, buoyancy and drag forces. The term dm/dt is the rate of change of size of the particle due to evaporation which can be estimated in terms of mass-transfer coefficient. The mass-transfer flux is written as following.[24]

$$j = k\frac{M_w}{RT}\left(P_w^{\text{sat}} - y_w\right) \qquad (3.49)$$

A suitable correlation to estimate heat-transfer coefficient can be used. Due to evaporation of water the temperature of the air also changes with axial position in the dryer. It can be taken into account by applying equation of change of energy. These equations can be solved to predict the performance of a spray dryer. The equations are valid for constant drying rate period. In the lower portion of the spray dryer the falling drying rate period may be observed. It requires an additional one-dimensional unsteady diffusion equation in spherical coordinates.

3.5.2 COOKING (BOILING)

Rice is consumed by a large population in the world. It is cooked by boiling. During cooking water is soaked by rice and gelatinization occurs at a particular temperature. Various attempts to model cooking of rice were reviewed by Shinde et al.[28] The models were classified into several categories. A group of model was called as shrinking core models. The group of models was simple in nature and provided analytical solutions relating time of cooking, θ, degree of cooking, and size of the rice grain, R. For example Suzuki's two models[28] are given by the following equations.

$$(1 - \alpha) = \exp(-k\theta) \tag{3.50}$$

$$R = 4\pi r_c^2 \frac{\rho}{M} \frac{dr_c}{dt} \tag{3.51}$$

Here, M is molecular weight of water and r_c is the radius of core, that is, uncooked rice. The shrinking core models assume that reaction rate controls the process.

Another group of models assume that the diffusion of water is slow and controls the process. The models based on this assumption were grouped as diffusion controlled models. Thus, the model were essentially one-dimensional unsteady-state diffusion expressed in cylindrical or spherical coordinates depending upon the assumption whether the rice grain was long or small. Engels et al.'s model[28] is given as

$$\frac{\partial W}{\partial t} = \frac{1}{r} D \frac{\partial}{\partial r}\left(r \frac{\partial W}{\partial r}\right) \tag{3.52}$$

Here, W is the water content. The surface was immediately saturated, that is at $t = 0$ and $r = R; W = W_e$; and at the surface: $D\left(\partial W / \partial t \frac{\partial W}{\partial t}\right) = k_1 \left(C_s - C_e\right)$

Engels modeled cooking of brown rice and considered rice consisting of two layers of different diffusivities.

$$D\frac{\partial W}{\partial r} = -\frac{D'}{\delta}\left(C_s - C_e\right) \tag{3.53}$$

Lin's model[28] is somewhat similar and is given as

$$\frac{\partial W}{\partial t} = D\left(\frac{\partial^2 W}{\partial x} + \frac{2}{r}\frac{\partial W}{\partial r}\right) - kW \tag{3.54}$$

The boundary conditions are

at $t = 0; W = W_0$; at $r = 0; \frac{\partial W}{\partial r} = 0$; and at $r = R; W = W_e$

The two-dimensional models for cooking of rice were also based on diffusion controlled models. These equations were solved numerically and were classified as numerical models.[28]

Model by Briffaz et al.[8,9] also considered no temperature gradient. It assumed that the water absorbed by rice reaches a critical value. After gelatinization, it can take excess water. The model considers swelling of rice during cooking. Model equations are one-dimensional diffusion of water into rice containing an uncooked core.

3.5.3 BAKING

Various food products are obtained by baking. It involves heating of dough with entrapped CO_2 in it. During baking heat transfer and mass transfer of water and CO_2 in the dough take place. Laws of conservation provide the model equations. At the boundary, heat transfer takes place by convection and radiation. This provides initial and boundary conditions. Zhou[14] reviewed the mathematical models for baking process. A couple of recent models are briefly discussed here to illustrate the nature of the baking processes.

Purlis[25] considered bread to be a cylindrical object and used equation for one-dimensional unsteady-state diffusion of water and heat transfer in cylindrical coordinates.

$$\rho C_p \frac{\partial T}{\partial t} = \frac{1}{r}\frac{\partial}{\partial r}\left(rk\frac{\partial T}{\partial r}\right) \tag{3.55}$$

$$\frac{\partial W}{\partial t} = \frac{1}{r}\frac{\partial}{\partial r}\left(rD_w\frac{\partial W}{\partial r}\right) \tag{3.56}$$

Here, k is thermal conductivity of dough, W is moisture content and D_w is diffusivity of water in the dough. It may be noted that k and D_w are not constant and have different values as the properties of the dough changes.

During baking the phase transition also takes place. The phases are dough, crumb, and crust. A jump in the properties was provided at the location of phase change which takes place at certain temperatures. The boundary conditions for Equations (3.55) and (3.56) are given as following[25]:

At the surface of bread:

$$-k\frac{\partial T}{\partial r} = h\left(T_s - T_\infty\right) + \epsilon\, \sigma\left(T_s^4 - T_\infty^4\right) \text{ and } -D_W\, \rho_s \frac{\partial W}{\partial r} = k_g\left(P_s\left(T_s\right) - P_\infty\left(T_\infty\right)\right)$$

And at the center of the bread:

$$\frac{\partial T}{\partial r} = \frac{\partial W}{\partial r} = 0$$

The model equations were solved using the properties as a function of temperature and incorporating phase transition.

A new evaporation–condensation–diffusion model was proposed for the baking of bread.[21] The model is based on the assumptions that the convective transport at the surface of the dough takes place due to convection and radiation with the oven air and radiative exchange with the oven walls. It leads to water evaporation at the surface of the product. Inside the bread heat transfer takes place by convection. Since water evaporates at the surface, a concentration gradient of water starts developing. Hence, water transfers from core toward the surface. As the temperature in the core increases, the gases present in bubble expands, evaporation of water at core–bubble interface increases and desolubilization of CO_2 (due to its reduced solubility) takes place. It results in increased pressure and hence deformation of the bread. It also results in opening of the pores and movement of gases through the pores toward the surface of the food item.

Transfer of water was assumed to take place by evaporation–condensation–diffusion mechanism.

Pores open as the pressure increases and changes in molecular structure related to the baking process takes place. The model considers the introduction of air into the dough through open pores. It considers all the variables, that is, temperature, concentration of water, CO_2, and dry matter to vary continuously in the product. The generation of CO_2 during baking was ignored. The model uses one-dimensional quasi steady-state form of laws of conservation of energy (enthalpy) and conservation of mass for water, air, and CO_2 for the food product. The 1D unsteady state diffusion equations in Cartesian coordinates are as following:

$$\frac{1}{V}\frac{\partial C_i}{\partial t} + \frac{\partial N_i}{\partial z} = -r_i \qquad (3.57)$$

where N_i is the diffusional flux of ith species (i = water, CO_2, and air) and r_i is the generation term for ith species. Only r_{CO_2} was nonzero term. The flux of water was assumed to be sum of two terms for liquid and vapor phases of water. The flux for gases (CO_2 and water vapor) was described by binary diffusion.

$$\frac{\partial N_i}{M_i} = N_i + \frac{\partial N_i}{\partial t} = -r_i \qquad (3.58)$$

The movement of water in the dough is shown in Figure 3.5. Dough was considered as successive layers of solid and bubble. Water moved in solid portion by diffusion. It crosses vapor layer by first getting evaporated one side and then getting condensed at the other side of the vapor layer.

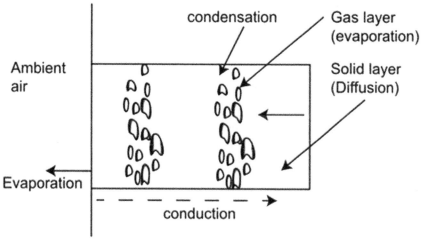

FIGURE 3.5 Mechanisms of heat and mass transfer during baking before pores are open.

The heat transfer was described by Fourier's law.

$$Q = k_h \frac{\partial T}{\partial z} \qquad (3.59)$$

Here, k_h is thermal conductivity of the aerated dough.

Total thermal energy transfer is sum of the heat transfer by conduction, transfer of sensible heat of all the species, and latent heat of evaporating and condensing vapor.[21] At later stages when the pores were open and interconnected, movement of vapors followed Darcy's law. For detailed model equations reader is referred to the work of Lucas et al.[21]

Assuming uniform temperature and concentration of each species at the beginning, the model equations were solved by assuming various terms estimated from steady-state flow of all species considered.

Papasidero et al.[23] proposed a simple model. Heat transfer due to convection of water was neglected and the model did not consider mass transfer of CO_2. Properties of the material involved were discussed in detail.[23]

3.5.4 DEEP FRYING

Deep frying is also an unsteady process. The heat-transfer coefficient which controls the variation of temperature during deep frying process is important to control the quality of the food.[26] There are four stages of frying.

i.   The food item gets heated up to the temperature of the boiling oil by natural convection.
ii. At the surface of the food item the water starts evaporating and leaves the food item in form of bubbles which causes enough turbulence. The heat transfer at the surface takes place by convection. The crust formation takes place. In both these stages, the heat transfer in the food item takes place by conduction.
iii. The temperature inside the food increases and the moisture content decreases. Several reactions, for example, oxidation, hydrolysis, etc. takes place during this phase. These reactions are reviewed in elsewhere.[34] As a result of these reactions, the structure of the food and its properties changes. The thickness of the crust increases with time.
iv. The water has already diminished. No more bubbles are observed.

The general model equations to describe deep frying consist of the following equations:

Equation of change for mass of water in form of liquid and vapor are for unsteady diffusion process and are same as Equations (3.43) and (3.44). Garrieri et al.[16] used the generation term as given in the following:

$$w = K_w C_w \tag{3.60}$$

Here, K_w is the mass-transfer coefficient and C_w is water concentration at the surface. The equation of change for energy is given by an equation similar to Equation (3.45).

Unlike the case of drying, the fluid phase surrounding the food item is not at constant temperature and concentration. The equation of change for water liquid and water vapor are described by unsteady convection–diffusion problem considering oil to be Newtonian fluid. Equations can be written in any coordinate system depending upon the shape of the food. These equations in Cartesian coordinates are same as Equation (3.19). The equation of change of thermal energy in Cartesian coordinate is same as Equation (3.11). Equations (3.11) and (3.19) require the knowledge of fluid velocity that can be obtained by using equation of change of momentum. In Cartesian coordinate the x-component of momentum balance is as the following equation[16]:

$$\rho\left(\frac{\partial v_x}{\partial t}+v_x\frac{\partial v_x}{\partial x}+v_y\frac{\partial v_x}{\partial y}+v_z\frac{\partial v_x}{\partial z}\right)=\left(\frac{\partial}{\partial x}\mu\frac{\partial v_x}{\partial x}+\frac{\partial}{\partial y}\mu\frac{\partial v_x}{\partial y}+\frac{\partial}{\partial z}\mu\frac{\partial v_x}{\partial z}\right)+\rho g_x\beta(T-T_\infty) \quad (3.61)$$

Equation (3.61) takes into account local variation of viscosity. The last term corresponds to the buoyancy force. Here, β is thermal expansion coefficient. The model equations after adding rate equations were solved using CFD software to study the acrylamide formation during deep frying.

A simple one-dimensional model for frying can also be used as reported elsewhere.[2,15] The model equations are one-dimensional diffusion of water and oil in the solid and heat transfer in the solid. To avoid duplication these equations are not presented. The model requires the position of the moving front which may be calculated as following.[15]

$$\overline{C_w}\frac{dx_l}{dt}=-D_w\frac{\partial C_w}{\partial x} \quad (3.62)$$

where $\overline{C_w}$ is the average concentration of water.

A two-dimensional model with similar features was proposed by Wu et al.[33] Swelling (puffing) of the food and shrinking of the food due to removal of moisture needs to be properly addressed in these models.

3.5.5 MEMBRANE PROCESSES

A number of membrane separation methods, including filtration, ultrafiltration, membrane distillation, etc. are used in food processing. The models for these processes are process specific. In general, the membrane separation involves the following steps.

i. Transport of ions from bulk to the surface of the membrane: The mass-transfer flux may be computed from using a correlation, film model or one-dimensional diffusion equation.
ii. Permeation of ions through the membrane: Mass-transfer flux is not described by diffusion. One can use Darcy's law. Expressions considering the porosity of membrane resembling the flow through capillaries have been used.[16]
iii. On the other side of the membrane a mechanism to remove the ions from the membrane is required. The film model or thermodynamic model can be used.

The resistance in the three steps can considered to be working in parallel.[27,29]

3.6 KINETICS OF FOOD MATERIALS

The quality of food items changes with time due to reactions. The changes in foods composition, which affect its quality, are described by the kinetic rate of these reactions. These reactions also have an effect on the equation of change of mass and appear as generation term. If the reaction is exothermic or endothermic then the generation term appears in the equation of change of thermal energy also (as heat of reaction). Thus, it is essential to the rate of reaction. The reactions related to food can be classified into three categories:

i. **Chemical reactions:** These are observed in almost all thermal processes. Due to a large number of chemical species present in the food it is difficult to correlate the kinetic data. However, the problem may be simplified if the major reactions can be identified. There may be several reactions in series or parallel producing more the one product. It has been often termed as 'multi-responsive' kinetics.
ii. **Enzymatic reactions:** These reactions involve enzymes as catalysts. However, the rate expressions are different than other catalyzed chemical reactions.
iii. **Microbial reactions:** The reactions involve microbial cells. The reaction is carried out by the enzymes released by the microbial cells.

3.6.1 ENZYME KINETICS MODELING

A large number of enzymatic reactions are explained by the Michaelis–Menten kinetics. It is based on the following reaction mechanism:

$$E + S \underset{k_2}{\overset{k_1}{\rightleftharpoons}} ES \overset{k_3}{\rightarrow} E + P$$

The enzyme, E, reacts with the substrate, S, and form an intermediate ES through a reversible reaction which then decomposes to give the product, P. The rate expression is given by

$$v = v_{max} \frac{[S]}{[S] + K_M} \qquad (3.63)$$

where v is the reaction rate and v_{max} is the maximum reaction rate observed at large values of concentration of substrate $[S]$ at which the reaction rate becomes constant. The constant K_M, known as Michaelis–Menten is given by

$$K_M = \frac{k_2 + k_3}{k_1} \qquad (3.64)$$

Other rate expressions are also in use. Detailed descriptions can be found in literature.[7]

3.6.2 KINETICS OF MICROBIAL PROCESSES

The microbial reactions, usually encountered in fermentations, have to be described by growth of microbial population as the microbial cells are living cells and multiply. They are governed by the birth-death process. The kinetics is usually described by Monod's equation.

$$\mu X = \frac{\mu_{max}[S]}{K_s + [S]} X \qquad (3.65)$$

where μ is the specific growth rate, μ_{max} is the maximum specific growth rate, and X is the biomass concentration.

It may be recalled that the cell growth may be limited by limited supply of the substrate acting as food for the cell. This situation is called as substrate limiting. Similarly, when the number of cells become too large, the waste generated by them acts as a poison to the reaction. It is called as product inhibition. The kinetics in both the cases can be described only by taking the type of the fermentor into account. A brief description is available in several text books.[19]

3.7   CONCLUSIONS

Food items are generally cooked using various processes, for example, baking, cooking by boiling, extrusion, deep frying, etc. During the processing, not only the properties of the food changes but also phase transition may also takes place. All these processes have been modeled using laws of conservation of momentum, mass, and energy.

It has been possible to model various processes by one-dimensional unsteady heat and mass transfer. Though, earlier models have been simple enough to get analytical solutions, recent models taking into account variation of food properties were solved numerically. It shows the importance of role of transport phenomena and kinetic studies in having an insight of the food processing.

The equation of change of chemical species especially water, and that for energy with appropriate boundary conditions and time varying properties of food material are sufficient to reveal the behavior of these processes.

KEYWORDS

- adsorption
- adsorption isotherms
- adsorption kinetics
- baking
- cooking
- deep frying
- diffusion
- dispersion
- drying
- drying model

- enzyme kinetics
- food
- food preservation
- food processing
- freeze drying
- freezing
- frying
- heat transfer
- kinetic modeling
- mass transfer
- mathematical model
- mathematical modeling
- membrane separation
- microwave heating
- momentum transfer
- non-Newtonian fluids
- pastes and dough
- porous materials
- rheological model
- rheology
- simultaneous heat and mass transfer
- spray drying
- thawing
- thermal processing
- transport processes
- unit operations
- viscoelastic fluids

REFERENCES

1. Alvis, A.; Velez, C.; Mendoza, M. R.; Villamiel, M.; Villada, H. S. Heat Transfer Coefficient During Deep-fat Frying. *Food Control* **2009**, *20*, 321–325.
2. Anderson, B. A.; Singh, R. P. Modeling the Thawing of Frozen Foods Using Air Impingement Technology. *Int. J. Refrig.* **2006**, *29*, 294–304.

3. Barkakati, P.; Begum, A.; Das, M. L.; Rao, P. G. Adsorptive Separation of Ginsenoside from Aqueous Solution by Polymeric Resins: Equilibrium, Kinetic and Thermodynamic Studies. *Chem. Eng. J.* **2010,** *161*, 34–45.

4. Basu, S.; Shivhare, U. S.; Mujumdar, A. S. Models for Sorption Isotherms for Foods: A Review. *Dry. Technol.: Int. J.* **2006,** *24*(8), 917–930.

5. Berk, J. *Food Process Engineering and Technology*. Academic Press: Berlington, 2009.

6. Bird, R. B.; Stewart, W. E.; Lightfoot, E. N. *Transport Phenomena*, 1st ed. John Wiley: New York, 1960.

7. Boekel, V.; Martinus, A. J. S. *Kinetic Modeling of Reactions in Foods*. CRC Press: Boca Raton, FL, 2009.

8. Briffaz, A.; Bohuon, P.; Méot, J. M.; Pons, B.; Matencio, F.; Dornier, M.; Mestres, C. Modeling of Brown Rice and Limited-water Cooking Modes and its Potential use for Texture Prediction. *J. Food Eng.* **2014,** *141*, 99–106.

9. Briffaz, A.; Bohuon, P.; Méot, J. M.; Dornier, M.; Mestres, C. Modeling of Water Transport and Swelling Associated with Starch Gelatinization during Rice Cooking. *J. Food Eng.* **2014,** *141*, 143–151.

10. Burkert, C. A. V.; Barbosa, G. N. O.; Mazutti, M. A.; Maugeri, F. Mathematical Modeling and Experimental Breakthrough Curves of Cephalosporin C Adsorption in a Fixed-bed Column. *Proc. Biochem.* **2011,** *46*, 1270–1277.

11. Chanes, J. W.; Ruiz, J. F. V.; Canovas, G. V. B. *Transport Phenomena in Food Processing*, CRC Press: Boca Raton, FL, 2003.

12. Chhabra, R. P.; Richardson, J. F. *Non-Newtonian Flow in the Process Industries Fundamentals and Engineering Applications*. Butterworth Heinemann: Oxford, 1999.

13. Curcio, S. A Multiphase Model to Analyze Transport Phenomena in Food Drying Processes. *Drying Technol.: Int. J.* **2010,** *28*(6), 773–785.

14. Farid, M. The Moving Boundary Problems from Melting and Freezing to Drying and Frying of Food. *Chem. Eng. Proc.* **2002,** *41*, 1–10.

15. Farid, M. M. *Mathematical Modeling of Food Processing*. CRC Press: Boca Raton, FL, 2010.

16. Garrieri, G.; De Bonis, M. V.; Pacella, C.; Pucciarelli, A.; Ruocco, G. Modeling and Validation of Local Acrylamide Formation in a Model Food During Frying. *J. Food Eng.* **2009,** *95*, 90–98.

17. Irudayaraj, J. *Food Processing Operations Modeling: Design and Analysis*. Marcel Dekker: New York, 2002.

18. Kondjoyan, A. A Review on Surface Heat and Mass Transfer Coefficients During Air Chilling and Storage of Food Products. *Int. J. Refrig.* **2006,** *29*, 863–875.

19. Levenspiel, O. *Chemical Reaction Engineering*, 3rd ed. John Wiley & Sons: New York, 1999.

20. Lim, A. P.; Aris, A. Z. Continuous Fixed-bed Column Study and Adsorption Modeling: Removal of Cadmium(II) and Lead(II) Ions in Aqueous Solution by Dead Calcareous Skeletons. *Biochem. Eng. J.* **2014,** *87*, 50–61.

21. Lucas, T.; Doursat, C.; Grenier, D.; Wagner, M.; Trystram, G.; Flick, D. Modeling of Bread Baking with a New, Multi-scale Formulation of Evaporation–Condensation–Diffusion and Evidence of Compression in the Outskirts of the Crumb. *J. Food Eng.* **2015,** *149*, 24–37.

22. Nissen, J. A.; Zammit, G. Ø. Modeling and Validation of Robust Partial Thawing of Frozen Convenience Foods During Distribution in the Cold Chain. *Proc. Food Sci.* **2011,** *1*, 1247–1255.

23. Papasidero, D.; Manenti, F.; Pierucci, S. Bread Baking Modeling: Coupling Heat Transfer and Weight Loss by the Introduction of an Explicit Vaporization Term. *J. Food Eng.* **2015,** *147,* 79–88.

24. Passos, M. L.; Ribeiro, C. P. *Innovation in Food Engineering: New Techniques and Products.* CRC Press: Boca Raton, FL, 2009.

25. Purlis, E. Baking Process Design Based on Modeling and Simulation: Towards Optimization of Bread Baking. *Food Contr.* **2012,** *27,* 45–52.

26. Rao, M. A.; Rizvi, S. S. H.; Datta, A. K. *Engineering Properties of Food.* CRC Press: Boca Raton, FL, 2005.

27. Sahu, J. K. *Introduction to Advanced Food Process Engineering.* CRC Press: Boca Raton, 2014.

28. Shinde, Y. H.; Vijayadwhaja, A.; Pandit, A. B.; Joshi, J. B. Kinetics of Cooking of Rice: A Review. *J. Food Eng.* **2014,** *123,* 113–129.

29. Singh, S. V.; Gupt, A. K.; Jain, R. K. Adsorption of Naringin on Non-ionic (neutral) Macroporus Adsorbent Resin from its Aqueous Solutions. *J. Food Eng.* **2008,** *86,* 259–271.

30. Skelland, A. H. P. *Non-Newtonian Flow and Heat Transfer.* John Wiley: New York, 1967.

31. Suzuki, M. *Adsorption Engineering.* Elsevier: Amsterdam, 1941.

32. Verma, A. K. *Process Modeling and Simulation in Chemical, Biochemical and Environmental Engineering.* CRC Press: Boca Raton, FL, 2014.

33. Wu, H.; Karayiannis, T. G.; Tassou, S. A. A Two-dimensional Frying Model for the Investigation and Optimisation of Continuous Industrial Frying Systems. *Appl. Therm. Eng.* **2013,** *51,* 926–936.

34. Zhang, Q.; Saleh, A. S. M.; Chen, J.; Shen, Q. Chemical Alterations Taken Place During Deep-fat Frying Based on Certain Reaction Products: A Review. *Chem. Phys. Lipids* **2012,** *165,* 662–681.

PART II
Review of Research Advances in Food Engineering

CHAPTER 4

ROLE OF ENCAPSULATION IN FOOD AND NUTRITION

ARPITA DAS[1] and RUNU CHAKRABORTY[2*]

[1]*Faculty of Chemistry and Chemical Engineering, Babes-Bolyai University, 400028 Cluj-Napoca, Romania. E-mail: arpita_84das@ yahoo.co.in*

[2]*Department of Food Technology and Biochemical Engineering, Jadavpur University, Kolkata 700032, West Bengal, India. E-mail: rchakraborty@ftbe.jdvu.ac.in, crunu@hotmail.com*

*Corresponding author.

CONTENTS

ABSTRACT

Microencapsulation is a useful tool to improve the delivery of bioactive compounds into food, particularly probiotics, minerals, vitamins, fatty acids, antioxidants, etc. Nowadays, aseptic microencapsulation is introduced for biodegradable material. New creations and future progress will be carried by double microencapsulation, improving strain and culture. Moreover, these technologies could promote the successful delivery of bioactive ingredients to the gastrointestinal tract.

4.1 INTRODUCTION

Today, manufacturers and others in the food industry are looking for newer and better ways to enhance the value of their products as well as differentiate themselves in the marketplace. Encapsulation is being seen by many as an effective technology for achieving these goals. Encapsulation is a process by which individual particles or droplets of solid or liquid material (the core) are surrounded or coated with a continuous film of polymeric material (the shell) to produce capsules in the micrometer to millimeter range, known as microcapsules.[45] Encapsulation involves solid, liquid, or gaseous component in a wall material, in order to form a particle which offer protection against oxygen, heat, humidity, and light. Such technologies are of significant interest to the pharmaceutical sector (e.g., for drug and vaccine delivery) and also have relevance for the food industry. The food industry continues to demonstrate a growing awareness in the use of microencapsulation. It is considered by many to be a successful delivery system which allows manufacturers to tap into new consumer/health trends. The industrial production of food frequently requires the addition of functional ingredients. Typically, these are used to control flavor, color, texture, or preservation properties, but increasingly ingredients with potential health benefits are also included. Adding bioactive ingredients to functional food presents many challenges,[12] particularly with respect to the stability of the bioactive compounds during processing and storage and the need to prevent undesirable interactions with the carrier food matrix. In this respect, encapsulation of the bioactive ingredients such as probiotics, minerals, vitamins, antioxidants, fatty acids, antioxidants, etc. could help to address some of these problems.

4.2   BENEFITS OF ENCAPSULATION

Encapsulation has a number of interesting advantages. One of the most important reasons for encapsulation of active ingredients is to provide superior stability to the final products and during processing along with less evaporation and degradation of volatile actives, such as aroma. Furthermore, encapsulation is used to mask unpleasant feelings during eating, such as bitter taste and astringency of different polyphenols. Another goal of employing encapsulation is to prevent reaction with other components in food products such as oxygen or water. In addition to the above, encapsulation may be used to immobilize cells or enzymes in food processing applications, such as fermentation process and metabolite production processes. They protect unstable sensitive materials from their environment prior to use, encapsulation increases better process ability, improving solubility, dispersibility, flowability; it increases shelf life by preventing these reactions (oxidation and dehydration).[8] The release of microparticle content at controlled rates can be triggered by shearing, solubilization, heating, pH, or enzyme action. This technology has different applications in the food, biomedical, pharmaceutical, and cosmetic industries as well as in agriculture and catalysis.[18] There is an increasing demand to find suitable solutions that provide high productivity and, at the same time, satisfy an adequate quality of the final food products.

4.3   DIFFERENT TECHNIQUES FOR ENCAPSULATION

There are various different types of techniques which are used for encapsulation, such as chemical (suspension, dispersion, emulsion, and polymerization); physicochemical (layer-by-layer assembly, sol–gel encapsulation, supercritical CO_2 extraction); physio-mechanical (spray drying, fluid-bed coating, electrostatic encapsulation). A lot of substances may be used to coat or encapsulate solids, liquids, or gases of different types and properties. However, regulations for food additives are more rigid than for pharmaceuticals. Different compounds, widely accepted for drug encapsulation, have not been approved for use in the food industry, because many of these substances have not been certified for food applications as "generally recognized as safe" materials. Actually, the whole food process should be designed in order to meet the safety requirements of governmental agencies such as the European Food Safety Authority or Food and Drug Administration in the USA.[46] Coating material must possess maximal protection of the active material against environmental conditions, to hold active material within capsules structure during processing or storage under various conditions, not to react with the encapsulated material, to have good rheological characteristics at high concentration, if it is needed and to have easy work ability during the encapsulation. Figure 4.1 represents schematic diagram of two representative types of microcapsules. Among all materials, the most widely used for encapsulation in food applications are polysaccharides. Starch and their derivate—amylose, amylopectin, dextrins, maltodextrins, polydextrose, syrups, and cellulose and their derivatives are commonly used. Plant exudates and extracts—gum arabic, gum tragacanth, gum karaya, mesquite gum, galactomannans, pectins, and soluble soybean polysaccharides are employed, too. Subsequently, marine extracts such as carrageenans and alginate are also present in foods. Microbial and animal polysaccharides like dextran, chitosan, xanthan, and gellan are also exploited. Apart from natural and modified polysaccharides, proteins and lipids are also appropriate for encapsulation. Examples of the most common milk and whey proteins are caseins, gelatine, and gluten. Among lipid materials suitable for food applications, there are fatty acids and fatty alcohols, waxes (beeswax, carnauba wax, candellia wax), glycerides, and phospholipids. In addition to the above, other materials are employed such as PVP, paraffin, shellac, inorganic materials. Coating material stabilizes core material, they are inert toward active ingredients, they control release under specific condition, and they are economical, flexible, non-hygroscopic, tasteless, stable and soluble in aqueous media or solvent.[11] But cost constraint always plays a key factor for choosing the most appropriate materials. No matter what is the material in question, the conversion of the physiochemical characteristics of the materials will be the precondition for successful food product development. So, it is prerequisite to study and analyze all properties of potential wall material in order to conclude and predict its behavior under conditions present in food formulations.[44]

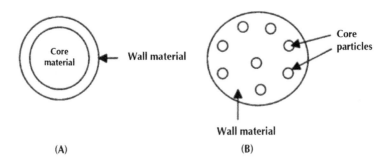

(A)   (B)

FIGURE 4.1   Schematic diagram of two representative types of microcapsules.[15]

4.4 NONMICROBIAL PRODUCTS

Spray drying has been traditionally used for the encapsulation of oil-based vitamins and fatty acids.[7,37] However, many emulsions, spray-chilling and liposome techniques show potential for the controlled release of bioactive compounds such as retinol, omega-3 fatty acids, yeasts, vitamins, and enzymes.[5,23,28,34,36] Many emulsions can yield delivery systems with novel encapsulation and delivery properties. The functional component can be encapsulated within the inner phase the oil phase or the outer water phase after drying; thus, a single delivery system can contain multiple functional components. Lesser used technologies for encapsulation of nonmicrobial bioactive ingredients include their incorporation into cyclodextrins[24] and coacervation.[6] Nanoemulsions, with droplets sizes between 100 and 500 nm, are produced by microfluidization or micelle formation techniques and are gaining popularity for both pharmaceutical and food applications.[20,42,47]

4.5 MICROBIAL PRODUCTS

Probiotic bacteria are defined as "live microorganisms which, administered in adequate amounts, confer a beneficial physiological effect on the host." Probiotics present two sets of problems when considering encapsulation: their size (typically between 1 and 5 mm diameter), which immediately excludes nanotechnologies, and the fact that they must be kept alive. This latter aspect has been crucial in selecting the appropriate encapsulation technology. Techniques and processes used for encapsulating probiotic microorganisms are presented in Table 4.1.

Microencapsulation by spray drying has been successfully used in the food industry for several decades.[22] Spray drying can be used to encapsulate active material within a protective matrix formed from a polymer or melt. Schematic diagram for spray drying encapsulation of nutraceuticals is presented in Figure 4.2. Although many techniques have been developed to microencapsulate food ingredients, spray-drying is the most common technology used in food industry due to low cost and available equipment. Aseptic microencapsulation is increasingly demanded because numerous biodegradable materials cannot be heat sterilized and sterilization by gamma rays may harm the encapsulated drug and degrade the polymer.[41] However, aseptic preparation of microspheres by conventional spray drying is difficult to achieve.

TABLE 4.1   Techniques and Processes Used for Encapsulating Probiotic Microorganisms.[2]

Microencapsulation techniques	Types of materials for coating	Major steps in processes
Spray-drying	Water-soluble polymers	Preparation of the solutions including microorganisms
		Atomization of the feed into spray
		Drying of spray (moisture evaporation)
		Separation of dried product form
Spray-congealing	Waxes, fatty acids, water-soluble and water-insoluble polymers, monomers	Preparation of the solutions containing core (e.g., probiotics)
		Solidification of coat by congealing the molten coating materials into nonsolvent
		Removal of nonsolvent materials by sorption, extraction or evaporation techniques
Fluidized-bed coating/air suspension	Water-insoluble and water-soluble polymers, lipids, waxes	Preparation of coating solutions
		Fluidization of core particles
		Coating of core particles with coating solutions
Extrusion	Water-soluble and water insoluble polymers	Preparation of coating solution materials
		Dispersion of core materials
		Cooling or passing of core-coat mixtures through dehydrating liquid
Coacervation/ phase separation technique	Water-soluble polymers	Core material is dispersed in a solution of coating polymer, the solvent for the polymer being the liquid manufacturing vehicle phase
		Deposition of the coating, accomplished by controlled, physical mixing of the coating and core materials in the vehicle phase
		Rigidifying the coating by thermal, cross-linking, or desolvation techniques, to form self-sustaining microcapsules
Electrostatic method	Oppositely charged polymers/compounds	Mixing of core and coating materials
		Extrusion of mixtures of core-coating materials in oppositely charged solutions
		Freeze-dry or oven-dry of microcapsules/ microspheres/beads

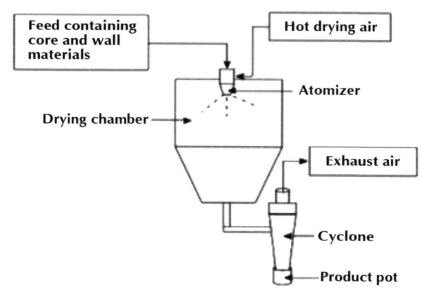

FIGURE 4.2 Schematic diagram for spray drying encapsulation of nutraceuticals.[19]

Extrusion is the simplest and most common technique used to produce probiotics capsules with hydrocolloids. The technique involves preparing a hydrocolloid solution, adding the probiotics ingredient to the solution, and dripping the cell suspension through a nozzle spray machine in the form of droplets which are allowed to fall freely into a hardening solution.[27]

Emulsion technique is based on the relationship between the discontinuous and continuous phase. Various supporting materials have been used to encapsulate probiotics by emulsion method including alginate, chitosan, and gelatin. This type of probiotics has been successfully applied to yoghurt cheddar cheese, ice-cream.[1] Figure 4.3 represent the preparation of capsules by the solvent-removal-induced encapsulation techniques. However, conventional emulsion-based processes bear some critical issues in relation to difficulty in the removal of an organic solvent, limitations in manufacturing facility, instability, and coalescence of emulsion droplets during hardening, and so on. This event led to the transformation of emulsion droplets to hardened microspheres in an efficient way. In the practice of this technique, halogenated ester organic solvents such as methyl chloroacetate and ethyl chloroacetate were chosen as dispersed solvents.[14,26]

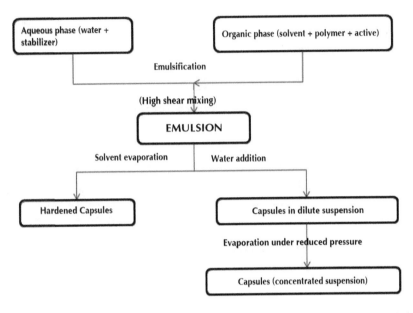

FIGURE 4.3 Preparation of capsules by the solvent removal induced encapsulation techniques.

4.6   APPLICATION OF ENCAPSULATION IN FOOD AND NUTRITION

One of the most important reasons for encapsulation of active ingredients is to provide improved stability in final products and during processing. Now-a-days, there is a trend toward a healthier way of living, which includes a rising awareness by consumers for what they eat and what benefits certain ingredients have in maintaining good health. Avoiding illness by diet is a unique offering of innovative so-called functional food, many of which are augmented with ingredients to promote health. However, simply adding ingredients to food products to improve nutritional value can compromise their taste, color, texture, and aroma. Sometimes they slowly degrade and lose their activity, or become hazardous by oxidation reactions. Ingredients can also react with components present in the food system, which may limit bioavailability and taste, odor, and color masking. The technology enables food companies to incorporate minerals, vitamins, flavors, and essential oils. Encapsulations also assist sensitive and fragile materials to survive processing and packaging conditions and stabilize the shelf life of the active ingredient.[40] Encapsulation is used to overcome all challenge by providing viable texture blending, appealing aroma releases. Disease preventing and health promoting properties of different nutrients and bioagents have been demonstrated.[3] Another benefit of encapsulation is less evaporation and degradation of volatile actives, such as aroma, which usually contains mixture of volatile and odorous organic molecules. Besides, flavors are usually expensive, and therefore food manufacturers are usually concerned about the preservation of aromatic additives.[35]

Encapsulation of probiotics for use in food application or biomass production can be achieved in several ways. The processes are spray drying, extrusion, emulsion, etc. The probiotics effect has been attributed to the production of acid, bacteriocins, competition of pathogens, and enhancement of immune system. Good probiotic viability and activity are considered essential for optimal functionality.[39] Calcium alginate has been widely used for the encapsulation of lactic acid and probiotic bacteria. Alginate capsules have some advantages. They easily form gel matrices around bacterial cells, they are not poisonous to the body (is safe or biocompatible), they are cheap, mild process conditions (such as temperature) are needed for their performance, and 0.5–4% concentration can easily be prepared and performed for experiment. Blending alginate with starch is a common practice, and it has been shown that encapsulation effectiveness of different bacterial cells especially lactic acid bacteria were improved by applying this method.[32] Microencapsulation of probiotics can be used efficiently for preparation of bacterial starter cultures with higher viability. Coating of the calcium chloride on sodium alginate capsules containing *Lactobacillus acidophilus* increased tolerance of the bacteria against harsh acidic (pH 2) and bile (1%) conditions.[13] Nowadays, by applying encapsulated starter culture bacteria, new innovations have been achieved in the manufacture of dairy probiotic products such as yogurt.

In addition to the above, encapsulation may be used to immobilize cells or enzymes in food processing applications, such as fermentation process and metabolite production processes. Immobilization of microbiological cells by entrapment within natural or synthetic polymers or by adsorption onto solid (in)organic carrier materials has become an increasing research area. Immobilization of cells provides ease of biomass separation and recovery, lower risk of microbial contamination, better use of equipment, and, as a consequence of these and other benefits, higher productivity and efficiency. Nowadays, immobilized cell technology is well established at commercial scales in secondary beer fermentation, alcohol-free, low-alcohol beer, and sparkling wine production. In order to meet the increasing demand for alcohol-free beer, several methods have been developed including alcohol removal from the product or limited fermentation of wart. In the second case, production is much better when immobilized cells are used.[30] Applications of immobilized yeast cells in wine production have been explored in a view to reduce labor requirements, to simplify time-consuming procedures, and thereby to reduce costs. To be convenient in wine production, the method has to be economical, easily performed in industrial conditions and not to cause oxidation and contamination of wine.[17]

Iron, calcium, and magnesium are the most important mineral which has been encapsulated and used in food industry. Iron deficiency produces anemia, which is one of the principal public health problems, affecting a quarter of the world's population in both industrialized and developing countries. The best way to prevent this problem is through the Fe fortification of food. However, the bioavailability of Fe is negatively influenced by interactions with food ingredients such as tannins, phytates, and polyphenols. Moreover, Fe catalyzes oxidative processes in fatty acids, vitamins, and amino acids, and consequently alters sensory characteristics and decreases the nutritional value of the food. Liposome technology is the method of choice for Fe fortification of fluid food products. An iron bioavailability study on milk enriched with $FeSO_4$ encapsulated in a lecithin liposome has been conducted.[9,43] Calcium is an essential mineral for bone growth and dental health. Growing concern about bone health in people of all ages has prompted the food industry to respond by adding calcium to foods and beverages. Magnesium plays a role as a physiological modulator affecting muscular contraction, cardiovascular function, and nerve impulse transmission. Magnesium can be encapsulated in the inner aqueous droplets of water-oil-water emulsions using triglyceride oils such as olive oil, rapeseed oil, and medium chain triacylglycerols (lipid phase) and polymeric surface-active species such as sodium caseinate (SC)/peptin (hydrophilic emulsifier) and polyglycerol polyricinoleate (PGPR) (lipophilic emulsifier) to prevent magnesium deficiency and associated clinical disorders, for example, hypertension, cardiovascular diseases, muscular weakness, and diarrhea.[10]

Vitamins can be incorporated in multiple emulsions in different ways depending on their solubility. While water-soluble (B-complex and C) vitamins have been encapsulated in water-oil-water emulsions, oil-water-oil emulsions can be more suitable in the case of lipid-soluble (A, D, E, and K) vitamins. Because of their known health implications, vitamin enrichment is a well-established strategy in the development of functional food.[16] Ascorbic acid (vitamin C), is added extensively to a variety of food products as either an antioxidant or a vitamin supplement. Its application as a vitamin supplement is impaired by its high reactivity and, hence, poor stability in solution.[29] For vitamin C encapsulation, both spray-cooling and spray chilling and fluidized-bed coating can be used when the vitamins are added to solid foods, such as cereal bars, biscuits, or bread. For application in liquid food systems, the best way to protect water-soluble ingredients is by encapsulation in liposomes. Liposomes are single or multilayered vesicles of phospholipids containing either aqueous-based or lipophilic compounds.[36] Lipid-soluble vitamins such as vitamin A, β-carotene, and vitamins D, E, or K are much easier to encapsulate than water-soluble ingredients. A commonly used procedure is spray-drying of emulsions.[31]

Among antioxidants, polyphenols are one of the most important groups. Resveratrol is a naturally occurring polyphenolic compound which has attracted much attention over recent years due to interest in its health benefits, including its antioxidant capacity, cardio-protection, anticancer activity, anti-inflammatory, and other effects. However, resveratrol has limited solubility in water, is sensitive to oxidation, and undergoes a light-induced conversion from the trans- to the cis-isomer. Due to the poor solubility of resveratrol, different inner aqueous phases containing select components were used to enhance the retention of resveratrol.[25] The primary emulsions were made with added canola oil containing PGPR and re-emulsified in water containing SC and NaCl. Less than 10% of the total encapsulated resveratrol was released to the external continuous aqueous phase, demonstrating the potential of multiple emulsions to encapsulate resveratrol for food applications. A new system for emulsion stabilization was developed based on formulations containing gum arabic, maltodextrin, and alginate as coating materials in water–oil–water emulsions containing phenolic mango seed kernel extract.[33] This extract is a good source of phenolic antioxidants with metal chelating and tyrosinase inhibiting activities. Anthocyanins belong to the most important group of hydrophilic plant pigments and have strong antioxidant, anticarcinogenic, and immune-modulating effects. Anthocyanins can be stabilized in the inner phase of double emulsions and released under gastrointestinal conditions. The release rate of free fatty acids during incubation is independent of the emulsifier used.[21] Carotenoids such as α-carotene, β-carotene, γ-carotene, lycopene, lutein, zeaxanthin, and β-cryptoxanthin, commonly found in vegetables and fruits, are being investigated as promising candidates for prevention of cancer, heart disease, and aging effects. Carotenoids have been suggested as candidate food components (in functional food development) for modulation of target functions related to defense against reactive oxidative species.[16] Incorporation of carotenoids in food has a wider range of uses, but there are some problems associated with instability (they are very sensitive to oxygen, light, and heat) and coloration of foods. Multiple emulsions can be used to overcome these limitations. Microcapsules containing both oil- and water-soluble carotenoids have been obtained by spray-drying of water–oil–water emulsions.[38]

Polyunsaturated fatty acids (PUFA) have been identified as essential to human subjects during the whole of their lifetime. Enrichment of food with fish oil has been studied intensively as an acceptable and effective means of increasing the levels of ω-3 PUFA in the general population. However, ω-3 PUFA are also prone to oxidation because of the high number of unsaturated double bonds in the fatty acyl chains. Encapsulation by emulsion spray-drying has been used successfully to increase the shelf-life of this type of ingredients and allows their use in a large variety of foods such as infant formulas and bread mixes. It was also found that the shelf-life of the fatty acids could be increased to more than 2 years by microencapsulation.[4]

4.7 CONCLUSIONS

Recent developments in food and nutrition sciences have highlighted the possibility of modulating some specific physiological functions in the organism through food intake. The beneficial effects of functional food derive from dietary active compounds, and therefore the design and development of these foods require strategies to control their presence. Encapsulation provides an effective method to cover those active compounds with a protective wall material and offers numerous advantages. These bioactive components include vitamins, fatty acids, antioxidants, minerals, and living cells. Encapsulated probiotic bacteria can be used in many fermented dairy products, such as yoghurt, cheese, cultured cream, and frozen dairy desserts, and for biomass production. Some of the main benefits are protection of various actives against evaporation, chemical reactions or migration in food, controlled delivery and preservation of stability of the bioactive compounds during processing and storage, prevention of undesirable interactions with other components in food products, etc. Materials used for design of protective shell of encapsulates must be food-grade, biodegradable, and able to form a barrier between the internal phase and its surroundings. Among all materials, polysaccharides are mostly used for encapsulation in food applications, whether the use of protein and lipid are less. Spray drying is one of the most extensively applied encapsulation techniques than extrusion and emulsion technique in the food industry because it is flexible, continuous, but more important an economical operation. Molecular inclusion in cyclodextrins and liposomal vesicles are more expensive technologies, and therefore, less exploited.

Finally, the paradigm of not wanting added bioactive compounds to alter food sensory properties or appearance might well be reconsidered. There are already commercial products available (yoghurt, breakfast cereal, kefir, candies, etc.) where particles containing encapsulated bioactive compound are clearly seen in the foods and even advertised on the label. Encapsulation is an important approach to meet all demands by delivering bioactive food components at the right time and right place. It may be forecasted that encapsulated bioactive compounds will play a significant role in increasing the efficacy of functional foods over the next period. With advanced strategies for stabilization of food ingredients and development of new approaches, we will be able to improve nutritional properties and health benefits of food compounds.

KEYWORDS

- **antioxidants**
- **emulsion technique**
- **extrusion**
- **fatty acid**
- **functional food**
- **microencapsulation**
- **probiotics**
- **spray-drying**
- **vitamins**

REFERENCES

1. Adhikari, K.; Mustapha, A.; Grun, I. U.; Fermando, L. Viability of Microencapsulated Bifidobacteria in Set Yoghurt during Refrigerated Storage. *J. Dairy Sci.* **2002,** *83*, 1946–1951.
2. Anal, A. K.; Singh, H. Recent Advances in Microencapsulation of Probiotics for Industrial Applications and Targeted Delivery. *Trends Food Sci. Technol.* **2007,** *18*, 240–251.
3. Ananta, E.; Volkert, M.; Knorr, K. Cellular Injuries and Storage Stabilities of Spray Dried *Lactobacillus rhamanous* GG. *Int. Dairy J.* **2005,** *15*, 399–409.

4. Andersen, S. Microencapsulated Omega-3 fatty Acids from Marine Sources. *Lipid Technol.* **1995,** *7*, 81–85.

5. Arnaud, J. P. Pro-liposomes for the Food Industry. *Food Technol. Eur.* **1995,** *2*, 30–34.

6. Arneado, C. J. F. Microencapsulation by Complex Coacervation at Ambient Temperature, 1996. Patent No. FR 2732240 A1.

7. Augustin, M. A.; Sanguansri, L.; Margetts, C.; Young, B. Microencapsulation of Food Ingredients. *Food Aust.* **2001,** *53*, 220–223.

8. Benita, S. *Microencapsulation: Methods and Industrial applications.* Marcel Dekker: New York, 1996.

9. Boccio, J. R.; Zubillaga, M. B.; Caro, R. A.; Gotelli, C. A.; Gotelli, M. J.; Weill, R. A New Product to Fortify Fluid Milk and Dairy Products with High-bioavailable Ferrous Sulfate. *Nut. Rev.* **1997,** *55*, 240–246.

10. Bonnet, M.; Cansell, M.; Placin, F.; David-Brian, E.; Anton, M.; Leal-Calderón, F. Influence of Ionic Complexation on Release Rate Profiles from Multiple Water in- oil-in-water (W/O/W) Emulsions. *J. Agric. Food Chem.* **2010,** *58*, 7762–7769.

11. Campos, C. A.; Gerschenson, L. N.; Flores, S. K. Development of Edible Films and Coatings with Antimicrobial Activity. *Food Bioprocess Technol.* **2011,** *4*, 849–875.

12. Champagne, C. P.; Roy, D.; Gardner, N. Challenges in the Addition of Probiotic Cultures to Foods. *Crit. Rev. Food Sci. Nutr.* **2005,** *45*, 61–84.

13. Chandramouli, V.; Kalasapathy, K.; Peiri, P.; Jones, M. An Improved Method of Microencapsulation and its Evaluation to Protect Lactobacillus spp. In Simulated Gastric Conditions. *J. Microbiol. Meth.* **2004,** *56*, 27–35.

14. Chung, Y.; Kim, J.; Sah, H. Reactivity of Ethyl Acetate and its Derivatives Toward Ammonolysis: Ramificatins for Ammonolysis-based Microencapsulation Process. *Adv. Polym. Technol.* **2009,** *20*, 785–794.

15. Desai, K. G. H.; Park, H. J. Recent Developments in Microencapulation of Food Ingredients. *Drying Technol.* **2005,** *23*, 1361–1394.

16. Diplock, A. T.; Aggett, P. J.; Ashwell, M.; Bornet, F.; Fern, E. B.; Roberfroid, M. B. Scientific Concept of Functional Foods in Europe. Consensus Document. *Br. J. Nutr.* **1999,** *81*, S1–S27.

17. Diviès, C.; Cachon, R.; Cavin, J. F.; Prévost, H. Theme 4: Immobilized Cell Technology in Wine Production. *Crit. Rev. Biotechnol.* **1994,** *14*, 135–53.

18. Dubey, R.; Shami, T. C.; Bhasker Rao, K. U. Microencapsulation Technology and Application. *Defence Sci. J.* **2004,** *59*, 82–89.

19. Fang, Z.; Bhandari, B. Spray Drying, Freeze Drying and Related Processes for Food Ingredient and Nutraceutical Encapsulation. In *Encapsulation Technologies and Delivery Systems for Food Ingredients and Nutraceuticals*; Garti, N., Julian McClements, D., Eds.; 2012; pp 71–102.

20. Flanagan, J.; Singh, H. Microemulsions: A Potential Delivery System for Bioactives in Foods. *Crit. Rev. Food Sci. Nutr.* **2006,** *46*, 221–237.

21. Frank, K.; Walz, E.; Gräf, V.; Greiner, R.; Köhler, K.; Schuchmann, H. P. Stability of Anthocyanin-rich W/O/W Emulsions Designed for Intestinal Release in Gastrointestinal Environment. *J. Food Sci.* **2012,** *77*, N50–N57.

22. Gouin, S. Micro-encapsulation: Industrial Appraisal of Existing Technologies and Trends. *Trends Food Sci. Technol.* **2004,** *15*, 330–347.

23. Guimberteau, F.; Dagleish, D.; Bibette. J. M. B. Emulsion multiple alimentaire compose´ e d'une emulsion primaire inverse disperse´ e auseind'unephaseaqueuse. *Patent FR* 828,378-A1, 2001.

24. Hedges, A. R.; Shieh, W. J.; Sikorski, C. T. Use of Cyclodextrins for Encapsulation in the Use and Treatment of Foods. *ACS Symp. Ser.* **1995**, *59*, 60–71.
25. Hemar, Y.; Cheng, L. J.; Oliver, C. M.; Sanguansri, L.; Agustin, M. Encapsulation of Resveratrol Using Water-in-oil-in-water Double Emulsions. *Food Biophys.* **2010**, *5*, 120–127.
26. Kim, J.; Hong, D.; Chung, Y.; Sah, H. Ammonolysis-induced Solvent Removal: Facile Approach for Solidifying Emulsion Droplets into PLGA Microspheres. *Biomacromolecules* **2007**, *8*, 3900–3907.
27. King, A. H. *Encapsulation of Food Ingredients: A Review of Available Technology, Focusing on Hydrocolloids in Encapsulation and Control Release of Food Ingredient*; American Chemical Society: USA, 1995; pp 213–220.
28. Kirby, C. J.; Whittle, C.; Rigby, N.; Coxon, D. T. Law BA: Stabilization of Ascorbic Acid by Microencapsulation. *Int. J. Food. Sci. Technol.* **1991**, *26*, 437–444.
29. Kirby, C. J.; Whittle, C.; Rigby, N.; Coxon, D. T.; Law, B. A. Stabilization of Ascorbic Acid by Microencapsulation in Liposomes. *Int. J. Food. Sci. Technol.* **1991**, *26*, 445–449.
30. Kosseva, M. R.; Panesar, P. S.; Kaur, G.; Kennedy, J. F. Use of Immobilised Biocatalysts in the Processing of Cheese Wheys. *Int. J. Biol. Macromol.* **2009**, *45*, 437–47.
31. Kowalski, R. E.; Mergens, W. J.; Scialpi, L. J. *Process for Manufacture of Carotenoid Compositions*, US Patent 6093348, 2000.
32. Krasaekoopt, W.; Bhandari, B.; Deeth, H. Evaluation of Encapsulation Techniques of Probiotics for Yoghurt. *Int. Dairy J.* **2003**, *13*, 3–13.
33. Maisuthisakul, P.; Gordon, M. H. Influence of Polysaccharides and Storage During Processing on the Properties of Mango Seed Kernel Extract (Microencapsulation). *Food Chem.* **2012**, *134*, 1453–1460.
34. McClements, D. J. Emulsion Stability. In Food Emulsions. Principles, Practices and Techniques, 2nd ed.; CRC Press: Washington DC; 2005; Chapter 7, pp 185–233.
35. Milanovic, J.; Manojlovic, V.; Levic, S.; Rajic, N.; Nedovic, V.; Bugarski, B. Microencapsulation of Flavors in Carnauba Wax. *Sensors* **2010**, *10*, 901–912.
36. Reineccius, G. A. Liposomes for Controlled Release in the Food Industries. In Encapsulation and Controlled Release of Food Ingredients. *ACS Symp. Ser.* **1995**, *590*, 113–131.
37. Re, M. I. Microencapsulation by Spray Drying. *Drying Technol.* **1998**, *16*, 1195–1236.
38. Rodríguez-Huezo, M. E.; Pedroza-Islas, R.; Prado-Barragán, L. A.; Beristain, C. I.; Vernon-Carter, E. J. Microencapsulation by Spray Drying of Multiple Emulsions Containing Carotenoids. *J. Food Sci.* **2004**, *69*, E351–E359.
39. Sandholm, M.; Myllarinen, T. P.; Crittenden, R.; Mogensen, G.; Fonden, R.; Saarela, M. Technological Challenges for Future Probiotic Food. *Int. Dairy J.* **2005**, *12*, 173–182.
40. Schrooyen, P.; De Ruiter, G.; De Kruif, C. Spray-dried Orange Oil Emulsions: Influence of Cold Water Dispersible Matrices on Retention and Shelf Life. In Proceedings of the International Symposium on the Controlled Release of Bioactive Materials; Controlled Release Society: Deerfield, 2000; pp 1317–1318.
41. Sergio, F.; Hans Peter, M.; Bruno, G. Ultrasonic Atomisation into Reduced Pressure Atmosphere-envisaging Aseptic Spray Drying for Microencapsulation. *J. Controlled Release* **2004**, *95*, 185–195.
42. Tan, C. P.; Nakajima, M. β-Carotene Nano-dispersions: Preparation, Characterization and Stability Evaluation. *Food Chem.* **2005**, *92*, 661–671.

43. Uicich, R.; Pizarro, F.; Almeida, C.; Diaz, M.; Bocchio, J.; Zubillaga, M.; Carmuega, E.; O'Donnell, A. Bioavailability of Microencapsulated Ferrous Sulfate in Fluid Cow's Milk. Studies in Human Beings. *Nut. Rev.* **1999,** *19,* 893–897.

44. Versic, R. J. The Economics of Microencapsulation in the Food Industry. In *Microencapsulation in the Food Industry*; Gaonkar, A. G., Vasisht, N., Khare, A. R., Sobel, R., Eds.; Ronald T. Dodge Company: Dayton, OH, 2014; pp 409–417.

45. Vidhyalakshmi, R.; Bhakyaraj, R.; Subhasree, R. S. Encapsulation "The Future of Probiotics"—A Review. *Adv. Biol. Res.* **2009,** *3*(3–4), 96–103.

46. Wandrey, C.; Bartkowiak, A.; Harding, S. E. Materials for Encapsulation. In *Encapsulation Technologies for Food Active Ingredients and Food Processing*; Zuidam, N. J., Nedovic, V. A., Eds.; Springer: Dordrecht, The Netherlands, 2009; pp 31–100.

47. Weiss, J.; Takhistov, P.; McClements, J. Functional Materials in Food Nanotechnology. *J. Food Sci.* **2006,** *71,* 107–116.

CHAPTER 5

FOOD PACKAGING TECHNOLOGIES

JOHN BROCKGREITENS[1] and ABDENNOUR ABBAS[2*]

[1]*Biosensors and Bionanotechnology Laboratory, Department of Bioproducts and Biosystems Engineering, University of Minnesota, Twin Cities, 2004 Folwell Hall, Falcon Heights, MN 55108, USA. Email: brock240@umn.edu*

[2]*Biosensors and Bionanotechnology Laboratory, Department of Bioproducts and Biosystems Engineering, University of Minnesota, Twin Cities, 2004 Folwell Ave., St. Paul, MN 55108, USA. E-mail: aabbas@umn.edu*

*Corresponding author.

CONTENTS

ABSTRACT

We began this chapter discussing food packaging and how it interacts both with food products and the outside environment. Then, we explored how these interactions could be used in functional "smart" packaging technologies. Smart packaging can be further defined as "active," "responsive," or both. We then concluded this chapter with a discussion of rising food packaging technologies followed by an overview of the challenges and obstacles to smart packaging technologies.

5.1 INTRODUCTION

Principally, food packaging is designed to contain food products and maintain food quality. The functions of packaging have been consolidated into four categories: containment, protection, communication, and convenience.[33] Containment is simply providing a vessel to store and transport food. Protection takes containment further in that it provides a sealed space that attempts to minimize food exposure to the environment. With the advent of printing, food packaging emerged as a medium to communicate product information (nutritional facts, ingredient lists, etc.) and marketing. Finally, convenience is concerned with the ease of using the packaging (portability, sealing, etc.). In this chapter, we will discuss the fundamentals of food packaging and introduce the concept of "smart" packaging systems that provide additional functionality to traditional packaging.

5.2 MATERIALS SCIENCE AND FOOD PACKAGING

In order to determine the properties of a food package one must first consider the materials that make up the package. Materials science combines physics, chemistry, engineering, and mathematics to understand and control material properties such as strength, composition, and conductivity.[3] The materials most commonly found in food packaging include paper/paperboard, polyethylene terephthalate plastic (PET), glass, tin, and other metals.[27] These materials are ideal for packaging in that they are cheap, widely available, and have favorable properties for packaging. As stated previously, the primary purpose of food packaging is to provide a barrier between food and threats from the outside environment. Figure 5.1 (adapted from Refs. [24,33]) shows an overview of the types of interactions that occur in packaging systems. As seen in Figure 5.1, a great deal of interaction can occur between food packaging and the environment. Water vapor, light, and external aromas can diffuse into food through the packaging material. This transfer can negatively impact food as it leads to dehydration, texture changes, and flavor/aroma loss. Additionally, gasses produced by microbes present in the food product (respiration) can diffuse out of the package. Finally, interactions with the packaging material itself are possible. In extreme cases, materials present in the package material can leach out into food products. A popular example is the plasticizer Bisphenol A that was used in plastic beverage containers and has been linked to endocrine disruption and birth defects.[34] Both of these interactions can be modeled mathematically. First, the behavior of gasses can be modeled using the **Ideal Gas Law**.

$$pV = nR \tag{5.1}$$

where p is the pressure in atm, V is the volume of the gas in L, n is the number of moles of the gas, R is the gas constant = 0.082057 L atm/mol K, and T is the temperature of the gas in Kelvin.

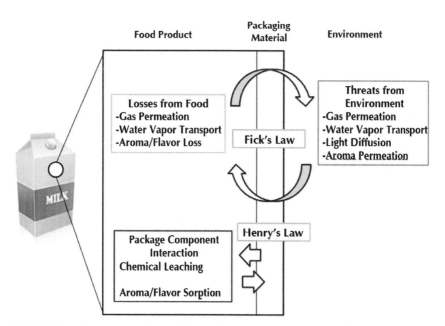

FIGURE 5.1   Overview of chemical interactions that occur through packaging material.

This law outlines external forces that drive the kinetic activity of theo-retical ideal gasses that only interact with each other elastically. Gasses that are involved with packaging, however, are often not under ideal condi-tions and have to contend with the material that compromises the package. The kinetic interaction between two or more different molecules is known as **diffusion**. Material properties like permeability and chemical composi-tion play important roles in food packaging. More permeable materials will allow greater transfer of gasses, food aromas, and water between the food and the environment. Chemical composition of packaging material is also a very important consideration as some materials and chemicals can impact the taste or texture of a food product. Mass transfer between the package and outside environment can be modeled using **Fick's Law of Diffusion**:

$$J = -D\frac{\Delta c}{\Delta x} \tag{5.2}$$

where J is the transport rate (flux) per unit area of permeant, c is the concen-tration of permeant, D is the diffusion coefficient, and $\Delta c/\Delta x$ is the concen-tration gradient of the permeant over a certain thickness (Δx).

Maximizing the properties of materials to limit light, vapor, odor, and water transfer into packaging has been the traditional focus of food package research. As materials technology has developed, food packaging has become more adept at accomplishing these primary functions and additional functionalities have been added to food packaging. This wave of advanced "smart" packaging has become popular both on the market and in research.

5.3 SMART PACKAGING

As the food packaging industry continues to evolve, new functionalities are being added. **Smart packaging** is defined as any packaging system that provides a specific service beyond the role of physical barrier between the food product and the outside environment.[4,6] This term is very broad and is often broken down into different categories as shown in Figure 5.1. In the greater realm of smart packaging, the term **intelligent packaging** is used to describe any packaging system that conveys information to the consumer about the enclosed product.[37,40] Within the industry, the terms smart and intelligent packaging are often used interchangeably, and there is much debate about their definitions.[37] To alleviate confusion surrounding these terms, the term responsive has been borrowed from the materials science field. **Responsive packaging** systems incorporate "stimuli-responsive" materials that react with heat, light, or a variety of chemical and biological triggers. Once a stimulus is applied, responsive materials (polymers, particles, surfaces, gels, etc.) change in shape, size, color, chemical functionality, or chemical structure.[1,5,36] Researchers take advantage of these predictable, tunable changes to make sensors and chemical release systems. Responsive systems can be either informative, corrective, or both. Informative responsive packaging incorporates a material that gives off a detectable signal in response to a specific **analyte** (or target). This term is most akin to the traditional term intelligent packaging (packaging that provides detailed information on the product), but lessens the confusion between the synonyms "smart" and "intelligent."

The other form of smart food packaging is **active packaging**. Active packaging systems directly interact with enclosed food products and alter food quality properties.[8,23] This definition is widely accepted in the industry and research in active packaging has grown substantially over the past two decades. Unlike responsive packaging, active systems are not triggered by specific target molecules. The third type of smart packaging system is **design-based smart packaging**. These packaging systems provide higher functionality based on their design. Examples include resealable packages or "easy-open" lids. We will not focus on design-based smart packaging but rather on advanced materials and their uses in active and responsive packaging systems.

Active and responsive packaging technologies are very important, advancing fields given their impacts on food safety. Putting sensors directly into food packaging supplies information directly to the consumer and prevents consumption of compromised food which in turn lowers the occurrence of foodborne illness. This consideration is important because food processing companies often have no way to ensure the quality of the products they sell once the food leaves the factory. Food packaging is designed first and foremost with food safety in mind. The United States Center for Disease Control and Prevention estimated that 1 in 6 Americans (approximately 48 million people) get sick and nearly 3000 people die each year from foodborne illnesses.[17] These illnesses are caused by a variety of biological pathogens (*Salmonella*, *Listeria*, Norovirus, etc.) and the most commonly reported illness is acute gastroenteritis. Active and responsive packaging systems have emerged to combat the growth and spread of these pathogens and other nonpathogenic spoilage organisms.

5.3.1 ACTIVE PACKAGING

Active packaging systems are used to improve food quality by incorporating materials or chemicals that directly interact with food products. This type of packaging has gained a lot of research attention due to recent advances in nanotechnology and materials science. Common techniques employed in active packaging include modified atmosphere conditions, antimicrobial films, and gas scavenging compounds.

5.3.1.1 MODIFIED ATMOSPHERE PACKAGING

Modified atmosphere packaging (MAP) introduces a controlled amount of gasses into the sealed environment of a packaging. This type of packaging is commonly used for sealed meat products and gasses like carbon dioxide and nitrogen are used. These gasses create an inhospitable environment for microbes like bacteria and fungi as they alter cell membranes, inhibit enzymatic activity, and block microbial nutrient transfer.[35] This type of active packaging is the most prevalent due to the fact that it does not require any advanced methods or technologies.

5.3.1.2 GAS SCAVENGERS

Gas scavengers remove gasses that degrade food quality as they build up in sealed packaging environments (Fig. 5.2). Oxygen, for example, causes browning of meats and facilitates the growth of aerobic bacteria. In production, ethylene acts as a ripening agent. In order to mitigate the impacts of these gasses and others, gas scavengers have been incorporated into packaging systems.[38] Common oxygen absorbing materials in food packaging include titanium dioxide, iron composites, and natural antioxidants.[15] One common example is oxygen-absorbing packets found in packaged cured meat products.

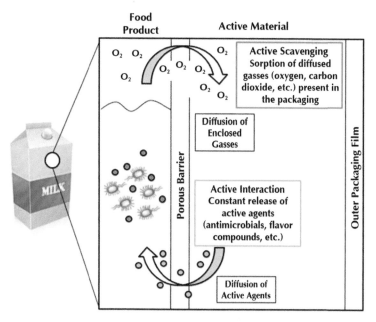

FIGURE 5.2   Overview of active food packaging interactions.

5.3.1.3   ANTIMICROBIAL PACKAGING

In order to combat the growth of both bacteria and fungi in food products, antimicrobial compounds can be embedded into packaging films or dispersed into food using a sachet. Compounds that directly interact with biological molecules (bacteria, fungi, etc.) are commonly referred to as **bioactive**.[7] Two major obstacles to this type of active packaging are the economic viability of antimicrobial systems and safety concerns related to the use of antimicrobials.[20]

5.3.2   RESPONSIVE PACKAGING

Responsive packaging systems incorporate triggering mechanisms. These triggers, based on stimuli responsive polymers, can be engineered to respond to specific stimuli like foodborne contaminants, spoilage microbes, pH, and temperature. Similar to active packaging, responsive systems that interact with biological molecules are known as **bioresponsive**.[16] This distinction is important for food packaging due to the fact that the majority of foodborne threats are either fungal or bacterial. Figure 5.3 shows a schematic of a responsive packaging system. Triggering agents diffuse and come into contact with the responsive material which in turn either gives off a signal (Informative Agent) and/or releases compounds to combat the threat (corrective agent).

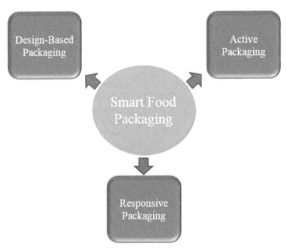

FIGURE 5.3   Hierarchy of smart packaging terminology.

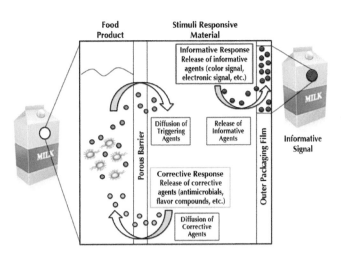

FIGURE 5.4   Schematic of responsive food packaging: Foodborne or external threats diffuse and interact with the responsive material yielding a detectable signal or release of corrective compounds.

5.3.2.1 STIMULI-RESPONSIVE MATERIALS

Stimuli-responsive materials change physical or chemical properties in response to external stimuli including temperature, pH, light, biological molecules, and more (Fig. 5.4). These materials can exist in the form of nanoparticles, gels, and layered films.[5] Common triggered reactions include assembly, collapse, shrinking, swelling, and change in chemical conformation. Within the last decade, research in these materials has boomed due their applicability in the biomedical industry.[1] Some of these materials have been utilized in food packaging and provided a foundation for a variety of responsive packaging systems.

5.3.2.2 INFORMATIVE RESPONSIVE PACKAGING

The major advantage that responsive packaging systems have over active systems is their ability to detect specific foodborne or external threats. This detection is accomplished through specific chemical modifications of stimuli-responsive materials. Signals can be color changes visible to the naked eye, changes in electronic properties measured by scanners, and a variety of other signals. Examples of informative responsive packaging systems include time–temperature indicators (TTIs) and radiofrequency identification (RFID) tags.[9,21,22]

TTIs have become common on packaged meat and dairy products. These color changing sensors indicate if a food product has been exposed to high temperatures (temperatures favorable for microbial growth) during transport. The indicator also registers the amount of time that the product was exposed to the high temperature.

RFID tags have emerged as a popular option for informative responsive packaging systems. RFID tags are printed electronic stickers that have come to replace barcodes on my products due to the fact that they can store more information and information can even be added to them. Like barcodes, they are read with simple scanners can provide information concerning shipping, handling, product origin, and more. RFIDs have yet to be incorporated into food packaging at a large scale, but recent developments have made these technologies cheaper and practical for food packaging applications. In addition to providing unique, modifiable information for a specific packaged product, RFIDs have recently been functionalized with food analyte sensors. By combining RFID tags with sensors, these labels can not only provide up to date information on the food package through the supply chain, but also can report on the quality of the food within the package.[9,13] Use of RFIDs in food packaging is expected to increase drastically as the costs associated with the tags decrease.

5.3.2.3 CORRECTIVE RESPONSIVE PACKAGING

Corrective responsive packaging systems utilize physical changes in stimuli-responsive materials. These systems act as compound laden "sponges" that shrink or swell when triggered and release their payloads into the food. This type of responsive technology is mainly used in the biomedical field for drug delivery systems. However, recent research has utilized these technologies for food packaging systems.[14]

5.4 RECENT TECHNOLOGICAL ADVANCES IN FOOD PACKAGING

Other significant advances in materials science, specifically nanotechnology, have had major impacts on the food packaging industry. These advances are far reaching and can potentially have major impacts in smart packaging in particular.

5.4.1 NANOCOMPOSITES

Nanocomposites are made up of base materials (metals, polymers, etc.) with nanomaterials dispersed within them. Nanomaterials are defined as materials that have at least one dimension in the nanoscale (10^{-9} m) and exist in many forms including films, particles, and fibers.[11] Nanomaterials can be added to current food packaging systems to improve structural and containment properties (improved protection from vapor and light diffusion). Food packaging nanotechnology can also act as an active agent. Silver nanoparticles, for example, are effective antimicrobial agents and use of these particles in food packaging has been explored.[2,10]

5.4.2 BIO-BASED MATERIALS

Other popular advances in food packaging are focused on the source of the packaging material. Recent global concerns with sustainability, persistent plastic pollution, and reliance on nonrenewable materials have fueled research and development for renewable packaging materials that can mimic current packaging materials.[39] To answer this challenge, packaging systems are being designed from bio-based, biodegradable, and/or compostable materials. **Bio-based** materials are simply derived from renewable, biological sources. These materials include bio-based plastics, edible films based on food materials, and paper products.[12,28,29] Smart food packaging technologies are no different and a great deal of research has been dedicated to functional, renewable packaging materials. Two technologies are at the forefront of this new wave of research: smart edible materials and bio-nanocomposites.

Smart edible materials are derived from food materials like proteins, lipids, and oils and possess higher functionalities like antimicrobial properties. Common materials include casein or whey protein films dosed with edible antimicrobials like cinnamon oil.[18,19,32] Production of these materials is currently limited due to their poor barrier properties and stability.

5.4.3   BIONANOCOMPOSITES

These are nanocomposites where the base material is derived from a renewable, biological source.[31] Common base materials include bioplastics like polylactic acid and thermoplastic starch.[30] These materials can then be combined with bio-based nanomaterials like chitosan, nanocrystalline cellulose, and carbon nanomaterials to produce biologically renewable composites. Bionanocomposite food packaging is still in its infancy as more research is needed to fully optimize the material properties, production methods, and disposal methods of these materials.

5.5   THE MARKET FOR SMART FOOD PACKAGING

The market for food packaging is expected to grow to over $300 billion by 2019.[25] Smart packaging is a small fraction of the this total with an estimated value of $23 billion.[26] This market is expected to grow to a value of $39.7 billion by 2020 as new materials and technologies become more industrially feasible. Currently, the smart packaging market is dominated by modified atmosphere packaging and TTIs. Another major contributing factor is that smart packaging systems are also used for medical products like vaccines and medications that must be monitored and controlled much like food products. This market will also see an increase as more major companies produce smart packaging products. Companies that currently produce smart packaging products include 3 M (US), BASF SE (Germany), and International Paper (US).

5.6 CHALLENGES AND PERSPECTIVES TO SMART FOOD PACKAGING

The adoption of smart food packaging technologies faces many challenges namely production, safety and health, and efficiency.

5.6.1 PRODUCTION

Food packaging systems are based on many simple, cheap materials like paper and polyethylene plastic. If smart food packaging is to be fully adopted by the industry, new technologies must be economically feasible. In order to achieve this feasibility, the packaging must be produced with minimal changes to existing production methods. Additionally, the source materials for these smart packages (antimicrobials, responsive materials, etc.) must be produced in a cost-effective manner.

5.6.2 SAFETY

Incorporating responsive materials into food packaging systems presents new threats from cross contamination and other risks. Great care must be taken to ensure that sensors are developed using food safe materials, or that the materials do not come into direct contact with the food. Functional materials within food packaging could have undesired impacts on food quality such as loss of the flavor or texture and even toxicity. Great care must be taken when putting new materials into contact with food and all smart food packages must be approved by the *United States Food and Drug Administration's Packaging and Food Contact Substances* (FCS) program. Safety is of paramount importance as novel materials like nanoparticles, which have not been fully studied for long term impacts, are incorporated into food packaging. With proper regulation, research, and monitoring safety concerns with smart food packaging can be managed and avoided.

5.6.3 EFFICIENCY

Smart food packaging must be efficient in terms of its desired goal. Active packaging and corrective responsive packaging must be able to prevent/attack food threats quickly and effectively. Triggers and responses must also be toward specific organisms or threats. For example, food safety sensors in milk production must only respond to pathogenic bacteria, not the beneficial bacteria. Also, sensors must be reliable. False positives could prove disastrous to food suppliers as food is unnecessarily recalled. False positives also diminish consumer trust in the sensing product. For active packaging, the response must be rapid and fully effective for an appreciable time frame.

5.7 CONCLUSIONS

Food packaging is poised to advance well beyond its primary functionality as new technologies and discoveries develop. New packaging functionalities are expected to reach the market over the next years, including food quality enhancement, quality indicators, and indicators of food spoilage or contamination. This development will be driven by the need to optimize product shelf life and quality, and by providing more efficient and timely control of food safety along the food supply chain. This wave of smart packaging will then impact many other industries as food-borne illness is prevented and food remains on shelves for longer periods of time.

KEYWORDS

- active packaging
- bioresponsive
- biosensors
- food packaging
- food safety
- foodborne illness
- nanocomposites
- nanotechnology
- responsive packaging
- smart packaging

REFERENCES

1. Bajpai, A. K.; Bajpai, J.; Saini, R.; Gupta, R. Responsive Polymers in Biology and Technology. *Polym. Rev.* **2011,** *51*, 53–97.
2. Becaro, A. A.; Puti, F. C.; Correa, D. S.; Paris, E. C.; Marconcini, J. M.; Ferreira, M. D. Polyethylene Films Containing Silver Nanoparticles for Applications in Food Packaging: Characterization of Physico-chemical and Anti-microbial Properties. *J. Nanosci. Nanotechnol.* **2015,** *15*, 2148–2156.
3. Bhandari, B. *Food Materials Science and Engineering.* Wiley, 2012.
4. Biji, K. B.; Ravishankar, C. N.; Mohan, C. O.; Gopal, T. K. S. Smart Packaging Systems for Food Applications: A Review. *J. Food Sci. Technol.* **2015,** 1–11.
5. Blum, A. P.; Kammeyer, J. K.; Rush, A. M.; Callmann, C. E.; Hahn, M. E.; Gianneschi, N. C. Stimuli-responsive Nanomaterials for Biomedical Applications. *J. Am. Chem. Soc.* **2014,** *137*, 2140–2154.
6. Brunazzi, G.; Parisi, S.; Pereno, A. The Instrumental Role of Food Packaging. In *The Importance of Packaging Design for the Chemistry of Food Products*; Springer, 2014; pp 57–86.
7. Cao, W.; Hench, L. L. Bioactive Materials. *Ceram. Int.* **1996,** *22*, 493–507.
8. Coma, V. Bioactive Packaging Technologies for Extended Shelf Life of Meat-based Products. *Meat Sci.* **2008,** *78*, 90–103.
9. Costa, C.; Antonucci, F.; Pallottino, F.; Aguzzi, J.; Sarriá, D.; Menesatti, P. A Review on Agri-food Supply Chain Traceability by Means of RFID Technology. *Food Bioprocess Technol.* **2013,** *6*, 353–366.
10. Dallas, P.; Sharma, V. K.; Zboril, R. Silver Polymeric Nanocomposites as Advanced Antimicrobial Agents: Classification, Synthetic Paths, Applications, and Perspectives. *Adv. Colloid Interface Sci.* **2011,** *166*, 119–135.
11. Duncan, T. V. Applications of Nanotechnology in Food Packaging and Food Safety: Barrier Materials, Antimicrobials and Sensors. *J. Colloid Interface Sci.* **2011,** *363*, 1–24.
12. Feichtinger, M.; Zitz, U.; Fric, H.; Kneifel, W.; Domig, K. J. An Improved Method for Microbiological Testing of Paper-based Laminates Used in Food Packaging. *Food Control* **2015,** *50*, 548–553.
13. Fiddes, L. K.; Chang, J.; Yan, N. Electrochemical Detection of Biogenic Amines During Food Spoilage Using an Integrated Sensing RFID Tag. *Sens. Actuators, B: Chem.* **2014,** *202*, 1298–1304.
14. Fuciños, C.; Fuciños, P.; Pastrana, L.; Rúa, M. Functional Characterization of Poly(*N*-isopropylacrylamide) Nanohydrogels for the Controlled Release of Food Preservatives. *Food Biopr. Technol.* **2014,** *7*, 3429–3441.
15. Gómez-Estaca, J.; López-de-Dicastillo, C.; Hernández-Muñoz, P.; Catalá, R.; Gavara, R. Advances in Antioxidant Active Food Packaging. *Trends Food Sci. Technol.* **2014,** *35*, 42–51.
16. Hendrickson, G. R.; Lyon, L. A. Bioresponsive Hydrogels for Sensing Applications. *Soft Matter* **2009,** *5*, 29–35.
17. http://www.cdc.gov/foodborneburden/estimates-overview.html, United States Centers for Disease Control and Prevention, Web, 2014.
18. Jo, H. J.; Park, K. M.; Na, J. H.; Min, S. C.; Park, K. H.; Chang, P. S.; Han, J. Development of Anti-insect Food Packaging Film Containing a Polyvinyl Alcohol

and Cinnamon Oil Emulsion at a Pilot Plant Scale. *J. Stored Prod. Res.* **2015**, *61*, 114–118.

19. Kechichian, V.; Ditchfield, C.; Veiga-Santos, P.; Tadini, C. C. Natural Antimicrobial Ingredients Incorporated in Biodegradable Films Based on Cassava Starch. *LWT—Food Sci. Technol.* **2010**, *43*, 1088–1094.

20. Kerry, J. P.; O'Grady, M. N.; Hogan, S. A. Past, Current and Potential Utilisation of Active and Intelligent Packaging Systems for Meat and Muscle-based Products: A Review. *Meat Sci.* **2006**, *74*, 113–130.

21. Kim, K.; Kim, E.; Lee, S. J. New Enzymatic Time–Temperature Integrator (TTI) that Uses Laccase. *J. Food Eng.* **2012**, *113*, 118–123.

22. Kim, M. J.; Jung, S. W.; Park, H. R.; Lee, S. J. Selection of an Optimum pH-indicator for Developing Lactic Acid Bacteria-based Time–Temperature Integrators (TTI). *J. Food Eng.* **2012**, *113*, 471–478.

23. Kruijf, N. D.; Beest, M. V.; Rijk, R.; Sipiläinen-Malm, T.; Losada, P. P.; Meulenaer, B. D. Active and Intelligent Packaging: Applications and Regulatory Aspects. *Food Additives Contam.* **2002**, *19*, 144–162.

24. Linssen, J. P. H.; Roozen, J. P. Food Flavour and Packaging Interactions. In *Food Packaging and Preservation*; Mathlouthi, M., Ed.; Springer US, 1994; pp 48–61.

25. Markets and Markets, *marketsandmarkets.com*, 2014.

26. Markets and Markets, *marketsandmarkets.com*, 2015.

27. Marsh, K.; Bugusu, B. Food Packaging—Roles, Materials, and Environmental Issues. *J. Food Sci.* **2007**, *72*, R39–R55.

28. Peelman, N.; Ragaert, P.; De Meulenaer, B.; Adons, D.; Peeters, R.; Cardon, L.; Van Impe, F.; Devlieghere, F. Application of Bioplastics for Food Packaging. *Trends Food Sci. Technol.* **2013**, *32*, 128–141.

29. Ramos, Ó. L.; Fernandes, J. C.; Silva, S. I.; Pintado, M. E.; Malcata, F. X. Edible Films and Coatings from Whey Proteins: A Review on Formulation, and on Mechanical and Bioactive Properties. *Crit. Rev. Food Sci. Nutr.* **2012**, *52*, 533–552.

30. Reddy, M. M.; Vivekanandhan, S.; Misra, M.; Bhatia, S. K.; Mohanty, A. K. Biobased Plastics and Bionanocomposites: Current Status and Future Opportunities. *Progr. Polym. Sci.* **2013**, *38*, 1653–1689.

31. Rhim, J. W.; Park, H. M.; Ha, C. S. Bio-nanocomposites for Food Packaging Applications. *Progr. Polym. Sci.* **2013**, *38*, 1629–1652.

32. Seydim, A. C.; Sarikus, G. Antimicrobial Activity of Whey Protein Based Edible Films Incorporated with Oregano, Rosemary and Garlic Essential Oils. *Food Res. Int.* **2006**, *39*, 639–644.

33. Singh, R. P.; Heldman, D. R. *Introduction to Food Engineering*. Gulf Professional Publishing, 2001.

34. Staples, C. A.; Dome, P. B.; Klecka, G. M.; Oblock, S. T.; Harris, L. R. A Review of the Environmental Fate, Effects, and Exposures of Bisphenol A. *Chemosphere* **1998**, *36*, 2149–2173.

35. Thompson, A. K. Modified Atmosphere Packaging. *Controlled Atmosphere Storage of Fruits and Vegetables*; 2010; pp 81–115.

36. Ulijn, R. V.; Bibi, N.; Jayawarna, V.; Thornton, P. D.; Todd, S. J.; Mart, R. J.; Smith, A. M.; Gough, J. E. Bioresponsive Hydrogels. *Mater. Today* **2007**, *10*, 40–48.

37. Vanderroost, M.; Ragaert, P.; Devlieghere, F.; De Meulenaer, B. Intelligent Food Packaging: The Next Generation. *Trends Food Sci. Technol.* **2014**, *39*, 47–62.

38. Vermeiren, L.; Heirlings, L.; Devlieghere, F.; Debevere, J. Oxygen, Ethylene and Other Scavengers. *Novel Food Packag. Technol.* **2003,** 22–49.

39. Weber, C. J.; Haugaard, V.; Festersen, R.; Bertelsen, G. Production and Applications of Biobased Packaging Materials for the Food Industry. *Food Additives & Contaminants* **2002,** *19,* 172–177.

40. Yam, K. L.; Takhistov, P. T.; Miltz, J. Intelligent Packaging: Concepts and Applications. *J. Food Sci.* **2005,** *70,* R1–R10.

PART III
Role of Food Engineering in Human Health

CHAPTER 6

APPLICATION OF PROBIOTIC AND PREBIOTIC FOR HUMAN HEALTH

MURLIDHAR MEGHWAL[1*] and HARITA R. DESAI[2]

[1]Food Science and Technology Division, Center for Emerging Technologies, Jain University, Jain Global Campus, Jakkasandra 562112, Kanakapura Main Road, Ramanagara District, Karnataka, India. E-mail: murli.murthi@gmail.com

[2]Department of Pharmaceutical Sciences and Technology, Institute of Chemical Technology, Nathalal Parekh Marg, Matunga, 400 019 Mumbai, Maharashtra, India. E-mail: hdesai27@gmail.com

*Corresponding author.

CONTENTS

ABSTRACT

Probiotics are live microbial strains whose human consumption confers health benefits. They have been studied to show versatile benefits on human health. They aid in stimulation of human immunity, decreases infectious, and antibiotic-mediated diarrheal incidences, decreases serum cholesterol, alleviate lactose intolerance, and repress life-staking tumors and cancers. Supplementation of *Lactobacilli* and *Bifidobacteria* in lactose-rich milk products can aid lactose intolerant individuals to tolerate lactose. Diarrhea has been found to elicit serious long-term effects affecting growth, nutrition, and cognition sometimes even causing mortality. Administration of a standard infant formula supplementation containing *B. bifidum* and *Streptococcus thermophilus* to human infants aged 5–24 months and suffering from chronic rotavirus associated diarrhea were found to show a lowering in the number of children suffering from diarrhea when compared to unsupplemented control groups. A supplementation of probiotics like *Lactobacillus reuteri*, *Lactobacillus acidophilus*, *Saccharomyces boulardii*, etc. have been found to show a lowering in the duration of diarrhea by 1–5 days and decreased shedding of rotavirus in rotaviral gastroenteritis. Probiotics have been found to show cholesterol lowering effects by versatile mechanisms like cholesterol assimilation enhanced cellular wall cholesterol binding, enhanced cholesterol incorporation in cell membrane, etc. A probiotic supplementation of *L. acidophilus* have been studied to show inhibitory effects on development of precancerous lesions and tumors owing to the ability of the probiotics to prevent growth of bacteria that produce enzymes like nitroreductase, azoreductase, etc. which catalyze conversion of procarcinogens to carcinogens. Prebiotics are selectively fermented components that cause specific alterations in composition and activity of gastrointestinal microflora, thus conferring benefits to host well-being and health. A supplementation of prebiotics mainly carbohydrates in nature like fructans, lactulose, etc. cause enhancement in beneficial microflora inherently present in human body as well as externally supplemented probiotic strains. Thus, a supplementation of probiotics and prebiotics has been found to show versatile benefits for human health not only in terms of enhancement of general well-being, but also in alleviation of diverse diseased conditions.

6.1 INTRODUCTION

Probiotics are nutritional supplements. Probiotics contains one or more cultures of living organisms and these microorganisms are typically bacteria or yeast. These probiotics are able to modify the endogenous microflora and they have a positive effect on the host. In recent years, there has been a lot of interest shown on probiotics for its health beneficial effects.[31,106]

The history of uses of probiotics is hidden in past but the systematic study of probiotics is not very long (Table 6.1).

TABLE 6.1　History of Uses of Probiotics

Researcher	Year	Observation and conclusions
Pasteur	1877	Observed antagonistic activity showing active opposition interaction between bacterial strains
		Suggested that nonpathogenic bacteria should be used to control pathogenic bacteria
Metchnikoff	1907	Observed that lactic fermentation of milk arrested putrefaction
		Suggested that consumption of fermented products would offer the same benefit to humans
		Felt that longevity in Bulgarian peasants was due to ingestion of "soured milks"
		Ingesting yogurt with *Lactobacilli* reduces toxic bacteria of the gut and prolongs life
Kipeloff	1926	Stressed importance of *Lactobacillus acidophilus* for good health
Rettger	1930	Early clinical application of *Lactobacillus*
Ferdinand Vergin	1950	Discussed the effects of antibiotics on beneficial intestinal bacteria
Parker	1974	First to use the term probiotics
Fuller	1980	Establishes first definition of probiotics

6.2 PROBIOTICS

Microbial colonization on Earth has been reported to occur 1 billion years after the complete formation of Earth's crust. The microbes have been cited to synthesize hydrocarbonated compounds that photosynthesize oxygen production. Oxygenic photosynthesis and aerobic respiration have been derived from microbial cell biochemistry.[66,128] Louis Pasteur is well known for his demonstrations of germ theory of disease which indicates the possible role of microorganisms in causing human infectious diseases.[4] In 1885, his laboratory experiments on feeding animals like rabbit, guinea pig, etc. with a diet deficient in microorganisms demonstrated the inevitable role played by microbiomes in digestion of food in the experimental animals. He also proposed the role of microorganisms in essential life survival.[61,119] A century later, Bernard Wostmann and his colleagues from the Lobund Laboratory at the university of Notre Dame in Indianapolis, Indiana gave insightful conclusions on the impact of microbiomes on host physiology. Based on their work on breeding experimental animals in germfree conditions, they illustrated that the absence of microbial colonization in animals should be supported by extraordinary nutritional supplementation for normal growth and development.[114]

In 1917, Alfred Nissle observed that a healthy soldier who failed to get shigellosis showed presence of *Escherichia coli* strains in his fecal matter. Based on these observations, he postulated that *E. coli* strains could be utilized to treat *Salmonella* and *Shigella*-mediated infections.[81]

Dr. Minoru Shirota in his medical research discovered that *Lactobacillus* strains were effective in suppressing harmful intestinal bacteria. Later, he successfully cultured the strain now known as *Lactobacillus casei* strain Shirota.[84] In 1935, he developed an edible tasty beverage fortified by the cultured strain that is now well known as "Yakult drink".[29] The term "Probiotic" was first established in 1960 by Lilly and Stillwell indicating substances secreted by some microbial organisms which stimulate growth of other microbial strains.[90] Fuller in 1989 defined Probiotics as live microfloral strains whose human supplementation aided in enhancing host health by restoring microfloral balance.[59]

6.3 FACTORS TO BE CONSIDERED FOR PREFERABLE PROBIOTICS AND PREFERENTIAL PROBIOTIC TRAITS

6.3.1 ACID TOLERANCE

A microbial strain, in order to impart its beneficial properties, is required to possess some inevitable properties like nonpathogenicity, ability to survive as colonies, acid, and bile resistance, healthy passage through the gastrointestinal tract in a viable state, ability to reproduce in the host anatomical conditions, adherence to host epithelial tissue, production of effective anti-microbial agents like acids, hydrogen peroxide, and bacteriocin, etc.[29,52] Desirable features of the probiotic strains depend on the anatomical region of the host where the strain needs to thrive and proliferate and the functions expected from the particular strain. Because several microbial strains differ in their cellular structure and composition, the strain ability to survive in host conditions vary considerably which are required to be taken into consideration while cultivating, storing, and transporting the microbial strain for human consumption.[101]

6.3.2   HUMAN GASTRIC JUICE TOLERANCE

Microbial potential to tolerate the gastrointestinal acidity and bile stress needs to be taken into prime consideration while selecting a particular strain from commercial sources and to ensure the positive health effects of the probiotics.[40] Presence of *Lactobacilli* strain in the human gastrointestinal milieu in sufficient number is an essential factor for maintenance of healthy equilibrium between beneficial and hazardous microflora in the gut by preventing the growth of notorious hazardous intestinal pathogens.[20] For a probiotic microorganism to impact the organ environment, a desirable population of 10^6–10^8 CFU/g of intestinal contents has been determined. The number may vary based on the strain type, processing conditions and gastric transit survival .[36] As per research on *Lactobacillus rhamnosus* GG strain striving in the gastrointestinal tract, its ability to survive the intestinal passage can be attributed to presence of a constant gradient between extracellular and cytoplasmic pH.[84] Acid resistance of several other Gram-positive probiotic strains is contributed to possession of F_0–F_1–ATPase enzyme which aids in proton translocation.[42] Gastric and intestinal survival of *Lactobacilli* is majorly determined by the fatty acid composition of the bacterial membrane which furthermore depends on several factors like growth, temperature, pH, growth phase, cultivation medium composition, etc.[37]

6.3.3   BILE TOLERANCE

Probiotic ability to hydrolyze bile salts has been found to be another relevant parameter for strain survival in host.[16,105] Bile salt hydrolysis has been contributed to possession of Bile Salt Hydrolases (BSHs) by probiotics but this enzyme possession has been detected to have detrimental effects on host health.[14]

6.3.4 ADHERENCE TO EPITHELIAL CELL SURFACE

Epithelial adhesion prevents probiotic elimination by peristalsis. Microbial adhesion to gut surfaces has been based on nonspecific physical interactions between adhesions and complementary receptors.[12,53] Epithelial adhesion property of probiotic strains has been attributed to auto-aggregation property, proteinaceous nature of surface components, and presence of surface layer proteins.[62]

6.3.5 IMMUNE-STIMULATORY BUT NOT PRO-INFLAMMATORY EFFECT

Immunoregulation by probiotics can be contributed to enhancement in nonimmunologic gut defense barrier which comprises of reduction in elevated intestinal permeability and normalization of altered gut microfloral ecology.[33,115] Probiotics have been studied to act as mediators of intestinal anti-inflammatory responses via inflammatory cytokine control mechanisms and hypersensitivity downregulation.[21]

6.3.6 ANTAGONISTIC ACTIVITY AGAINST PATHOGEN

Probiotics elevate host defense mechanisms by enhancing nonspecific host resistance to invasive microbial pathogens and altered gut microfloral stabilization.[9]

6.4 MAJOR EXAMPLES OF PROBIOTICS AND THEIR UNIQUE FEATURES

L. acidophilus has been frequently used as a probiotic in recent times for maintenance of intestinal microfloral balance.[18,84,134] It shows several benefits like stabilization of intestinal microflora, preventing colonization of enteropathogenic bacteria by competing for intestinal wall adhesion, reduction of lactose intolerance, prevention of lactose-induced diarrhea and colon cancer, immunostimulation, etc.[46,54,111] It shows distinct properties like production of antimicrobials, adhesion to mucosal surface, acid and bile tolerance, etc. which contributes to its above benefits in human system. Its antiinfective functions are mainly associated to production of lactic acid which further acts as pH stabilizing and antimicrobial compound.[85,107]

L. rhamnosus GG is indicated in prevention and treatment of gastrointestinal infections, diarrhea, vaccine potentiation by immune stimulation, and antiallergic effects.[115] It shows acid and bile tolerance, excellent growth and colonization property, and good intestinal adhesion characteristics. Its excellent adherence and immunomodulation properties can be contributed to its pili-like appendages. It promotes survival of intestinal epithelial cells by secretion of proteins like Major Secreted Protein Msp1 and Msp2 which prevent cellular cytokine-induced apoptosis.[8,88,84] Its host interaction property can be attributed to presence of an extracellular polysaccharide layer.

Various health benefits of *L. casei Shirota* (LcS) are mainly contributed to its immunomodulation properties. It induces immunoregulation by release of Th1 associated cytokines, interleukin (IL)-12, interferon (IFN)-γ, and tumor necrosis factor (TNF)-α. It stimulates immune responses by promotion of phagocytosis and enhancement of natural killer cell cytotoxicity.[35,72,82,115]

Lactobacillus plantarum species has been mainly associated for its antimicrobial activity in female urogenital infections. Its anti-infective property can mainly be contributed to production of Hydrogen peroxide and induction of coaggregation of vaginal pathogens.[55,135]

Lactobacillus reuteri is a normal inhabitant of gastrointestinal tract of healthy humans and is preferred as a probiotic for its antimicrobial properties. *L. reuteri* metabolizes glycerol resulting in production of 3-hydroxy propionaldehyde which shows strong and selective antimicrobial activity against *Salmonella*, *Clostridium*, *Listeria*, and *Escherichia* species. It is used to prevent and treat various gastrointestinal infections.[41,51]

The genus *Bifidobacterium* is present in colon and forms 25% of cultivable fecal bacteria in adults and 80% in infants. It comprises of various species like *Bifidobacterium bifidum*, *Bifidobacterium infantis*, *Bifidobacterium longum*, *Bifidobacterium breve*, etc. It shows safe survival but does not colonize in colon. It produces lactic acid in synergism to *L. acidophilus* species. It catalyzes fermentation reaction in colon resulting in production of short-chain fatty acids (SCFAs) like acetate, propionate, and butyrate. Additionally they also aid to synthesize vitamins like Vitamin B.[56,57,73]

The antimicrobial activity of *B. infantis* can be contributed to secretion of antimicrobial compounds other than acids. *B. longum* SBT2928 produces a protein factor which inhibits adhesion of enterotoxigenic *E. coli* strain. The antimicrobial activity of many Bifidobacterial strains can be contributed to production of antibacterial lipophilic factors with high molecular weight, competition with pathogens for epithelial binding sites, etc. *B. breve* has been studied to enhance antigen specific IgA-antibody directed against rotavirus. Bifidobacterial species has been indicated in several gastrointestinal infectious rotavirus induced and antibiotic-associated diarrhea. *B. longum*, *B. breve*, and *B. infantis* have been associated with prevention of relapse of chronic pouchitis after induction of antibiotic remission.[56,117]

S. thermophilus belongs to Gram-positive bacterial group and is closely related to *Lactococcus lactis*. It is highly adapted to grow on lactose and converts lactose to lactate during growth. It possesses antibacterial activity against intestinal pathogens and hence used to ameliorate gastrointestinal diseases. It is widely used in association with *L. casei* for prevention of antibiotic-associated diarrhea and in association with *B. bifidum* in reducing the incidences of Rotavirus induced diarrhea in infants. It positively acts on ceramide and sphingolipids in human skin and hence can be explored for several skin ailments. Consumption of *S. thermophilus* has been associated with reduction in serum cholesterol levels, exert antioxidant activity, act as a folic acid generator and serve as an anti-infective in Bacterial vaginosis.[7,99]

Saccharomyces boulardii is a yeast strain and is indicated in antibiotic-associated diarrhea and recurrent *Clostridium difficile* intestinal infections. Its probiotic property include gastric survival, maintenance of optimum 37°C conditions during its use in vitro and in vivo and antimicrobial efficacy against a host of pathogens. It is widely indicated as a probiotic drug to strengthen host physiology.[15,100]

6.5 PROPERTIES OF PROBIOTICS

The success of probiotic activity depends on possession of certain critical attributes by the probiotic species which aid the probiotic strain to exhibit the desirable activity in human body.[4,71] Figure 6.1 shows properties of probiotics.

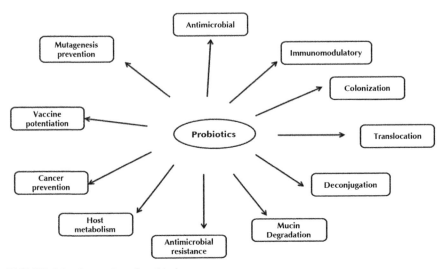

FIGURE 6.1   Properties of probiotics.

6.5.1   ANTIMICROBIAL PROPERTY

The antimicrobial property of probiotics is contributed to secretion of anti-infective products like bacteriocins and organic acids like lactic acid, acetic acid, butyric acid, and hydrogen peroxide.[3] The antimicrobial property of *L. acidophilus* can be contributed to production of bacteriocin Lactacin B and acidolin.[122] The antimicrobial action of major probiotic strains can be due to exertion of either single or multiple of following properties:

    i.   pH modulation,
   ii.   antimicrobial peptide secretion,
  iii.   inhibition of bacterial invasion, and
   iv.   blockade of bacterial epithelial adhesion.[26]

6.5.2   IMMUNOMODULATORY PROPERTY

Stimulation of immune system is another probiotic property by which they exert their activity. The reduction in intensity and severity of diarrhea on *L. reuterii* consumption is due to immunomodulation of human intestinal tract.[113,115] The immunomodulatory activity of probiotics can be due to their probiotic effect on epithelial cells, dendritic cells, macrophages, B-lympho-cytes, natural killer cells, T cells, etc.[6]

6.5.3 INTESTINAL COLONIZATION

The basic mechanism of gastrointestinal colonization by probiotics is adhe-sion to intestinal epithelial cells.[47] The extent of beneficial effect exerted by any probiotic strain depends on their ability to colonize the intestine in suffi-cient numbers.[80] Execution of all relevant probiotic properties like immu-nomodulation, antimicrobial activity, metabolism, etc. is dependent on the effective colonization of the concerned probiotic strain. The colonization of probiotic strains is induced by its hydrophobic nature and aggregation property.[49]

6.5.4 TRANSLOCATION IN INFECTIOUS DISEASE

Probiotic translocation in human body can result in severe nosocomial infec-tions like Endocarditis, Bacteraemia, Pneumonia, septicaemia, etc. Platelet aggregation by probiotic can result in endocarditis. *Lactobacillus* consump-tion over a 5-year period by 45 patients in the United States was found to show occurrence of Bacteraemia. Consumption of *S. boulardii* containing probiotics has been associated with *Saccharomyces cerevisiae* fungemia in critical care patients.[91] Liver abscesses in immunocompromised patients with Crohn's disease have been linked to *L. acidophilus* consumption.[28]

6.5.6 DECONJUGATION

Conjugation is a mechanism by which several endogenous and exoge-nous substances in human body are metabolized, recirculated, and finally excreted.[125] It forms a major metabolic pathway responsible for ensuring excretion of toxic compounds from human body. Glycine, taurine, gluc-uronic acid, and sulphate are four major molecules used in conjugation reac-tions. In presence of deconjugating enzymes, conjugated substances can be deconjugated by activity of some probiotic strains thus causing changes in physicochemical properties of the conjugates, thus affecting the pathophysi-ologies.[124] Glycine and taurine conjugates have been found to be deconju-gated in presence of *Lactobacilli* and *Bifidobacterial* species resulting in adverse effects like decreased bile acid recirculation, steatorrhea, decreased efficacy of contraceptive drugs, etc.[13]

6.5.7 ANTIMICROBIAL RESISTANCE

Probiotic strains like *Lactobacillus* and *Bifidobacterium* have been found to be resistant to clinically relevant antimicrobials like Vancomycin. Thus administration of probiotics with antimicrobial resistant properties could of value in patients on antibiotics.[70]

6.5.8 METABOLIC ACTIVITY

Cholesterol lowering property of probiotics can mainly be contributed to assimilation, binding and degradation of cholesterol by metabolic ways mediated by probiotics.[58] Cholesterol is assimilated by probiotics and used for self-metabolism. Furthermore, cholesterol is catabolized by probiotics.[60]

6.5.9 ANTIMUTAGENIC AND ANTICARCINOGENIC PROPERTIES

Anticarcinogenic properties of probiotics can be contributed to varied mechanisms like prevention of growth of bacteria producing enzymes responsible for converting procarcinogens to carcinogens, tumor-growth inhibition by production of immunostimulating metabolites, and production of TNF.[39,103]

6.5.10 VACCINE—ADJUVANT PROPERTY

Probiotics enhance vaccination effects by affecting antibody responses following vaccination and improving the seroconversion rates. *L. acidophilus* has been found to increase effectivity of mucosal vaccines like malaria, HIV, and infant diarrhea.[89]

6.6 MECHANISM OF ACTION OF PROBIOTICS

Several theories have been proposed on varied mechanisms by which probiotics exert their beneficial effects in human system indicate that colonization may not be the only means of probiotic activity (Fig. 6.2).

For example, *B. longum* exerts its effects by becoming a part of human intestinal microflora, whereas *L. casei* species exerts its probiotic effect in a subtle manner as it transits through by remodeling the existing microfloral community. Four major pathways by which most probiotics exert their effect on human body include modulation of barrier function, production of antimicrobial contents, competition for intestinal adherence, and immunomodulation.[132]

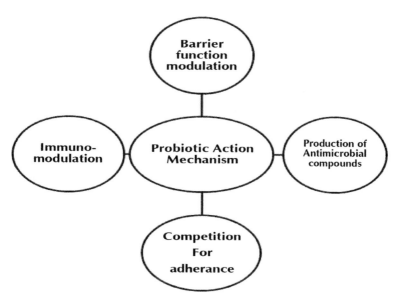

FIGURE 6.2   Mechanism of action for probiotics.

Probiotics modulate the epithelial barrier function by influencing the intestinal cell apoptosis or elevating mucosal production. Antimicrobial activity of probiotics can be contributed to their ability to induce host cells to produce pathogen mitigating peptides like defensins (hBD protein) and cathelicidins. Additionally, probiotics stimulate production of antimicrobial factors like Bacteriocins, hydrogen peroxide, nitric oxide and SCFAs like lactic acid and acetic acid which lower the lumen pH. SCFAs alleviate Gram-negative pathogens by disruption of outer membrane whereas bacteriocins inhibit Gram-negative bacteria by pore formation and disruption of inner membrane.[34] Competition for adherence to epithelial binding sites is another mechanism of pathogenic inactivation by probiotics. Probiotics like *S. boulardii* secret heat labile factors which decrease bacterial adherence to epithelium. Probiotic strains like *L. casei* augment secretion of pathogen specific Immunoglobin A thus preventing invasion. Interference with pathogenic quorum sensing signaling (cell-to-cell communication) is another relevant mechanism observed for probiotics like *L. acidophilus* that inhibits *E. coli* colonization.[67]

6.7 ADVANTAGES OF PROBIOTICS

Out of the several advantages conferred by probiotics on human health, some advantages are far-weighted in terms of general human routine survival and growth. These include benefits on human intestinal gut, epithelium, immune system, lactose intolerance, cholesterol assimilation, infectious diarrhea, preventive mechanisms for life-staking disease like cancer, general improvement of body mechanisms against asthma, allergies, etc.[24,115]

6.7.1 *INTESTINAL GUT*

Studies on gut microbiota of infants have indicated that establishment of microbes in the gastrointestinal tract starts immediately after delivery. Microbial composition of an adult gets established after 2 years of age.[63] Microorganisms present in the maternal vagina and fecal matter along with microbes in environment (air, water, etc.) form the important sources of bacteria. Moreover, the gut microbial composition can be changed by altering the diet fed to infants, that is, infants fed with breast milk show majorly *Bifido-bacterial* spp., whereas infants fed with formula milk show more complex microbial composition.[79] Human GIT comprises of about 10^{14} microfloral cells which include 1000 diverse bacterial types. Major microbial colonization is observed in colon. Large intestine mainly shows facultative anaerobic strains such as *E. coli* and *Streptococcus* species.[74] Adult gut shows nonsporing anaerobes which comprise of *Bacteroides* spp., *Bifidobacterium* spp., *Eubacterium* spp., *Clostridium* spp., *Lactobacillus* spp., *Fusobacterium* spp., and *Gram-positive cocci*. *Enterococcus* spp. *Enterobacteriaceae*, *methanogens*, dissimilatory sulphate-reducing bacteria are present in lower numbers. Colonic microbes readily ferment and degrade substrates available from diet and endogenous secretions.[120] Growth substrate comprise of carbohydrates, proteins, bacterial secretions, lysis products, sloughed epithelial cells, and mucins. Some of the several beneficial effects of gut microbiota comprise of immunostimulation, improved digestion and absorption, synthesis of essential vitamins, pathogenic inhibition, cholesterol lowering, and antigas distension activity.[5]

6.7.2   EPITHELIUM

Epithelium lining the gastrointestinal mucosa acts as a protective barrier against pathogenic intrusion. Probiotics beneficially act on the barrier function of epithelium by production of protective cytokines which mediate regeneration of damaged epithelial cells and prevent epithelial cell apoptosis occurring due to exposure to derogatory agents and pathogens.[10] Probiotics aid to counteract adverse effects of inflammatory cytokines on the epithelial barrier function. Probiotic protective action in inflammatory bowel disease can be contributed to probiotic mediated correction of imbalances between the effector and regulatory mechanisms of epithelium. Effectivity of *L. rhamnosus* GR-1, *L. reuteri* RC-14, etc. on vulvovaginal candidiasis (Fig. 6.3) can be contributed to protection of the vaginal epithelium against pathogenic intrusion by production of hydrogen peroxide.[11]

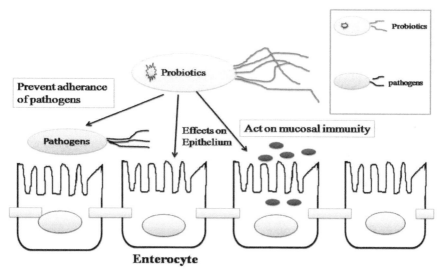

FIGURE 6.3   Effect of probiotics on epithelium.

6.7.3   IMMUNE SYSTEM

Probiotic supplementation during weaning has found to cause elevation in antibody response of infants to hazardous infections like influenza, diphtheria, meningitis, etc.[38] *L. sakei* have been found to modulate in vitro expression of pro-inflammatory and anti-inflammatory molecules like IL-1β, IL-8, and TNF-α. Interaction of probiotics with Peyers Patch M cells has been studied to be an important contributing factor to elevation in antigenic transport across intestinal epithelium.[76,115] Immunomodulatory activities of probiotics have been found to be species-specific. Human blood leucocytic capacity was found to be stimulated by *L. acidophilus* Lal. Probiotic supplementation of *L. rhamnosus* HN001 and *Bifidobacterium lactis* HN109 has been studied to enhance the cytotoxic potential of natural killer (NK) cells.[135] Use of *L. casei Shirota* causes elevation in IL-2 production causing elevation in NK cell activity. Probiotic strains like *B. bifidum* and *L. acidophilus* Lal have been studied to elevate production of IgA by B cells thus aiding to maintain intestinal humoral immunity.[97,115]

6.7.4   CANCER

Probiotics have been studied to have an inhibitory effect on development of precancerous lesions and tumors in animal models. As per studies by Hosono et al., milk fermented with *L. delbrueckii* subspecies *bulgaricus* or *Enterococcus faecalis* was found to show antimutagenic activity against NQO.[54] Anticancerous effects of probiotics like *L. acidophilus* have been contributed to their ability to prevent the growth of bacteria that produce enzymes like Beta-glucuronidase, nitroreductase, and azoreductase which catalyze conversion of precarcinogens to carcinogens. Other mechanisms by which probiotics aid in cancer prevention include metabolite production that stimulate immune system leading to tumor growth inhibition and production of TNF by macrophages.[104,115]

6.7.5   RESPIRATORY INFECTIONS

A large number of clinical trials have indicated that probiotic therapy can be considered critical for prevention and management of respiratory ailments and infections. Effectivity of probiotics in respiratory problems can be contributed to several mechanisms like (i) survival competition against pathogens, (ii) enhancement of respiratory epithelial barrier function, (iii) enhancement of cellular immunity by increasing activity of natural killer cells (NK) cells and macrophages in airways, (iv) modulation of intestinal microbiota and thus regulation of the amount of allergen exposed to lungs by circulatory route, (v) anti-inflammatory cytokine induction, (vi) increased production of secretory IgA, etc.[102] Studies on a murine model of asthma demonstrated that administration of *L. reuterri* LGG strain and *B. breve* declined hyperresponsiveness of airways by alleviation of number of inflammatory cells in bronchoalveolar lavage (BAL) fluid and decreased inflammation of lung tissue.[133] Administration of *E. faecalis* FK-23 orally was found to suppress asthmatic response by attenuation of Th17 cell development. An important role of probiotics like *L. reuterri* LGG *L. casei* strain shirota (LcS) and DN114001 have been implicated in prevention and treatment of respiratory infections like influenza. Garaiova et al.[60] have reported that *L. reuterri* LGG protects mice from H1N1 influenza virus infection by immunoregulation. Probiotic use has been demonstrated to show lower incidences of ventilator –associated pneumonia, reduced respiratory infections in healthy and hospitalized children and decreased duration of common cold infection.[60]

6.7.6 ALLERGIES

Development of allergies has been contributed to an alteration in gut microflora composition. Probiotics have been found to exert a beneficial effect on allergies by elevation of mucosal barrier function. Human consumption of *Lactobacillus GG* strain has been associated with alleviation of food allergies induced by milk protein.[112] A 6-month administration of probiotics to pregnant women has been found to lower incidence of childhood eczema by 50% when compared to placebo. The elimination of eczema has been contributed to a two-fold increase in transforming growth factor β_2 (an inflammatory cytokine) in the breast milks of women.[92]

6.7.7 LACTOSE INTOLERANCE

The presence of lactase enzyme in the world population depends on the ethnicity and level of consumption of dairy products. Lactase deficiency is seen in 80% Hispanic people and almost 100% of Asian and Native American people.[110] Absence of lactase enzyme causes lactose to be metabolized by colonic microflora producing methane, carbon dioxide, and hydrogen gas leading to symptoms of abdominal cramps, flatulence, bloating, and diarrhea. Supplementation of *Lactobacilli/Bifidobacterium* strains in lactose rich milk products can aid lactose intolerant individuals to tolerate lactose. This is mainly contributed to presence of microbial beta-galactosidase enzyme in these fermented products which aids in lactose hydrolysis.[48]

6.7.8 CHOLESTEROL ASSIMILATION

Oral administration of strains of *L. johnsonni* and *L. reuteri* in pigs has been studied to show decrease in serum cholesterol levels.[126,130] Probiotics have been found to show cholesterol lowering effects by varied mechanisms, such as

i. deconjugation of bile acids by BSHs enzymes produced by probiotics,
ii. cholesterol assimilation,
iii. cholesterol coprecipitation with deconjugated bile,
iv. enhanced cholesterol binding to cell walls,
v. cholesterol incorporation into cellular membrane during growth,
vi. cholesterol conversion into coprostanol,
vii. cholesterol fermentation into short-chain fatty acids, etc.

6.7.9 INFECTIOUS AND ANTIBIOTIC-MEDIATED DIARRHEA

Diarrhea has been found to elicit serious long-term effects affecting growth, nutrition, and cognition sometimes even causing mortality.[77,78] Administration of a standard infant formula supplementation containing *B. bifidum* (1.9×10^8 CFU/g) and *Streptococcus thermophilus* (1.4×10^7 CFU/g) to human infants aged 5–24 months and suffering from Chronic Rotavirus associated diarrhea was found to show a lowering in the number of children suffering from diarrhea when compared to unsupplemented control groups.[32] Probiotic supplementation of *L. reuteri*, *L. acidophilus*, *S. boulardii*, etc. have been found to show a lowering in the duration of diarrhea by 1–5 days and decreased shedding of rotavirus in Rotaviral gastroenteritis. Ingestion of probiotic strains of *Lactobacilli* GG alone or in combination with other genera have been found to reduce frequency, duration, and severity of antibiotic-induced diarrheal changes mainly induced by *Clostridium difficile*.[75]

6.7.10 HEALTH BENEFITS FOR HUMANS OF PROBITOICS

- Elimination of lactose intolerance
- Antidiarrheal
- Immunomodulatory
- Antidiabetic
- Anticarcinogenic

- Hypocholesterolemic
- Antihypertensive.

6.8   HOW TO INCREASE THE SURVIVAL OF PROBIOTICS

Probiotic activity in human body mainly depends on good retention of its characteristic features. Versatile procedures adopted to stimulate probiotic growth, its maintenance, storage, and packaging can render the strain inactive by adversely affecting its colony counts. Probiotic stabilization and preservation techniques like drying (as a means of probiotic dehydration technique), use of cell protectants, etc. have been widely studied.[22] Drying techniques like spray-drying and freeze-drying are the two most commonly used techniques, however, in recent times vacuum oven drying, fluidized bed drying and a combination of both have also been widely applied for probiotic stabilization. Retention of probiotic activity after drying is dependent on several parameters like probiotic type, species used, drying conditions, composition of inoculums and culture media, cryoprotectant used, etc.[43] Freeze drying in synergism with judicial optimization of composition suspension medium and cryoprotectant type has been found to be a highly effective stabilization methodology. Preservation by atomization (spray-drying) has been a widely adopted stabilization technique for lactobacillus cultures. Vacuum drying at low temperatures ranging between 30 and 80°C has also been explored as a probiotic stabilization technique. Commonly known as controlled low-temperature vacuum dehydration (CLTV), this technique can be used for drying *L. acidophilus* species. Low-temperature drying has been found to be more beneficial to atomization techniques. Freeze-drying has been widely used to for *B. lactis* βb-12 whereas CLTV has been found to be efficacious for *L. delbruekii* species *bulgaricus*. Loss of probiotic viability during storage can be contributed to temperature fluctuations mainly rise in temperatures.[17] The extent of loss of activity is dependent on temperature, moisture content, protectors, and oxidative stress. Mechanism of probiotic death during storage has been studied to be different from mechanisms causing probiotic death during thermal stresses. Use of cell protectants like whey, polysaccharides, glycerol, skimmed milk powder, betaine, trehalose, sugars (lactose, glucose), etc. have been studied to enhance probiotic survival during adverse processing and storage conditions. Protective mechanisms of these substances include probiotic enrichment against hyperosmotic stress, enhanced cellular adaptability to stresses, etc.[129] Use of sugars for probiotic preservation by drying and in other processing steps has been demonstrated to enhance probiotic survival to higher temperatures stresses. Currently, microencapsulation techniques have been studied to protect probiotic strains. Entrapment of probiotic strains in a matrix confers the probiotic cell resistance against external temperature, oxidation, and other stresses by the barrier properties of the encapsulation matrix. Examples of microencapsulating agents include cellulose acetate phthalate, gelatin, vegetable gums, alginates, chitosan, fats, etc. Alginates have been widely used in probiotic stabilization by microencapsulation. Furthermore, addition of prebiotic like inulin, fructans, etc. to probiotics have been found to confer enhanced survival and activity of probiotics.[50]

6.9   QUALITY ASSURANCE OF PROBIOTICS

The performance of probiotics in human body is mainly dependent on its ability to thrive and stabilize in the human anatomical conditions. Hence, retention of the conventional probiotic traits on the basis of which the probiotic is selected is an inevitable requirement. Some of these traits include mucosal adhesion, acid, and bile stability, survival during gastrointestinal transit and colonization in sufficient amounts. Selection of a probiotic strain during manufacturing and storage requires the need to confirm the possession of these properties. Screening and selection criteria for probiotics comprise of testing for phenotype and genotype stability, plasmid stability, carbohydrate, and protein-utilization patterns, acid and bile tolerance, survival and growth, bile metabolism, intestinal adhesion properties, antimicrobial production, antibiotic resistance patterns, gut pathogenic inhibition potential, and immunogenicity mucosal adherence (relevant for colonization) can be evaluated using enterocyte like Caco-2 tissue culture cells and intestinal mucus as substrates. Additionally, HT29-MTX tissue culture cells and human ileostopy glycoproteins are some other substrate s utilized for adhesion studies. Probiotic colonization can be studied on biopsies from intestinal mucosa and rectal mucosa. Probiotic ability to produce antimicrobial substances like growth-inhibiting metabolites like organic acids, hydrogen peroxide and bacteriocins, adhesion inhibitors also need to be evaluated.[45]

6.10   REGULATORY SPECIFICATIONS

As per US regulatory classifications, probiotics intended for dietary supplementation need to be regulated by Dietary Supplement Health and Education

Act and a probiotic intended for therapeutic benefits comparable to drug activity need to be regulated by Food and Drug Administration.[116] Probiotic use in India is regulated by Food Safety and Standard Act. Use of probiotics in foods is regulated by guidelines laid down by Indian Council of Medical Research and Department of Biotechnology.[19]

6.11 SAFETY OF PROBIOTICS FOR HUMAN USE

Use of probiotics in human supplementation has been found to show enormous advantages. Moresoever, unjustified indiscriminate use of probiotics can lead to hazardous side effects on human health. These side effects can be manifested as systemic infections, perilous metabolic activities, abnormal immunostimulation in susceptible individuals and induction of gene transfer mechanism. Volunteers supplemented with *Lactobacilli, Bifidobacterium* species, etc. have been studied to exhibit local or systemic infections like septicemia and endocarditis. Humans treated with *S. boulardii* were found to show fungemia.[30] Adverse manifestations of probiotic supplementation include metabolic stimulation inducing diarrhea and intestinal lesions through deconjugation and dehydroxylation of bile salts. Oversupplementation of probiotics can lead to have adverse effects similar to those observed due to overgrowth of inherently present gastrointestinal strains. Probiotic mucosal degradation has been demonstrated to be detrimental to human health. Parenteral administration of bacterial cell wall components like peptidoglycan polysaccharides from Gram-positive bacteria like *Lactobacilli* species have been studied to elicit cytokine secretion causing fever, arthritis, and autoimmune diseases. Externally supplemented probiotics can transfer resistant genes to endogenous flora or pathogens. Extent of gene transfer depends on properties of genetic material transferred (e.g., plasmids, transposons): nature of donor and receptor strains, concentration, and extent of contact. The safety of probiotics can be determined by in depth analysis of intrinsic features of strain, strain pharmacokinetics, and probability of interaction between the host and the strain.[2,115]

6.12 PREBIOTICS

The World Health Organization (WHO) defines a prebiotic to be a nondigestible substance that aids in benefiting the host physiology by selective stimulation of growth of indigenous bacteria.[95]

- Nondigestible food supplements or ingredients
- Not absorbed or degraded
- Alter the balance of intestinal flora and by acting as substrates stimulate the growth of beneficial bacteria (i.e., Lactobacillus and Bifidobacterial)
- Nondigestible food ingredients
  - Fructo-oligosaccharides (FOS) (chicory, inulin)
  - Lactulose
- Positively affect the endogenous flora
- Stimulate the growth of one or a limited number of bacterial species
  - FOS—*Bifidobacteria*
  - Lactulose—*Lactobacilli*

Prebiotics promote beneficial bacteria already present in colonic milieu as well as enhance efficacy of externally supplemented probiotic bacteria. A prebiotic comprises of a fermented ingredient that stimulates alternation in composition and activity of microbiomes present in gastrointestinal tract thus indirectly promoting host well-being. Prebiotics are obtained from fruits, vegetables, dietary fibers present in polysaccharides like starch by various processing steps like extraction, concentration, hydrolysis, or enzymatic modulation. Prebiotics comprise of indigestible oligosaccharides or mixtures of FOS and polysaccharides.[44,65] They are commonly composed of carbohydrates mainly fructans (inulin), galacto-oligosaccharides (GOS) and lactulose (synthetic disaccharide). A prebiotic supplementation to infant formula aids in conferring benefits of human milk oligosaccharides which comprise of oligosaccharide mixtures found in mammalian milk which have innumerable advantages on neonatal development.[86] FOS (aka oligofructose), isomaltooligosaccharide, xylooligosaccharide (XOS), inulin, fiber, oligomate, palatinose, pyrodextrin, and raftiline are the some of the major examples of the prebiotics.

6.13 PREBIOTIC SUPPLEMENTATION AND HUMAN HEALTH

6.13.1 EFFECT ON GUT

Studies investigating the beneficial effects of prebiotic supplementation on bifidobacterial composition have indicated that oligosaccharide supplementation causes increase in number of fecal *Bifidobacterium* concentration. Similarly, supplementation of disaccharides like lactulose, soyabean oligosaccharide, and GOS induce elevation in bifidobacterial species. The extent of elevation depends on the duration for which the prebiotic was supplemented. Furthermore a dose-related enhancement in fecal lactobacilli ($P < 0.001$) was observed in 12 subjects fed with GOS for a week. The prebiotic supplementation led to enhancement in concentration of beneficial bacteria and decrease in hazardous or less-desirable microbes. Supplementation of 3 g/day of lactulose to eight subjects was found to show decrease in fecal clostridial and bacteroidal concentration.[25,98]

6.13.2 HOST HEALTH AND PHYSIOLOGY

6.13.2.1 PROMOTION OF BACTERIAL FERMENTATION

The colonic microbiota plays a very relevant role in promoting host health and well-being. They aid in fermenting and metabolizing colonic dietary contents leading to production of vital vitamins, antioxidants, defensins, etc. They potentiate activity of beneficial microbiomes in human gut and reduce metabolic activity of hazardous bacteria by alleviating the bacteria by various mechanisms.[95]

6.13.2.2 DIETARY SUBSTRATE UTILIZATION

Human diet comprising of nondigestible oligosaccharides, fibers, proteins, etc. acts as growth substrates for gut microflora. Anaerobic fermentation forms the important mechanism of energy production for versatile anaerobic microbial strains residing in small intestine. Main growth substrates comprise of nondigestible carbohydrates like resistant starch, dietary fibers, resistant dextrins, non-starch polysaccharides (NSPs0 like pectins, arabinogalactans, gum Arabic, guar gum, and hemicelluloses and nondigestible oligosaccharides like raffinose, stachyose, ITF, galactans, and mannans.[118]

6.13.2.3 BENEFICIAL METABOLITE PRODUCING ACTIVITIES

Intestinal microbiota metabolize consumed prebiotics to produce metabolism products like acetate, propionate, butyrates, lactates, pyruvates, ethanol, succinate, and gases like hydrogen, carbon dioxide, methane, and hydrogen sulphide.[123]

6.14 PREBIOTIC TYPES AND HEALTH BENEFITS

6.14.1 FRUCTANS

Fructans comprise of polymeric fructose linked to glucose moiety at the terminal position. They pass through the upper gastrointestinal tract without metabolism reaching the colon. In colon, they are fermented to SCFAs by colonic microflora.[93]

6.14.2 INULIN

Inulin is found in leeks, onions, wheat, asparagus, garlic, and chicory. It is a naturally occurring nondigestible fructan type. A general lowering of insulin and triacylglycerol levels has been indicated in a 8-week double-blind randomized placebo controlled trial on the effect of 10 g inulin on fasting blood lipid, glucose, and insulin levels of healthy middle-aged men and women with moderately raised plasma cholesterol and triacylglycerol levels. Inulin has been demonstrated to promote digestive health, lipid metabolism, increase calcium absorption, thus decreasing osteoporosis and reducing risks of cancerous activity in colon and breast.[27]

6.14.3 FRUCTO-OLIGOSACCHARIDES

FOS comprise of mixtures consisting of glucose linked to fructose units by β-(1,2) links with a degree of polymerization between 1 and 5. It is commonly found in plants like Asparagus, wheat, Jerusalem chokes, rye, and onion (25–40% of dry matter). Commercially, FOS are obtained by using immobilized cells in calcium alginate gel or enzyme loaded on an insoluble carrier or by sucrose conversion catalyzed by fungal fructosyltransferase enzyme. In colon, FOS are metabolized by fermentation to lactate, SCFAs (acetate, propionate, and butyrate), and gas. FOS have been found to potentiate Bifidobacterial growth in colon along with suppression of hazardous bacterial growth. A decrease in fecal pH, increase in fecal/colonic organic acids, decrease in fecal bacterial enzymatic activities, modification in fecal neutral sterols, enhanced magnesium absorption, reduction in colonic tumor development (by enhancement of colon butyrate concentrations and local-immune system effectors) are some of the other benefits of FOS on human health.[69,87,115]

6.14.4   GALACTO-OLIGOSACCHARIDES

GOS collectively comprise of compounds belonging to a carbohydrate group involving oligo-galactose with lactose and glucose. They are commercially produced from lactose by β-galactosidase enzyme.[127] The protection of breast-fed infants from intestinal pathogenic bacteria has been contributed to presence of oligosaccharides resembling. GOS naturally found in human milk. Commercially, GOS comprise of mixtures of oligosaccharides formed by linking one or more galactosyl moieties to a terminal glucose/galactose units like galactobioses or galactotrioses. The physicochemical and physiological properties of GOS are dependent on mixture composition. Their properties like sweetness, solubility, osmolarity, crystal formation ability, and reactivity decreases as molecular size increases. They have been demonstrated to stimulate growth of *Bifidobacterium* species, improve calcium absorption, provide constipative relief, stimulate human immune system, prevent colonic cancer, etc.[83,115,121]

6.14.5   LACTULOSE

Lactulose is a synthetic disaccharide comprising of galactopyranosyl-D-fructofuranose. A significant increase in *Lactobacilli* and *Bifidobacterial* colony counts and a decrease in Bacteroidal concentration was observed in studies of fecal cultures obtained from healthy human volunteers maintained on a lactulose-rich diet. A decline in fecal *Clostridium perfringens*, Bacteroidaceas, fecal indoles, skatol, and phenols and activities of β-glucuronidase, nitroreductase, and azoreductase activities was observed after lactulose consumption by healthy human volunteers.[109,131]

6.14.6   EMERGING AND NOVEL PREBIOTICS

XOS, *resistant starch*, and *pectic oligosachharides* are some of the emerging and novel prebiotics. *XOS* comprise of xylose chains linked by β (1–4) bonds produced enzymatically by hydrolysis of xylan obtained from Birch woods, oats, corn, etc.; these emerging prebiotics have been found to show prebiotic potential due to their selectivity for *Bifidobacterial* species A randomized parallel placebo-controlled double blind study of XOS supplementation in healthy humans was found to show increased fecal levels of Bifidobacteria, butyrates, activities of α-glucosidase, and β-glucuronidase and decreased amounts of acetate and *p*-cresol *resistant starch* can be defined as the starch fraction that escapes upper gastrointestinal digestion and is consequently fermented by colonic microbial colonies. Studies on human flora-associated rats colonized with microfloral populations was found to show a 10–100-fold enhancement in *Lactobacilli* and Bifidobacterial concentration and a predominant decline in enterobacteria, thus confirming its prebiotic potential oligosaccharides obtained from enzymatic or physical hydrolysis of pectin (a complex galacturonic acid-rich polysaccharide) have been found to promote growth of Bifidobacterial species as indicated by studies on fecal cultures of healthy human volunteers supplemented by the same.[96,136]

6.15 SYNBIOTICS

A probiotic organism in combination with its prebiotic food is defined as a synbiotic. Providing both the organism and substrate at the time of ingestion may offer improved chance of survival in GI tract. A synergistic application of a probiotic strain and its supporting prebiotic food can potentiate the health benefits derived. A probiotic supplementation without its prebiotic food from which it derives its nutrition is found to show low survival with greater intolerance for oxygen, acidic pH, and temperature modulations. Prebiotics play an important role in probiotic preservation.[64] The efficacy of a symbiosis can be contributed to (i) improvement in probiotic viability and (ii) imparting of specific health benefits. Synbiotic food consumption predominantly aids in enriching health benefits by stimulation of gut metabolism and maintenance of gut physiological features. The health benefits derived from a synbiotic supplementation is mainly due to elevation in SCFA, ketones, carbon disulphide, and methyl acetate. A supplementation of *L. acidophilus* CHO-220 and inulin to hypercholesterolemic subjects in a randomized, double-blind, placebo-controlled parallel design study was found to positively influence the modulation of irregularly shaped red blood cells (RBCs). An increase in RBC membrane fluidity with reduction in cholesterol enrichment of RBC membrane was observed.[1,23]

6.16 CARE, CANCERS, AND PRECAUTIONS IN USE OF PREBIOTICS AND PROBIOTICS

Mucin produced by Goblet cells and enterocytes in human gastrointestinal tract forms a first line of defense for underlying cells. Probiotics have been studied to regulate balance between production and degradation of mucin leading to adverse pathophysiological consequences. Hence, caution should be taken while administering probiotics with mucin degrading properties to critically ill patients.[68]

6.17 CONCLUSIONS

Probiotics or beneficial microbiomes show innumerable effects on diverse aspects of human well-being. Based on the strain used, a range of benefits can be derived from probiotic use. These include immunostimulation, cancer prevention, cholesterol assimilation, infection control, respiratory regulation, etc. The selection of a probiotic strain with preferable traits like optimized acid tolerance, bile tolerance, gastric transit, epithelial adherence, etc. forms an inevitable criterion for ensuring positive probiotic activity in human pathophysiology. The in-vivo activity of probiotic can be stimulated by administration of prebiotics like fructans, lactulose, resistant starch, etc. in synergism with the probiotic strain.

KEYWORDS

- bifidobacterial
- cholesterol-lowering
- dietary
- enzyme
- food
- gut
- health
- human health
- L. acidophilus
- lactobacilli
- lactose intolerance
- prebiotics
- probiotics
- strain

REFERENCES

1. Akdis, M.; Frei, R.; O'Mahony, L. Prebiotics, Probiotics, Synbiotics, and the Immune System. *Curr. Opin. Gastroenterol.* **2015,** *31*(2), 153–158.
2. Akkermans, L. M.; Haller, D.; Sanders, M. E. Safety Assessment of Probiotics for Human Use. *Gut Microbes* **2010,** *1*(3), 164–185.
3. Ali, F. S.; Hussein, S. A.; Saad, O. A. O. Antimicrobial Activity of Probiotic Bacteria, *Egypt. Acad. J. Biol. Sci.* **2013,** *5*(2), 21–34.
4. Amiot, J. Immunomodulatory Effects of Probiotics in the Intestinal Tract. *Curr. Issues Mol. Biol.* **2008,** *10*(1–2), 37–54.
5. Angelakis, E.; Merhej, V.; Raoult, D. Related Actions of Probiotics and Antibiotics on Gut Microbiota and Weight Modification. *Lancet, Infect. Dis.* **2013,** *13*(10), 889–899.
6. Arvilommi, H. Probiotics: Effects on Immunity. *Am. J. Clin. Nutr.* **2001,** *73*(2), 444–450.
7. Aryana, K. J.; Mena, B. Influence of Ethanol on Probiotic and Culture Bacteria *Lactobacillus bulgaricus* and *Streptococcus thermophilus* Within a Therapeutic Product. *Open J. Med. Microbiol.* **2012,** *2*, 70–76.
8. Auclair, J.; Frappier, M.; Millette, M. *Lactobacillus acidophilus* CL1285, *Lactobacillus casei* LBC80R, and *Lactobacillus rhamnosus* CLR2 (Bio-K+): Characterization, Manufacture, Mechanisms of Action, and Quality Control of a Specific Probiotic Combination for Primary Prevention of *Clostridium difficile* Infection. *Clin. Infect Dis.* **2015,** *60*(2), S135–S143.
9. Barlow, J.; Costabile, A.; Gibson, G. R.; Rowland, I.; Tejero-Sarinena, S. *In Vitro* Evaluation of the Antimicrobial Activity of a Range of Probiotics Against Pathogens: Evidence for the Effects of Organic Acids. *Anaerobe* **2012,** *18*(5), 530–538.
10. Barrett, K. E.; Resta-Lenert, S. Live Probiotics Protect Intestinal Epithelial Cells from the Effects of Infection with enteroinvasive *Escherichia coli* (EIEC). *Gut* **2003,** *52*(7), 988–997.
11. Batish, V. K.; Duary, Raj, K.; Grover, S.; Rajput, Y. S. Assessing the Adhesion of Putative Indigenous Probiotic *Lactobacilli* to Human Colonic Epithelial Cells. *Indian J. Med. Res.* **2011,** *134*, 664–671.
12. Batish, V. K.; Grover, S. Probiotic *Lactobacilli* to Human Health. *Indian J. Med. Res.* **2015,** *13*(3), 67–91.
13. Begley, M.; Hill, C.; Gahan, C. G. M. Bile Salt Hydrolase Activity in Probiotics. *Appl. Environ. Microbiol.* **2006,** *72*(3), 1729–1738.
14. Begley, M. Hydrolase Activity in Probiotics. *Environ. Microbiol.* **2010,** *2*(9), 769–778.
15. Bernasconig, P.; Mcfarland, L. V. *Saccharomyces boulardii*: A Review of an Innovative Biotherapeutic Agent, Microbial Ecology. *Health Dis.* **1993,** *6*, 157–171.
16. Besendorfer, V.; Kos, B.; Matosic, S.; Suskovic, J. The Effect of Bile Salts on Survival and Morphology of a Potential Probiotic Strain *Lactobacillus acidophilus* M92. *World J. Microbiol. Biotechnol.* **2000,** *16*(7), 673–678.
17. Bezkorovainy, A. Probiotics: Determinants of Survival and Growth in the Gut. *Am. J. Clin. Nutr.* **2001,** *73*, 399–405.
18. Bhadekar, R.; Dixit, G.; Samarth, D. T. V. Comparative Studies on Potential Probiotic Characteristics of *Lactobacillus acidophilus* Strains. *Eur.-Asian J. Biosci.* **2013,** *7*, 1–9.
19. Bhattacharya, S. K.; Ganguly, N. K.; Sesikeran, B. ICMR-DBT Guidelines for Evaluation of Probiotics in Food. *Indian J. Med. Res.* **2011,** *134*(1), 22–25.

20. Bhutada, S. A; Tambekar, D. H. Acid and Bile Tolerance, Antibacterial Activity, Antibiotic Resistance and Bacteriocins Activity of Probiotic *Lactobacillus* Species. *Recent Res. Sci. Technol.* **2010**, *2*(4), 94–98.

21. Biedrzycka, E. Immunostimulative Activity of Probiotic Bifidobacterium Strains Determined in vivo using Elisa Method. *Pol. J. Food Nutr. Sci.* **2003**, *12*(53), 20–23.

22. Bigetti, G. K. Dried Probiotics for Use in Functional Food Applications. In *Food Industrial Processes—Methods and Equipment*, 2012, ISBN: 978-953-307-905-9.

23. Blaauw, R.; Lombard, M.; Mugambi, M. N.; Musekiwa, A.; Young, T. Synbiotics, Probiotics or Prebiotics in Infant Formula for Full Term Infants: A Systematic Review. *Nutr. J.* **2012**, *11*, 81.

24. Boby, V. U.; Suvarna, V. C. Probiotics in Human Health: A Current Assessment. *Curr. Sci.* **2005**, *88*(11), 1744–1788.

25. Bodinier, M.; Denery, S.; Gourbeyre, P. Probiotics, Prebiotics, and Symbiotics: Impact on the Gut Immune System and Allergic Reactions. *J. Leukoc. Biol.* **2011**, *89*, 685–695.

26. Boirivant, M.; Strober, W. The Mechanism of Action of Probiotics. *Curr. Opin. Gastroenterol.* **2007**, *23*(6), 679–692.

27. Boon, N.; Jacobs, H.; Possemiers, S.; Verstraete, W.; van de Wiele, T. Inulin-type Fructans of Longer Degree of Polymerization Exert more Pronounced In Vitro Prebiotic Effects. *J. Appl. Microbiol.* **2007**, *102*, 452–460.

28. Borriello, S. P. Safety of Probiotics that Contain *Lactobacilli or Bifidobacteria. Clin. Infect. Dis.* **2003**, *36*(6), 775–780.

29. Bothe, E.; Gyorgy, E. Acid and Bile Tolerance, Adhesion to Epithelial Cells of Probiotic Microorganisms. *U.P.B. Sci. Bull., B* **2010**, *72*(2), 199–201.

30. Boyle, R. J.; Robins-Browne, R. M.; Tang Mimi, L. K. Probiotic Use in Clinical Practice: What are the Risks?. *Am. J. Clin. Nutr.* **2006**, *83*, 1256–1264.

31. Brummer, R. J.; de Vos, W. M.; Rijkers, G. T. *Br. J. Nutr.* **2011**, *106*(9), 1291–1296.

32. Canani, R.; Berni, G. Alfredo, Vecchio Andrea Lo, Probiotics as Prevention and Treatment for Diarrhea. *Curr. Opin. Gastroenterol.* **2009**, *25*(1), 18–23.

33. Carmuega, E.; Galdeano, C.; Maldonado, Alejandra, M.; Perdigón, G.; Weil, R. Mechanisms Involved in the Immunostimulation by Probiotic Fermented Milk. *J. Dairy Res.* **2009**, *76*(4), 446–454.

34. Carvalho, Isabel, S.; Hemaiswarya, S.; Raja, R.; Ravikumar, R. Mechanism of Action of Probiotics. *Braz. Arch. Biol. Technol.* **2013**, *56*(1), 78–82.

35. Cats, A.; Kuipers, E. J. Effect of Frequent Consumption of a *Lactobacillus casei*-containing Milk Drink in *Helicobacter pylori*-colonized Subjects. *Aliment. Pharmacol. Ther.* **2003**, *17*, 429–435.

36. Champagne, C.; Gardner, N. Challenges in the Addition of Probiotic Cultures to Foods. *Crit. Rev. Food Sci. Nutr.* **2005**, *45*, 61–84.

37. Chin, J.; Kailasapathy, K. Survival and Therapeutic Potential of Probiotic Organisms with Reference to *Lactobacillus acidophilus* and *Bifidobacterium* spp. *Immunol. Cell Biol.* **2000**, *78*, 80–88.

38. Chin, J.; Matsuzaki, T. Modulating Immune Responses with Probiotic Bacteria. *Immunol. Cell Biol.* **2000**, *78*, 67–73.

39. Choi, J.; Choi, Y. J.; Elise, L.; Ma, Im, E.; Pothoulakis, C.; Rhee, S. H. The Anti-cancer Effect of Probiotic *Bacillus polyfermenticus* on Human Colon Cancer Cells is Mediated through ErbB2 and ErbB3 Inhibition. *Int. J. Cancer J. Int. Cancer* **2010**, *127*(4), 780–790.

40. Chua, K. H.; Leong, S. F.; Sahadeva, R. P. Survival of Commercial Probiotic Strains to pH and Bile. *Int. Food Res. J.* **2011,** *18*(4), 1515–1522.

41. Connolly, E. *Lactobacillus reuteri* ATCC 55730 a Clinically Proven Probiotic. *Nutrafoods* **2004,** *3*(1), 15–22.

42. Corcoran, B. M.; Fitzgerald, G.; Ross, R. P.; Stanton, C. Survival of Probiotic *Lactobacilli* in Acidic Environments is enhanced in the Presence of Metabolizable Sugars. *Appl. Environ. Microbiol.* **2005,** *71*(6), 3060–3067.

43. Corcoran, B. M.; Fitzgerald, G. F.; Ross, R. P.; Stanton, C. Survival of Probiotic *Lactobacilli* in Acidic Environments is enhanced in the Presence of Metabolizable Sugars. *Appl. Environ. Microbiol.* **2005,** *71*(6), 289–299.

44. Crandall, P. G.; Lee, S. O.; Bryan, C. A.; Pak, D.; Ricke, S. C. The Role of Prebiotics and Probiotics in Human Health. *J. Prob. Health* **2013,** *1*, 108.

45. Crittenden, R.; Isolauri, E.; Playne, M.; Salminen, S.; Tuomola, E. Quality Assurance Criteria for Probiotic Bacteria. *Am. J. Clin. Nutr.* **2001,** *73*(1), 393–398.

46. Czarnecki, Z.; Goderska, K.; Nowak, J. Comparison of the Growth of *Lactobacillus acidophilus* and *Bifidobacterium bifidum* species in Media Supplemented with Selected Saccharides including Prebiotics. *Acta Sci. Pol., Technol. Aliment* **2008,** *7*(2), 5–20.

47. De Champs, C.; Balestrino, D.; Forestier, C.; Maroncle, N.; Rich, C. Persistence of Colonization of Intestinal Mucosa by a Probiotic Strain, *Lactobacillus casei* subsp. *rhamnosus* Lcr35, after Oral Consumption. *J. Clin. Microbiol.* **2003,** *41*(3), 1270–1273.

48. De Vrese, M. Probiotics Compensation for Lactase Insufficiency. *Am. J. Clin. Nutr.* **2001,** *73*, 421–429.

49. Del Piano, M. Evaluation of the Intestinal Colonization by Microencapsulated Probiotic Bacteria in Comparison with the Same Uncoated Strain. *J. Clin. Gastroenterol.* **2010,** *44*(1), 42–46.

50. Desai, A. R.; Powell, I. B.; Shah, N. P. Survival and Activity of Probiotic *Lactobacilli* in Skim Milk Containing Prebiotics. *J. Food Sci.* **2004,** *69*(3), FMS57–FMS60.

51. Dimaguila, M. A.; V. T.; Hunter, C. Effect of Routine Probiotic, *Lactobacillus reuteri* DSM 17938, use on Rates of Necrotizing Enterocolitis in Neonates with Birthweight < 1000 grams: A Sequential Analysis. *Pediatrics* **2012,** *12*(142), 1471–2431.

52. Ding, W. K.; Shah, N. P. Acid, Bile, and Heat Tolerance of Free and Microencapsulated Probiotic Bacteria. *J. Food Sci.* **2007,** *72*(9), 446–450.

53. Dunnel, C.; Kelly, P.; Halloran, S. Mechanisms of Adherence of a Probiotic *Lactobacillus* Strain During and After In Vivo Assessment in Ulcerative Colitis Patients. *Microbial Ecol. Health Dis.* **2004,** *16*, 96–104.

54. Elise, L. M. Anticancer effect of Probiotic *Bacillus polyfermenticus* on Human Colon Cancer Cells is Mediated through ErbB2 and ErbB3 Inhibition. *Int. J. Cancer* **2010,** *127*, 780–790.

55. Epton, H. A. S.; Wilson, A. R.; Sigee, D. Anti-bacterial Activity of *Lactobacillus plantarum* Strain SK1 against *Listeria monocytogenes* is due to Lactic Acid Production. *J. Appl. Microbiol.* **2005,** *99*, 1516–1522.

56. Faid, M.; Zinedine, A. Isolation and Characterization of Strains of Bifidobacteria with Probiotic Proprieties In Vitro. *World J. Dairy Food Sci.* **2007,** *2*(1), 28–34.

57. Fioramonti, J. Review Article: Bifidobacteria as Probiotic Agents—Physiological Effects and Clinical Benefits. *Aliment Pharmacol. Ther.* **2005,** *22*, 495–512.

58. Fitzgerald, G. F.; Ross, R. P.; Russell, D. A.; Stanton, C. Metabolic Activities and Probiotic Potential of bifidobacteria. *Int. J. Food Microbiol.* **2015,** *149*(1), 88–105.

59. Fuller, R. Probiotics in Human Medicine. *Gut* 1991, *32*(4), 439–442.

60. Garaiova, I. Probiotics and Vitamin C for the Prevention of Respiratory Tract Infections in Children attending Preschool: A Randomised Controlled Pilot Study. *Eur. J. Clin. Nutr.* **2015,** *69,* 373–379.

61. Geison, G. L. *The Private Science of Louis Pasteur.* Princeton University Press: Princeton, 1995.

62. Gibas-Dorna, M.; Olejnik, A.; Piątek, J. The Viability and Intestinal Epithelial Cell Adhesion of Probiotic Strain Combination—*In Vitro* Study. *Ann. Agric. Environ. Med.* **2012,** *19*(1), 99–102.

63. Gibson, G. Human Gut Microbiota and its Relationship to Health and Disease. *Nutr. Rev.* **2012,** *69*(7), 392–403.

64. Gibson, G. R.; Kolida, S. Synbiotics in Health and Disease. *Ann. Rev. Food Sci. Technol.* **2011,** *2,* 373–393.

65. Gibson, G. R.; Roberfroid, M. B. Dietary Modulation of the Human Colonic Microbiota: Introducing the Concept of Prebiotics. *J. Nutr.* **1995,** *125,* 1401–1412.

66. Gillen, A.; Sherwin, F. Louis Pasteur's Views on Creation, Evolution, and the Genesis of Germs. *Ans. Res. J* **2008,** *1,* 43–52.

67. Gogineni, V. K.; Morrow, Lee, E.; Maleske, M. A. Probiotics: Mechanisms of Action and Clinical Applications. *J. Probiotics Health* **2013,** *1,* 101.

68. Guarner, F. Prebiotics and Mucosal Barrier Function. *J. Nutr.* **2006,** *136*(8), 2269.

69. Gudmand-Hoye, E.; Olesen, M. Efficacy, Safety, and Tolerability of Fructooligosaccharides in the Treatment of Irritable Bowel Syndrome. *Am. J. Clin. Nutr.* **2000,** 72, 1570–1575.

70. Gueimonde, M.; de los Reyes-Gavilán, G.; Margolles, A.; Sánchez, B. Antibiotic Resistance in Probiotic Bacteria. *Frontiers Microbiol.* **2013,** *4,* 202.

71. Halvorsen, R. The Use of Probiotics for Patients in Hospital: A Benefit Risk Assessment. Opinion of the Steering Committee of the Norwegian Scientific Committee for Food Safety. *Norwegian Scientific Committee for Food Safety (VKM)*—Doc. no version 09-100, 2009.

72. Havenaar, R.; Schaafsma, G.; Spanhaak, S. The Effect of Consumption of Milk Fermented by *Lactobacillus casei* Strain Shirota on the Intestinal Microflora and Immune Parameters in Humans. *Eur. J. Clin. Nutr.* **1998,** 52, 899–907.

73. Hayasawa, H.; Ishibashi, N.; Yaeshima, T. Bifidobacteria: Their Significance in Human Intestinal Health. *Mal. J. Nutr.* **1997,** *3,* 149–159.

74. Hemarajata, P.; Versalovic, J. Effects of Probiotics on gut Microbiota: Mechanisms of Intestinal Immunomodulation and Neuromodulation. *Ther. Adv. Gastroenterol.* **2013,** *6*(1), 39–51.

75. Hempel, S. Probiotics for the Prevention and Treatment of Antibiotic-associated Diarrhea. *Syst. Rev. Meta-anal., JAMA* **2012,** *307*(18), 1959–1969.

76. Herich, R.; Levkut, M. Lactic Acid Bacteria, Probiotics and Immune System. *Vet. Med.—Czech* **2002,** *47*(6), 169–180.

77. Hickson, M. Probiotics in the Prevention of Antibiotic-associated Diarrhoea and *Clostridium difficile* Infection. *Ther. Adv. Gastroenterol.* **2011,** *4*(3), 185–197.

78. Horosheva, T. V.; Sorokulova, I.; Vodyanoy, V. Efficacy of *Bacillus* Probiotics in Prevention of Antibiotic-associated Diarrhoea: A Randomized, Double-blind, Placebo-controlled Clinical Trial. *JMM Case Rep.* **2014,** *1*(3), 1–6.

79. Isolauri, E. Probiotics: A Role in the Treatment of Intestinal Infection and Inflammation. *Gut* **2002,** *50*(3), 54–59.

80. Isolauri, E.; Salminen, S. Intestinal Colonization, Microbiota, and Probiotics. *J. Paediatrics* **2013**, *149*(5), 115–120.
81. Jurgen, S.; Sonnenborn, U. The Non-pathogenic *Escherichia coli* Strain Nissle 1917. *Feat. Versatile Probiot.* **2009**, *21*(3–4), 122–158.
82. Kaga, C.; Takagi, A. *Lactobacillus casei* Shirota Enhances the Preventive Efficacy of Soymilk in Chemically Induced Breast Cancer. *Cancer Sci.* **2013**, *104*(11), 1508–1514.
83. Kajander, K.; Korpela, R.; Niittynen, L. Galacto-oligosaccharides and Bowel Function. *Scandinavian J. Food Nutr.* **2007**, *51*(2), 62–66.
84. Kekkonen, R. A.; Lahti, L.; Salonen, A. Associations between the Human Intestinal Microbiota, *Lactobacillus rhamnosus* GG and Serum Lipids Indicated by Integrated Analysis of High-throughput Profiling Data. *Peer J.* **2013**. DOI:10.7717/peerj.32.
85. Kermanshahi, R. K.; Tahmourespour, A. The Effect of a Probiotic Strain (*Lactobacillus acidophilus*) on the Plaque formation of Oral Streptococci. *Bosnian J. Basic Med. Sci.* **2011**, *11*(1), 38–40.
86. Kunz, C.; Kuntz, S.; Rudloff, S. *Bioactivity of Human Milk Oligosaccharides, Food Oligosaccharides: Production, Analysis and Bioactivity*. First Edition. John Wiley and Sons, Ltd., 2014.
87. Larqué, E.; Sabater-Molina, M.; Torrella, F.; Zamora, S. Dietary Fructooligosaccharides and Potential Benefits on Health. *J. Physiol. Biochem.* **2009**, *65*(3), 315–328.
88. Lebeer, S.; Segers, M. E. Towards a Better understanding of *Lactobacillus rhamnosus* GG—Host Interactions. *Microbial Cell Factories* **2014**, *13*(S1–S7), 2–16.
89. Licciardi, P. V.; Tang, M. L. Vaccine Adjuvant Properties of Probiotic Bacteria. *Discov. Med.* **2011**, *12*(67), 525–33.
90. Lilly, D. M.; Stillwell, R. H. Probiotics-growth Promoting Factors Produced by Micro-organisms. *Science* **1965**, *147*(3659), 747–248.
91. Liong, M. T. Safety of Probiotics: Translocation and Infection. *Nutr. Rev.* **2008**, *66*(4), 192–202.
92. Liu, Z. Q.; Yang, G.; Yang, P. C. Treatment of Allergic Rhinitis with Probiotics: An Alternative Approach. *North Am. J. Med. Sci.* **2013**, *5*(8), 465–468.
93. López, Mercedes, G.; Urías-Silvas, J. E. Agave Fructans as Prebiotics. In *Recent Advances in Fructooligosaccharides Research*; López, Mercedes, G., Urías-Silvas, J. E., Eds.; 2007. ISBN:81-308-0146-9.
94. López-Caballero, M. E.; López de Lacey, A. M.; Montero, P.; Pérez-Santín, E. Survival and Metabolic Activity of Probiotic Bacteria in Green Tea. *LWT–Food Sci. Technol.* **2014**, *55*(1), 411–414.
95. Macfarlane, G. T.; Steed, H.; Macfarlane, S. Bacterial Metabolism and Health-related effects of Galacto-oligosaccharides and Other Prebiotics. *J. Appl. Microbiol.* **2008**, *104*(2), 305–344.
96. Mäkeläinen, H.; Ouwehand, A. C.; Rautonen, N.; Saarinen, M.; Stowell, J. Xylo-oligosaccharides and Lactitol Promote the Growth of *Bifidobacterium lactis* and *Lactobacillus* species in Pure Cultures. *Benef. Microbes.* **2010**, *1*(2), 139–148.
97. Maldonado, G. C. Proposed Model: Mechanisms of Immunomodulation Induced by Probiotic Bacteria. *Clin. Vaccine Immunol.* **2007**, *14*(5), 485–492.
98. Marcel, R. Prebiotic Effects: Metabolic and Health Benefits. *Br. J. Nutr.* **2010**, *104*(S2), 1–63.
99. Massaguer-Roig, S.; Zacarchenco, P. B. Properties of *Streptococcus Thermophilus* Fermented Milk Containing Variable Concentrations of *Bifidobacterium longum* and *Lactobacillus Acidophilus*. *Brazilian J. Microbiol.* **2006**, 37, 338–344.

100. McFarland, L. V. Systematic Review and Meta-analysis of *Saccharomyces boulardii* in Adult Patients. *World J. Gastroenterol.* **2010**, *16*(18), 2202–2222.
101. Mirlohi, M.; Soleimanian-Zad, S. Investigation of Acid and Bile Tolerance of Native Lactobacilli Isolated from Fecal Samples and Commercial Probiotics by Growth and Survival Studies. *Iranian J. Biotechnol.* **2009**, *7*(4), 233–240.
102. Mortaz, E. Probiotics in the Management of Lung Diseases. *Mediat. Inflamm.* **2013**, Article ID 751068, 10.
103. Nami, Y. Probiotic Potential for Anticancer Activity of Bacterium *Lactobacillus plantarum*, **2008**, *65*(3), 67–87.
104. Nami, Y. Assessment of Probiotic Potential and Anticancer Activity of Newly Isolated Vaginal Bacterium *Lactobacillus plantarum* 5BL. *Microbiol. Immunol.* *58*(9), 492–502.
105. Neville, B. A.; O'Toole, P. W. Probiotic Properties of *Lactobacillus salivarius* and Closely Related *Lactobacillus* species. *Future Microbiol.* **2010**, *5*(5), 759–774.
106. Nino, B. Probiotics, Prebiotics and Gut Microbiota. In *International Life Sciences Institute Monograph Series*, 2013, ISSN 2294-5490.
107. Oise, B. C. Antibacterial Effect of the Adhering Human *Lactobacillus acidophilus* Strain LB. *Antimicrobial Agents Chemother.* **1997**, *41*(5), 1046–1052.
108. Okumura, K.; Takeda, K. Effects of Fermented Milk Drink Containing *Lactobacillus casei* Strain Shirota on the Human NK-Cell Activity, Effects of Probiotics and Prebiotics. *J. Nutr.* **2007**, *137*, 3791–793.
109. Oliveira, R.; Pinheiro, S. Use of Lactulose as Prebiotic and its Influence on the Growth, Acidification Profile and Viable Counts of Different Probiotics in Fermented Skim Milk. *Int. J. Food Microbiol.* **2011**, *145*(1), 22–27.
110. Ouwehand, A. C. Antiallergic Effects of Probiotics. *J. Nutr.* **2007**, *137*, 794S–797S.
111. Oyetayo, E. Safety and Protective Effect of *Lactobacillus acidophilus* and *Lactobacillus casei* Used as Probiotic Agent In Vivo. *Afr. J. Biotechnol.* **2003**, *2* (11), 448–452.
112. Ozdemir, O. Various effects of Different Probiotic Strains in Allergic Disorders: An Update from Laboratory and Clinical Data. *Clin. Exp. Immunol.* **2010**, *160*(3), 295–304.
113. Patel, A. R.; Shah, N. P.; Prajapati, J. B. Immunomodulatory Effects of Probiotics in the Treatment of Human Immunodeficiency virus (HIV) Infection. *Biomed. Preventive Nutr.* **2014**, *4*(1), 81–84.
114. Pleasants, J. History of Germfree Animal Research at Lobund Laboratory. In *Proceedings of the Indiana Academy of Science*, 1965.
115. Rabia, A. S.; Nagendra, P. Immune System Stimulation by Probiotic Microorganisms. *Crit. Rev. Food Sci. Nutr.* **2014**, *54*(7), 938–956.
116. Reid, G. Regulatory and Clinical Aspects of Dairy Probiotics. In *FAO/WHO Expert Consultation on Evaluation of Health and Nutritional Properties of Powder Milk with Live Lactic Acid Bacteria*, 2015. ftp://ftp.fao.org/es/esn/food/Reid.pdf.
117. Reuter, G. The Lactobacillus and Bifidobacterium Microflora of the Human Intestine: Composition and Succession. *Curr. Issues Intest. Microbiol.* **2001**, *2*(2), 43–53.
118. Roberfroid, M. B. Prebiotics: Preferential Substrates for Specific Germs? *Am. J. Clin. Nutr.* **2001**, *73*, 406–409.
119. Sams, E.; Keith, T.; Marvin, W. The Battle for Life: Pasteur, anthrax, and the First Probiotics. *J. Med. Microbiol.* **2014**, *63*(Part 11), 1573–1574.
120. Shanahan, F. Molecular Mechanisms of Probiotic Action: It's all in the Strains. *Gut* **2011**, *60*, 1026–1027.

121. Shoaf, K.; Mulvey, G. L.; Armstrong, G. D.; Hutkins, R. W. Prebiotic Galactooligo-saccharides Reduce Adherence of Enteropathogenic *Escherichia coli* to Tissue Culture Cells. *Infect. Immun.* **2006,** *74*(12), 6920–6928.
122. Shokryazdan, P. Probiotic Potential of *Lactobacillus* Strains with Antimicrobial Activity against some Human Pathogenic Strains. *Biol. Med. Res. Int.* **2014,** 16, Article ID 927268.
123. Slavin, J. Fiber and Prebiotics: Mechanisms and Health Benefits. *Nutrients* **2013,** *5*(4), 1417–1435.
124. Snel Wilco, C. A.; Pang, K. S.; Mulder, G. J. Glutathione Conjugation of Bromosul-fophthalein in Relation to Hepatic Glutathione Content in the Rat *In Vivo* and in the Perfused Rat Liver. *Hepatology* **1995,** *21*(5), 1387–1394.
125. Testa, B.; Kramer, S. D. The Biochemistry of Drug Metabolism—An Introduction: Part 4. Reactions of Conjugation and their Enzymes. *Chem. Biodivers.* **2015,** *12*(4), 685–696.
126. Tomaro, D. C. Cholesterol Assimilation by *Lactobacillus* Probiotic Bacteria: An In Vitro Investigation. *Bio. Med. Res. Int.* **2014,** *380316*, 9.
127. Torres, D. P. M.; Maria, G. P. F.; Teixeira, J. A.; Rodrigues, L. R. Galacto-Oligosaccha-rides: Production, Properties, Applications, and Significance as Prebiotics. *Comprehensive Rev. Food Sci. Food Safety* **2010,** *9*, 438–454.
128. Trevors, J. T. The Subsurface Origin of Microbial Life on the Earth. *Res. Microbiol.* 2002, *153*(8), 487–491.
129. Tripathi, M. K.; Giri, S. K. Probiotic Functional Foods: Survival of Probiotics during Processing and Storage. *J. Functional Foods* **2014,** *9*, 225–241.
130. Tsai, C. C. Cholesterol-lowering Potentials of Lactic Acid Bacteria Based on Bile-salt Hydrolase Activity and Effect of Potent Strains on Cholesterol Metabolism In Vitro and In Vivo. *Sci. World J.* **2014,** *690752*, 10.
131. Tuohy, K. M. A Human Volunteer Study to Determine the Prebiotic Effects of Lactu-lose Powder on Human Colonic Microbiota. *Microbial. Ecol. Health Dis.* 2002, *14*, 165–173.
132. Walker, A. W. Mechanisms of Action of Probiotics. *Clin. Infect. Dis.* **2008,** *46*(2), 87–91.
133. West, N. P. Probiotic Supplementation for Respiratory and Gastrointestinal Illness Symptoms in Healthy Physically Active Individuals. *Clin. Nutr.* **2014,** *33*(4), 581–587.
134. Wheater, D. The Characteristics of *Lactobacillus plantarum*, *L. helveticus* and L. *casei*. *J. Gen. Microbiol.* **1955,** *12*, 133–139.
135. Wold, A. E. Immune Effects of Probiotics. *Scand. J. Nutr.* **2001,** *45*, 76–85.
136. Zaman, S. A.; Sarbini, S. R. The Potential of Resistant Starch as a Prebiotic. *Crit. Rev. Biotechnol.* **2015,** *13*, 1–7.

CHAPTER 7

FUNCTIONAL FOOD, NUTRACEUTICALS, AND HUMAN HEALTH

MURLIDHAR MEGHWAL[1*] and TRIDIB KUMAR GOSWAMI[2]

[1]*Food Science and Technology Division, Center for Emerging Technologies, Jain University, Jain Global Campus, Jakkasandra 562112, Kanakapura Main Road, Ramanagara Dist., Karnataka, India. Email: murli.murthi@gmail.com*

[2]*Agricultural and Food Engineering Department, Indian Institute of Technology, Kharagpur 721302, West Bengal, India. E-mail: tkg@agfe.iitkgp.ernet.in*

Corresponding author.

CONTENTS

ABSTRACT

There is a link between nutrition and medicine. Functional foods are best nutrition in present scenario. There are a lot of food sources for nutraceuticals and they have enhanced bioavailability of nutrition. The chemistry and structure of nutraceuticals can be helpful to understand the bioavailability of these food-based nutraceuticals. Micronutrients, vitamins, isoflavones, flavanoids, carotenoids and lycopene are some of the major nutraceuticals. Soy protein, soy isoflavones, cardiovascular and bone health are also relevant to the same. The study on using GCBs as a functional additive has dominant compound 5-caffeoylquinic acid in GCB; phenolics released during simulated digestion were found highly bioavailable in vitro. Results indicate that phenolic compounds from bread enriched with powdered GCB were highly mastication-extractable, which may predict their high bioaccessibility and bioavailability; sensory characteristics results indicated that a partial replacement of wheat flour in bread with up to 3% ground GCB powder gives satisfactory overall consumer acceptability. This chapter includes information on the following:

Phytoestrogens: Mechanism of action, menopause, breast, and prostate cancer. Citrus flavanoids and other natural cholesterol lowering agents; carotenoids: metabolism and disease; lycopene: source, properties and nutraceuticals potential, nutraceuticals—garlic, grape, tea; Omega-3 fatty acids, antioxidant, chemoprevention, and Functional Food; olive oil and plant sterols, Omega-3 fatty acids and eicosanoids; Omega-3 fatty acids and lipoprotein metabolism; Omega-3 fatty acids, insulin resistance, and rheumatoid arthritis; and free radicals and oxidative stress—antioxidant mechanisms, the biochemical basis for nutraceuticals for the chemoprevention of disease.

Herbs and spices have lot of application to functional foods. Recent studies on examining the consumers' willingness to buy functional foods showed that consumers with higher health motivation and more trust in the food industry reported higher willingness to buy functional foods than the participants with lower health motivation and less trust in the industry and study suggests that cultural factors play a significant role in the acceptance of functional foods. Therefore caution should be exercised in generalizing research findings from Western countries to others.

7.1   INTRODUCTION

The food industry from many years has been interested in tracing out the components in foods which have health benefits and can be used in the development of functional food and nutraceutical products, such as fiber, phytosterols, peptides, proteins, isoflavones, saponins, phytic acid, probiotics, prebiotics, and functional enzymes.[1] Currently, the popularity of functional foods,[10] which consist of dietary supplements,[30] fruits and vegetable-based products,[12] and beverages, are in great demand due to their health promoting effect, prevention, and therapy of diseases.[8,25] Herbs and spices[27] are also receiving great attention in medicinal uses and are used all over the globe.[27] There is need for research on efficacy and safety determination of these commodities.[35] The functional food manufactures distributes products satisfying specific needs for those involved in sports at the professional level and for those who are interested in keeping their own healthy lifestyle.[9,10,22] The functional and its related food products aim at creating novel high quality ready-to-eat foods with functional activity which can be useful for promoting public health.[9,21] Some of the latest projects going on these aspect and their objectives is to include the application of plant cell and in vitro culture systems to create very large amounts of high-value plant secondary metabolites with recognized anticancer,[25,31] antilipidemic, anticholesterol,[29] antimicrobial, antiviral, antihypertensive, and anti-inflammatory properties and to include them in specific food products.[21] The functional food and nutraceuticals has found their application in social research context.[22] Functional food and nutraceuticals aim at the realization of ready-to-eat food for breakfast and sport with a high content of compounds endowed with functional activity useful for promoting public health.[22] There is already a wide range of foods available to today's consumer but now the challenge is to identify functional foods that have the potential to improve health and wellbeing, reduce the risk from or delay the onset of major diseases such as cardiovascular disease, cancer, osteoporosis, obesity, diabetes, blindness, malnutrition, anemia, and so on.[9,10,21] Along with a healthy lifestyle, functional food[22] can provide good health and wellbeing. Irregular, unbalanced, untimely, non-hygienic and non-healthy eating habits unequivocally and causally associated with the risk of obesity, cardiovascular disease, type 2 diabetes, stroke, cancers,[25,31] and neurodegenerative disorders.[10,18] A functional food can be defined as

- Foods or dietary components that may provide a health benefit beyond basic nutritional requirement.[28]

- Foods that have biologically active components which impart health benefits or physiological effects.
- Food that includes both functional attributes of conventional foods and new food products redesigned new components.[10]

This chapter presents basic information on functional food and nutraceuticals, and how these are related to human health.

7.2 CLASSIFICATION OF FUNCTIONAL FOODS

Various attributes of functional food are indicated in Figure 7.1. Functional food can be classified as follows:

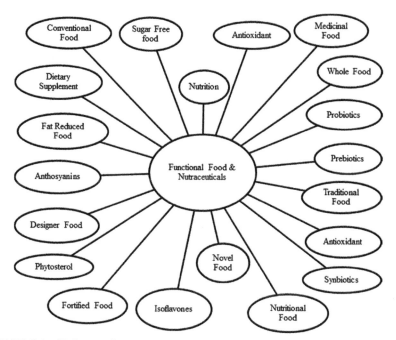

FIGURE 7.1   Various attributes of functional food.

7.2.1 CONVENTIONAL OR WHOLE FOODS

Conventional foods are mainly consumed for two purposes, namely: keeping good general health, and for keeping oneself live and for sustenance.

7.2.2 FUNCTIONAL FOODS

The functional food contains probiotics (life organisms) and prebiotics.[20] A few examples of probiotics are lactose intolerance, diarrhea, disease resistance, and healthy gut.[16,20,23]

7.2.3 DESIGNER FOODS (NUTRACEUTICALS INCORPORATED FOODS)

Designer foods are taken for the treatment of specific disease for the prevention of: Cancer,[31] heart disease,[14] diabetes mellitus, and osteoporosis.[25]

Functional Food Products

a. Conventional product marketed with new information: oats soluble fiber.
b. Conventional product with added or enhanced ingredient: spread or added plant sterols; chocolate with enhanced flavanols.
c. "New" product to deliver functional ingredient: Beverages—functional teas, water.[10,32]
d. Removing negative functionality: Oils with heart—healthy profile.

Unique features of functional food are

a. Being a conventional or everyday food to be consumed as part of the normal or usual diet.
b. Composed of natural (as opposed to synthetic) components sometimes in increased concentration or present in foods that would not normally supply them.
c. Having positive effect beyond nutritive value/basic nutrition that may enhance well—being and health and/or reduce the risk of disease or provide health benefit so as to improve the quality of life including physical, psychological, and behavioral performances, and have scientifically based claims.

From a practical point of view, a functional food can be (1) a natural food, (2) a food to which a component has been added, (3) a food from which a component has been removed, (4) a food where the nature of one or more components has been modified, (5) a food in which the bioavailability of one or more components has been modified, or (6) any combination of these possibilities.[10,22,24]

The consumption of midday meal or midnight snack food is becoming one of the most popular eating-habit trends, despite its close correlation with several metabolic disorders. Eating small amounts of food on a regular basis is considered healthy, but often available snacks in the market provide excess calories and fats and little or no nutrient value to the diet. In addition to that most snacks contain preservatives, sweeteners, and flavoring agents that can have negative effects on health. Healthy diets include factors such as making proper food, keeping track of meal times, and regulating the amount of food intake.[24]

7.3   EMERGENCE OF FUNCTIONAL FOOD AND MEDICINAL NUTRIENTS

The origin of the functional food[22] lies in the roots of human civilization origin.[10] Many past societies viewed sickness and disease as punishment sent down from the gods. The treatment of ailments usually involved prayer to appease angry gods in conjunction with the consumption of "magic potion." The early writing of the medicinal application of plants may be viewed as the basis of modern pharmacological medicine such as Aurveda, Hunani, and homeopathy. The origin of functional foods is probably a combination of at least two things: (1) Our distant ancestors probably noticed that when animals were ailing, they often ate certain plants that they would not other- wise eat; and (2) our distant ancestors thought that the activities of plants and other aspects of nature probably seemed magical.[10]

7.4   INFLUENCE OF DIET ON DISEASE

The recent studies show that food habit has lot of influence on our diet. The improper food habit has known to cause 30% cardiovascular disor- ders, 35% cancers,[31] 70% constipation, 50% obesity, 25% diabetes type-II, and 30% dental caries. Tremendous amounts of money, human death, and human sufferings may be saved by teaching the population about healthy eating habits and by producing modified foods to reduce the risk of these diseases.

7.5   GLOBAL NUTRACEUTICALS AND FUNCTIONAL FOOD INDUSTRY

Different factors are responsible for the rapid global growth of the nutraceuticals and functional food industry[34] and some of the major reasons are mentioned below:

- An increase in public health consciousness.
- Increased access to the information through education and an enquiring medium has resulted in a rapidly emerging self-care movement among consumers.
- Change in perspective from food as sustenance to food as having nutritional benefit.
- An aging population.
- Escalating health-care costs.
- Recent advances in research and technology.
- Change in government regulations and accountability.
- Expansion of the global marketplace.[24]

7.6   REASONS FOR GROWING INTEREST IN FUNCTIONAL FOODS

- Collective health consciousness grows with more-advanced age.
- With age comes an increased incidence of disease.
- More newspaper and magazine articles are dedicated to the relationship between diet and health, nutraceuticals concepts.[34]
- More television programs address topics of disease prevention and treatment.
- Public awareness via the Internet regarding the etiology, prevention, and treatment of various diseases is most significant.
- Dissatisfaction with current western medical services.
- Side effects of current western medical services.
- General dietary recommendations for disease risk reduction, for example, more fruits, vegetables[12], and whole grains are well accepted.[8,28]
- The emerging concept is that intact foods are probably more powerful that individual components.
- Functional food is cheaper compared to western medicines.
- Availability of functional foods will continue to grow because it is sought after and very profitable.

7.7 CONSUMER INTEREST IN FUNCTIONAL FOOD

Functional food and nutraceuticals are not intended only to satisfy hunger and provide the necessary human nutrients, but also to prevent nutrition-related diseases and increase the physical and mental wellbeing of their consumer.[10,34] Specially, among participants in space science and missions, recognition of nutraceuticals and dietary supplements[28,30] is growing for their potential in reducing health risks and to improve health quality and eating habits during long-term flights and missions.[33] India and China are the two major countries known for their production of traditional functional food products and nutraceuticals.[5,10,34]

- Demand for characteristics based on household model suggests human capital.
- Nutrition knowledge, ability to find and process new information: health status.
- Potential benefits of functional foods higher when predisposed to disease. Health risks (substitution possibilities).[22]
- How well functional foods fit into current diet, food habits: cost.
- How much more costly than alternative: includes hidden cost of lost utility if not as tasty.

7.8 INDUSTRY INTEREST IN FUNCTIONAL FOODS

Companies working in food industry have high expectations for food products that can fulfill the requirements of consumers for good health and healthy life style.[33]

- Market demand for particular characteristics.[24]
- Add value or market share to existing product or product ingredient, for example: soy-protein branding for addition to cereals, beverages.[11]
- Create new market for innovative products, for example: new beverages.
- Added value upstream in supply chain when commodity becomes differentiated, for example: soybeans with low linoleic (less trans-fats).

7.9 IMPACT OF FUNCTIONAL FOODS ON MARKET

Estimates of functional food market are having huge potential and depend on what are included [24]:

- Clear impact in some markets: calcium-enhanced market; soy beverages rapid growth; functional breads and grains 10% of that market.[10]
- Investments in new brands focus on reliability of composition, taste: Solae soy protein joint venture by Dupont and Bunge.
- Investments in new products tailor packages of functional components: Quaker Oats "Nutrition for Women" combines vitamins,[2] soy, calcium.

Effective scientific research alone does not make a product successful in the marketplace. The product must be in an adequate form so that the consumers can easily accept it.[24] Therefore, as one of the first steps of product development, it is necessary to explore which diseases consumers are concerned about so that the product could be successful in the market. According to surveys, primary health concerns among consumers are cardio-vascular diseases, stress, high blood pressure, malignant tumor, diseases of the digestive system, arthritis, and obesity. However, it is more interesting, to identify those diseases that consumers consider as curable with the help of nutrition consumption. Another important success factor for the marketing of Functional Food is the price for this type of food in comparison to conventional food products.[10] Examples of recently launched Functional Food products indicate that consumers are only willing to accept limited price increments for such products. In general, price increments of 30–50% are observed in high-volume Functional Food segments like functional dairy products. It seems most likely that higher price increments will be accepted by consumers only for such Functional Food products that have a proven health effect related to a disease which directly influences them in the near future.[22] However, as yet, such products have been rarely launched in the market. In this sense, relatively high price increments are probably the main reason for the limited market success of several Functional Food products introduced over recent years. It appears that concepts for Functional Food products should be based on food products with a positive health image and avoid a distinct medical or clinical perspective.[10] Consumers expect Functional Food in such retail outlets and most of them are not willing to go to specific shops just to buy Functional Food products. Such a strategy does not exclude serving specific distribution channels (pharmacies, health food shops) either with the same or a modified product. In addition, it should be Functional Food products can be made available for impulse buying (e.g., in specific convenience-oriented shops) as well.[33] Siegrist et al.[26] did internet survey based study on examining the consumers' willingness to buy functional foods for German and Chinese people and showed that consumers in

China were much more willing to buy functional foods, compared with their German counterparts, in both countries, the participants with higher health motivation and more trust in the food industry reported higher willingness to buy functional foods than the participants with lower health motivation and less trust in the industry; food neophobia had a negative impact on acceptance of functional foods in the Chinese people but same was observed for the German people and finally study suggest that cultural factors play a significant role in the acceptance of functional foods; therefore caution should be exercised in generalizing research findings from Western countries to others.

7.10 MICRONUTRIENTS AND MACRONUTRIENTS

There are two types of nutrients which we need for our survival, micronutrients and macronutrients. This division is based on the quantity of a nutrient the body needs. Human needs micronutrients in small amounts and macronutrients in large amounts.

Micronutrients are nutrients that the human body needs in minute amounts so that it can function properly. Although, they are needed only in small amounts, their deficiency leads to critical health problems. Most of the diseases and conditions that people face today are due to deficiency of micronutrients. The World Health Organization (WHO) says that if we eliminate this deficiency, labor efficiency will increase multifold. Here is a list of micronutrients: vitamins: Vitamin A, Vitamin B, Vitamin C, Vitamin D, Vitamin E, Vitamin K, and carotenoids;[2,28] minerals: boron, calcium, chloride, chromium, cobalt, copper, fluoride, iodine, iron, magnesium, manganese, molybdenum, phosphorous, potassium, selenium, sodium, and zinc; organic acids: acetic acid, citric acid, lactic acid, malic acid, choline, and taurine. It is better to get these from natural sources such as fruits and vegetables,[12] as this seems to increase their usefulness.[8]

Macronutrients constitute the bulk of the food we eat. The macronutrients are proteins, carbohydrates, and fats.

a. **Proteins** are called building blocks of life. It is what most of our body is made up of. Proteins themselves are made of amino acids. Some good sources of protein include legumes, soy, milk, and milk products. Standard amino acids: arginine, aspartic acid (aspartate), asparagine, cystine, glutamic acid (glutamate), glutamine, glycine, histidine, isoleucine (branched chain amino acid), leucine (branched chain amino acid), lysine, methionine, phenylalanine, proline, serine, threonine, tryptophan, tyrosine, and valine (branched chain amino acid).

b. **Carbohydrates** are made up of sugar or starches. They are the main energy providers for our body. Excess carbohydrates are converted into fat and stored in our body. All foods have carbohydrates in some measure. Major carbohydrates: Fructose, glucose, sucrose, ribose, amylose, amylopectin, maltose, lactose, and galactose.

c. **Fats** are substances that your body stores for future use. Although, most people think that fats are to be avoided altogether, there is a distinction to be made here. There are good fats and bad fats. Bad fats are to be avoided and good fats are to be eaten. Good fats are very much necessary for the proper functioning of the body. Proteins, carbohydrates, and fats are to be eaten in proper proportion. If they are not, lifestyle diseases will affect health.

(i) **Saturated fats:** Butyric acid, caproic acid, caprylic acid, capric acid, lauric acid, myristic acid, pentadecanoic acid, palmitic acid, margaric acid, stearic acid, arachidic acid, behenic acid, lignoceric acid, and cerotic acid.

(ii) **Monounsaturated fats:** Myristol, pentadecenoic, palmitoyl, heptadecenoic, oleic acid, eicosen, erucic acid, and nervonic acid. Polyunsaturated fats: Linoleic acid (LA)—an essential fatty acid, α-linolenic acid (ALA)—an essential fatty acid, stearidonic acid, arachidonic acid, timnodonic acid (EPA), clupanodonic acid, and cervonic acid (DHA).

(iii) **Essential fatty acids**: These two essential fatty acids are the starting point for other important omega-acids[6] (e.g., DHA, EPA) (1) ALA (18:3) Omega-3 fatty acid[6] and (2) LA (18:2) Omega-6 fatty acid.[6] Macronutrients that support metabolism are water and air and air with water vapor which also contains water.

7.11 SOYFOODS: A POTENT FUNCTIONAL FOOD ADJUNT

Soybean is one of the world's most valuable crops. Soybean is the most widely grown and utilized legume in the world. Soybean is native crop of Eastern Asia. Soybean grows in pods featuring edible seeds. Soy seeds comprise of 8% seed coat or hull, 90% cotyledons, and 2% hypocotyls axis.

7.11.1   NUTRITIONAL ASPECTS OF THE SOY FOOD

- High in fiber.
- Low in calories.
- Low in fat.
- Cholesterol-free.[29]

7.11.2   FUNCTIONAL ASPECTS OF THE SOY FOOD

- Emulsify fat and bind water.
- Elastic gel texture, imparting an interesting mouth feel.
- Low in viscosity.

7.11.3   COMPOSITION AND MAJOR COMPONENTS OF SOYBEAN

Soy protein is excellent source of vegetable protein. The most well balanced of all vegetable proteins are rich source eight essential amino acids. High lysine content and hence nutritionally complementary to cereal protein possesses protease inhibitors.[11] Soybean helps in maintaining protein balance in diet replacing animal protein. The soybean composition is as follows: oil—18%; protein—38%; soluble carbohydrate—15%; insoluble carbohydrate—15%; and moisture, ash, and others—14%. The major soy proteins are glycinin, β-conglycinin. The major soy isoflavones are genistein and diadzein.

7.11.4   HEALTH BENEFITS OF SOY CONSUMPTION

- Polyunsaturated, Omega-6 and Omega-3 are two essential fatty acids[6] and, saturated fatty acids of soy reduces cholesterol levels, heart attacks,[14] and stroke.
- The isoflavones or phytoestrogens of soy: relieves certain menopausal symptoms, prevents cancer,[17] and reverses osteoporosis.[25]
- In non-insulin-dependent diabetes promotes serum insulin production and maintains potassium to sodium (3/1–11/1).

7.12   FUNCTIONAL FOOD INGREDIENTS FROM TEA

7.12.1   GREEN TEA: CATECHINS AND MAIN THE AFLAVINS IN BLACK TEA

Currently, green tea has been recognized as a functional food for the prevention human diseases and consumed by a large proportion of the world's population.[15,32,36] Demand for tea consumptions along with other drugs has been found to be increasing.[32] Thus, there is a need for finding scientific evidences for possible drug interaction and reactions are needed to be established to ensure the safety and avoid side effects. Catechins comes under flavanol compounds and are present in green tea, chocolate, cocoa, pomegranate juice, and red wine.[4,15,32] Green tea (*Camellia sinensis*) is abundant source of catechins consisting of[13,28]: epigallocatechin-3-gallate (EGCG), epigallocatechin, epicatechin, and epicatechin-3-gallate.[15,32]

7.13   HEALTH BENEFITS OF THEAFLAVINS OF TEA: AS INTERMEDIATES OF MEDICINE

- Numerous studies have found that theaflavins have effects of antioxidation and cancer,[17] scavenging, or inhibiting free radicals.[25]
- Theaflavins are also shown to slowdown tumor spreading, decreasing blood thickness, and preventing heart diseases.[14]
- Theaflavins are more efficient than tea catechins in regulating fat metabolism and improving blood circulation.[15,32]
- Theaflavins have also been shown to improve immune system function and have anti-inflammation effect.
- In addition, theaflavins can help repair the damage caused by smoking by increasing the body's ability to degrade and eliminate the harmful components in smoking.
- Theaflavins and gallates are also inhibitors of HIV (human immunodeficiency virus) reverse transcriptase and other DNA and RNA polymerases.

Theaflavins can be used as natural pigments and antioxidation agent in natural food. The antioxidation effect of theaflavins is higher than tea polyphenols, BHA, and BHT. Owing to the functions of deodorization sterilization, pigmentation, etc., theaflavins have been widely used in chemical commodity products such as shampoo, toothpaste, and natural dyes.[32]

7.14   HEALTH BENEFITS OF TEA POLYPHENOLS

- To cure cardiovascular diseases tea polyphenols showed good effects on fibrinolysis and antiblood agglutination.[32]
- Especially the free catechins which possess the ability to prevent platelet from agglutination.[15]
- To inhibit reducing the possibility of hypertensive and coronary heart disease.[14]
- EGCG kill leukemia cell.
- Weight loss.
- Prevent brain damage.
- Lowers cholesterol.
- Anticarcinogenic: It possesses the ability of blocking the formation of endogenous nitrosoamine, which is a carcinogen.
- To strengthen the capacity of human immunity and antisenescent activity.
- To enhance the tenacity and permeating ability of blood capillary and improve the resistance of blood vessel.[32]

The study on green coffee beans (GCBs) from Ethiopia, Kenya, Brazil, and Colombia using as a functional additive found to have dominant compound 5-caffeoylquinic acid in GCB. Phenolics released during simulated digestion were found highly bioavailable in vitro. Simulated digestion released phytochemicals acts as chelating and reductive agents, free radical scavengers, and lipid-preventers; results indicate that phenolic compounds from bread enriched with powdered GCB were highly mastication-extractable, which may predict their high bioaccessibility and bioavailability. Sensory characteristics results indicated that a partial replacement of wheat flour in bread with up to 3% ground GCB powder gives satisfactory overall consumer acceptability and bread enriched with GCB possessed higher anti-radical activity than control samples. This study showed that powdered GCB may be used directly, without extract preparation, for food supplementation.[7]

7.15   PHYTOPROTECTANTS IN VEGETABLES

Raw garlic[3,29] and onions have allylsulphides, and these work against *Helicobacter pylori* and stomach ulcers. Alliinandallicin work as anticancer,[17] hypocholesterolemic, and hypotriglyceride.[29] It is, however, not certain if cooked garlic[3,29] and onions work as well as raw. Tomatoes have carotenoids

chiefly lycopene, and it works as anticancer[25] and also works against prostate cancer and other cancers.[25] Lycopene is helpful against heart disease.[14]
Lady's finger, puisakmucinous vegetables covered with mucus have soluble
fiber pectin that has hypoglycemic effects.[12] It reduces blood cholesterol.[29]
Phytoprotectants (gingerolinginger, curcumin) have anti-inflammatory
action and carminative action.[19] They relieve flatulence and accumulation
of gas in the alimentary canal. Phytoprotectants in whole cereals (such as
brown rice, whole wheat, and whole grains; those are rich in fiber) help in
reducing risk of colon cancer[25] and neutralize cancer causing agents in the
intestine.[11] Terpenoids, phytic acid, ellagic acid, phytoestrogens (lignans)
reduce risk of heart disease and cancer risk, particularly breast cancer.[25]

7.16 FUNCTIONAL CEREALS AND BAKERY PRODUCTS

Cereals (in particular oat and barley) offer another alternative for the production of functional foods.[11] Additionally, cereals can be applied as sources
of nondigestible carbohydrates that, besides promoting several beneficial
physiological effects, can also selectively stimulate the growth of lactobacilli and bifidobacteria present in the colon and act as prebiotics. Cereals
contain water soluble fiber, such as beta-glucan and arabinoxylan, oligosaccharides, such as galacto- and fructo-oligosaccharides[23] and resistant starch,
which have been suggested to fulfill the prebiotic concept.[11] While functional
foods are rapidly increasing in popularity in sectors such as dairy products or
confectionery, in bakery products they are still relatively underdeveloped.[10]
These products however provide the ideal matrix by which functionality
can be delivered to the consumer in a highly acceptable food. In late 2003,
Unilever innovated the bakery sector by introducing a white bread called
Blue Band Goede Start, which was the first white bread containing the nutritional elements normally available in brown bread including fibers, vitamins
(B1, B3, and B6), iron, zinc, and inulin (a starch that comes from wheat).[2,22]

7.17 LIMITATIONS AND CHALLENGES BEFORE FUNCTIONAL
FOOD AND WAYS TO OVERCOME

Due to the limited consumers' awareness of the health effects of newly developed functional ingredients, there is a strong need for specific information
and communication activities to eliminate these evident deficits. This relates
in particular to pioneering companies opening a specific market segment,

for which targeted information activities to consumers and opinion leaders (like medical doctors, nutritional advisers) are regarded as crucial success factor.[24]

7.18 CONCLUSIONS

Functional food and nutraceuticals have many attributes and major of them have been mentioned in this chapter. The health benefits of some the major functional food ingredients are discussed like tea, soybeans, sprouted grains, vegetables, and spices. Many dietary constituents provide ingredients for possible use in functional foods. Functional food properties of tea and soy have been discussed in this chapter. There are many health disorders, diseases, and complexities which can be managed with the help of functional food and nutraceuticals. With proper support from appropriate government agencies and research from food scientists, there is a tremendous potential for processed functional foods in the future.

KEYWORDS

- bioavailability
- cancer
- carotenoids
- flavonoids
- functional food
- human health
- isoflavones
- lycopene
- medicinal plants
- micronutrients
- nutraceuticals
- phytotherapy
- vitamins

REFERENCES

1. Boye, J. I. Nutraceuticals and Functional Food Processing Technology. *Food Res. Int.* **2015**, *72*, 110–170.

2. Broekmans, W. M. R.; Klopping-Ketelaars, I. A. A.; Schuurman, C. R. W. C.; Verhagen, H.; van den Berg, H.; Kok, F. J.; van Poppel, G. Fruits and Vegetables Increaseease Plasma Carotenoids and Vitamins and Decrease Homocysteine in Humans. *J. Nutr.* **2000**, *130*, 1578–1583.

3. Das, T.; Choudhury, A. R.; Sharma, A.; Talukder, G. Effects of Crude Garlic Extract on Mouse Chromosome In Vivo. *Food Chem. Toxicol.* **1996**, *34*, 43–47.

4. De Pascual-Teresa, S.; Santos-Buelga, C.; Rivas-Gonzalo, J. C. Quantitative Analysis of Flavan-3-ols in Spanish Foodstuffs and Beverages. *J. Agric. Food Chem.* **2000**, *48*, 5331–5337.

5. Devasagayam, T. P.; Tilak, J. C.; Singhal, R. Functional Foods in India: History and Scope. *Anti-angiogenicfunct. Med. Foods* **2007**, *1*, 70–90.

6. Din, J. N.; Newby, D. E.; Flapen, A. D. Omega 3 Fatty Acids and Cardiovascular Disease-fishing for a Natural Treatment. *Br. Med. J.* **2004**, *328*, 30–35.

7. Dziki, D.; Gawlik-Dziki, U.; Pecio, L.; Różyło, R.; Świeca, M.; Krzykowski, A.; Rudy, S. Ground Green Coffee Beans as Functional Food Supplement-preliminary Study. *LWT—Food Sci. Technol.* **2015**, *63*(1), 691–699.

8. Galaverna, G.; Sforza, S.; Di Silvestro, G.; Marchelli, R. Variation of the Antioxidant Activity in Fruit Juices during Technological Treatments. In *Biologically-active Phytochemicals in Food*; Pfannhauser, W., Fenwick, G. R., Khokhar, S., Eds.; Royal Society of Chemistry: Cambridge, 2001; pp 474–476.

9. Giardi, M. T.; Rea, G.; Berra, B. *Bio-farms for Nutraceuticals, Functional Food and Safety Control by Biosensors.* Library of Congress Cataloging-in-Publication Data: USA, 2010; pp 6–363.

10. Gibson, G. R.; Williams, C. M. *Functional Foods: Concept to Product.* CRC Press: Boca Raton, FL, 2000.

11. Gopaldas, T. Iron-deficiency Anemia in Young Working Women can be Reduced by increasing the Consumption of Cereal-based Fermented Food or Gooseberry Juice at the Workplace. *Food Nutr. Bull.* **2002**, *23*, 94–105.

12. Gordon, M. H.; Walker, A. F.; Roberts, W. G. A Human Study Investigating the Effects of Increased Fruit and Vegetable Consumption in Smokers. In *Biologically-active Phytochemicals in Food*; Pfannhauser, W., Fenwick, G. R., Khokhar, S., Eds.; Royal Society of Chemistry: Cambridge, 2001; pp 21–23.

13. Graham, H. N. Green Tea Composition, Consumption, and Polyphenol Chemistry. *Prev. Med.* **1992**, *21*, 334–350.

14. Hertog, M. G. L.; Feskens, E. J. M.; Hollman, P. C. H.; Katan, M. B.; Kromhout, D. Dietary Antioxidant Flavonoids and Risk of Coronary Heart Disease. The Zutphen Elderly Study. *Lancet* **1993**, *342*, 1007–1011.

15. Higdon, J. V.; Frei, B. Tea Catechins and Polyphenols: Health Effects, Metabolism and Antioxidant Functions. *Crit. Rev. Food Sci. Nutr.* **2003**, *43*, 89–143.

16. Kaur, I. P.; Chopra, K.; Saini, A. Probiotics: Potential Pharmaceutical Applications. *Eur. J. Pharm. Sci.* **2002**, *15*, 1–9.

17. Krishnaswamy, K. Indian Functional Foods: Role in Prevention of Cancer. *Nutr. Rev.* **1996**, *54*, 127–131.

18. Mattson, M. P. Awareness of Hormesis will Enhance Future Research in Basic and Applied Neuroscience. *Crit. Rev. Toxicol.* **2008**, *38*, 633–639.
19. Polasa, K.; Nirmala, K. Ginger—Its Role in Xenobiotic Metabolism. *Nutr. News* **2002**, *23*, 1–6.
20. Rani, B.; Khetarpaul, N. Probiotic Fermented Food Mixtures: Possible Applications in Clinical Antidiarrhoea Usage. *Nutr. Health* **1998**, *12*, 97–105.
21. Rea, G.; Antonacci, A.; Lambreva, M.; Margonelli, A.; Ambrosi, C.; MariaGiardi, T. *The Nutra-snacks Project: Basic Research and Biotechnological Programs on Nutra-ceutics.* Library of Congress Cataloging-in-Publication Data: USA, 2010; pp 6–363.
22. Roberfroid, M. B. Concepts in Functional Foods: The Case of Inulin and Oligofructose. *J. Nutr.* **1999**, *129*(7), 1398–1401.
23. Roberfroid, M. B. Functional Effects of Food Components and the Gastrointestinal System: Chicory Fructo-oligosaccharides. *Nutr. Rev.* **1996**, *54*, 38–42.
24. Sanders, M. E.; Huis-in't-Veld, J. Bringing a Probiotic Containing Functional Food to the Market: Microbiological, Product, Regulatory and Labeling Issues. *Ant. Van Leeuw.* **1999**, *76*, 293–315.
25. Shukla, Y.; Taneja, P. Antimutagenic Effects of Garlic Extract on Chromosomal Aberrations. *Cancer Lett.* **2002**, *176*, 31–36.
26. Siegrist, M.; Shi, J.; Giusto, A.; Hartmann, C. Worlds Apart: Consumer Acceptance of Functional Food and Beverages in Germany and China. *Appetite* **2015**. DOI:10.1016/j.appet.2015.05.017.
27. Singh, J.; Bagchi, G. D.; Srivastava, R. K.; Singh, A. K. Pharmaceutical Aspects of Aromatic Herbs and their Aroma Chemicals. *J. Med. Aromat. Plant Sci.* **2000**, *22*, 732–738.
28. Skinner, M.; Hunter, D. *Bioactives in Fruit: Health Benefits and Functional Foods.* John Wiley and Sons, Ltd.: NY, 2013.
29. Spigelski, D.; Jones, P. J. H. Efficacy of Garlic Supplementation in Lowering Serum Cholesterol Levels. *Nutr. Rev.* **2001**, *59*, 141–236.
30. Starling, S. India Rethinks Supplements Sector. *Funct. Foods Neutraceut.* **2004**, *10*, 138–150.
31. Surh, Y. J. Cancer Chemoprevention with Dietary Phytochemicals. *Nat. Rev. Cancer* **2003**, *3*, 768–780.
32. Trevisanato, S. I. and Kim, Y. I. Tea and Health. *Nutr. Rev.* **2000**, 58, 1–10.
33. Vergari, F.; Tibuzzi, A.; Basile, G. *An Overview of the Functional Food Market: From Marketing Issues and Commercial Players to Future Demand from Life in Space.* Library of Congress Cataloging-in-Publication Data: USA, 2010; pp 308–343.
34. Wildman, R. E. C. *Handbook of Neutraceuticals and Functional Foods.* CRC Press: Boca Raton, FL, 2001.
35. Yamada, S.; Taki, Y.; Misaka, S.; Okura, T.; Deguchi, Y.; Umegaki, K.; Watanabe, H.; Watanabe, Y.; Skinner, M. In *Pharmacokinetic and Pharmacodynamic Interaction of Functional Foods with Medicines Bioactives in Fruit: Health Benefits and Functional Foods*, 1st ed.; Skinner, M., Hunter, D., Eds.; John Wiley & Sons: New York, 2013.
36. Yang, C. S.; Wang, X.; Lu, G.; Picinich, S. C. Cancer Prevention by Tea: Animal Studies, Molecular Mechanisms and Human Relevance. *Nat. Rev. Cancer* **2009**, *9*, 429–439.

GLOSSARY

Functional food: A modified food or food ingredient that provides a health benefit beyond satisfying traditional nutrient requirements.

Nutraceutical: Food or part of a food that offers medicinal and/or health benefits including prevention or treatment of disease.

Prebiotic: A colonic food which encourages the growth of favorable intestinal bacteria (e.g., bifidobacteria and lactobacilli).

Probiotic: A mono- or mixed culture of microorganisms which when applied to animal or man affects the host beneficially.

CHAPTER 8

PHYTOCHEMICALS, FUNCTIONAL FOOD, AND NUTRACEUTICALS FOR ORAL CANCER CHEMOPREVENTION

SATARUPA BANERJEE[1*] and JYOTIRMOY CHATTERJEE

School of Medical Science and Technology, Indian Institute of Technology, Kharagpur 721302, West Bengal, India. E-mail: satarupabando@gmail.com, jchatterjee@smst.iitkgp.ernet.in

*Corresponding author.

CONTENTS

ABSTRACT

Delays in detection, increased mortality rate and 5-year survival rate, chemoresistance to conventional therapies, chances of secondary tumor formation leading to field cancerization after cure are the major concerns faced today in oral cancer diagnosis and treatment. Physical and physiological side effects produced by the conventional treatment option in oral cancer are also remarkable problem forcing people to opt for the unconventional alternative and complementary medicines and chemopreventive strategies of herbal medications. This overview highlights on the natural chemopreventive components reported in the recent literatures which have shown to exert anticancer potential against oral precancerous and cancerous lesions, as well as possess chemopreventive and modulating efficacy emphasizing the major molecular mechanisms beneath the process. Implementation of these dietary chemopreventive approaches in daily routine, discussed in this chapter, may further aid in reduction of global oral cancer burden.

8.1 INTRODUCTION

Oral cancer is one of the deadliest diseases of world being sixth cause of death. A few study reports that occurrence of oral cancer is decreasing in the USA, some other reports suggest that due to increased immigration of South Asian people to the USA, prevalence of oral cancer still exists there as threat. Recent finding suggests that now the case of oral cancer is increasing in patients of younger age. Despite all the advances and studies, 5-year survival rate for oral cancer is still lower than other types of cancer.[78] Dramatic rise of oral cancer within the population below 60 in the USA and the Europe is also of great concern.[29] The main reason of oral cancer in Asian countries is thought to be due to increased exposure to carcinogen and region specific epidemiological reasons like use of tobacco and/or areca nut,[110] while alcohol consumption play the major role in western countries.[2] Adaptation to genetic changes like p53 mutations or bcl-2 overexpression is the major underlying causes of such association. Conventional radiotherapy, gene therapy, CO_2 laser surgery, or other surgical excision like electrosurgery, cryotherapy, chemotherapy, as well as combined immunotherapy and chemotherapy are the major options for treatment.[9] In most of the case they are associated with immense physical and physiological side effects and chances of secondary tumor formation or field cancerization often exists. Again, problem of early diagnosis is still unsolved.[79] Here, lies the effectiveness of preventive medicine as well as complementary or alternative intervention strategies. Rising cost of health care and limited resources for health prevention effort in the USA is also alarming. Low-cost herbal remedy for oral cancer prevention may solve or alleviate the problem. Natural products having, low cost, and lesser side effects are widely accepted among people for chemoprevention nowadays.

This chapter reviews the role of natural products for chemoprevention and treatment purposes in oral cancer.

8.2   ORAL CARCINOGENESIS AND CHEMOPREVENTION

Any cancerous condition inside oral cavity is considered as oral cancer, 95% of which attributes to squamous cell carcinoma. Oral carcinogenesis is a multistep process. It is often preceded through dysplastic or non-dysplastic oral premalignant disorders, where multiple genetic alterations, individual predisposition to environmental toxins are the major underlying causes,[20] which can be therefore be modulated, arrested, and even reversed with specific components called chemopreventive agents.[30] The process of chemoprevention can be defined as the use of natural, synthetic, or biologic agents for prevention by blocking necessary genetic steps, or progression reversal during carcinogenesis.[88]

Till 2000, out of more than 1000 chemopreventive studied agents, 13-cis-retinoic acid, a vitamin A derivative was most potent but simultaneously toxic too. Due to the failure of clinical trial with retinoids because of their systemic toxicity, major researches were initiated in area of chemoprevention by natural agents considering their better safety profile. Dietary carotenoids, fruits, and vegetables were regarded as protective factors against oral cancer, while carotenoids, retinoids, vitamin E (α-tocopherol), and polyphenols (green tea) and protease inhibitors (soy) were major chemopreventive agents found till date.[5] Intake of vitamin C, E, beta carotene, curcumin from turmeric, black raspberry, and strawberry showed considerable primary chemopreventive results in primary clinical trials, but in spite of early positive significant results, Bowmin Birk Inhibitor from soybeans did not showed further longtime potential. Much clinical studies are therefore needed.[87] Previous studies suggested that fruits and vegetables were important for anticancer activity, but a recent study suggested mainly micronutrients and phytochemicals are responsible for that.[109] The important components which showed marked evidence in oral cancer include polyphenols, phenolic acids, flavonoids, coumarins, etc. Few other important polyphenols studied are resveratrol, black tea polyphenol, green tea polyphenols like EGCG, curcumin, quercetin, gossypol, etc.,[22,24] phenolic acids include ellagic acids, ferulic acids while few flavonoids are kaempferol, quercetin, naringenin, etc.[25]

8.3 MECHANISM OF NATURAL CHEMOPREVENTION

Dietary phytochemicals has recently received attention as cancer chemopreventive agents due to their pleiotropic effects on multiple carcinogen-activated oncogenic pathways including ROS-induced carcinogenesis and also for their increased safety profile.[115,124] The major mechanisms involved in chemoprevention of *in vivo* oral cancer by natural agents mainly includes improvement of inherent antioxidant mechanism against oxidative damages, restoration of xenobiotic metabolizing enzymes, modulation of lipid peroxidation activities, inhibition of cell proliferation, and induction of apoptosis.[98] These processes in turn lead to significant reduction in tumor volume or decrease in dysplasia. Few studies highlighted the role of angiogenesis or epithliomesenchymal transition on chemoprevention mechanism too.[19,20] Chemoprevention is generally exerted by two major mechanisms. The rationales behind tumor blocking agents generally includes free radical scavenging, induction of DNA repair, exerting antioxidant activity, blocking carcinogen uptake, and finally inducing phase II drug-metabolizing and inhibiting phase I drug metabolizing enzymes. Tumor suppressors acts by altering gene expression, inhibiting cell proliferation, induction of apoptosis, and modulating different signal transduction pathways. Recent studies were focused on the dosage optimization for phase 0 clinical trials and toxicity reduction as well and improvements on delivery systems of few selected natural chemopreventive agents.[99] Chemopreventive activity is also exerted by combination of several cellular activities including antioxidant, anti-inflammatory effect, NF-κB is associated inflammation reduction and metastasis inhibition too.[103]

8.4 ROLE OF NATURAL CHEMOPREVENTIVE AGENTS IN IN VITRO AND IN VIVO NATURAL CHEMOPREVENTION MODELS WITH THEIR PROBABLE MECHANISM OF ACTION

Chemoprevention of natural components against oral cancer was mainly studied in vitro in different cell lines like KB, HSC-9, etc. or in vivo animal models like 7,12-dimethylbenz[a]anthracene (DMBA) induced hamster buccal pouch carcinogenesis. Efficacy of the components alone or in combination is assessed with respect to their antiproliferative, antiapoptotic, or regression of dysplastic condition. DMBA induced carcinoma in Syrian hamster buccal mucosa (DMBA-HBM) is considered as an excellent model for carcinogenesis and in vivo chemoprevention potentiality assessment as well due to its morphological, histopathological, oxidant–antioxidant status, and molecular marker expression profile similarity with human models.[7,107] Table 8.1 depicts the role of different phytochemicals, natural, and dietary components alone or in combination on DMBA-HBM model with their probable mode of action, while Table 8.2 depicts the role of crude plant extracts, where component-based analysis are still needed. Table 8.3 presents the role of chemicals in different normal and cancerous epidermoid and immortalized keratinocytes like SCC-9, SSC-4, FaDu, IHOK, SAS, CAL27, etc. and their probable mechanism of action.

TABLE 8.1 Phytochemicals, Natural and Dietary Components on DMBA-HBM Model with their Probable Mode of Action.

Constituent	Source	Probable mode of action	References
Bovine milk lacto-ferrin and black tea polyphenols	–	Modulation of carcinogen metab-olizing enzymes and cellular redox status	[13]
Dietary turmeric (1%)	Turmeric	Decreased cell proliferation, inflammation, enhanced apop-tosis, aberrant expression of differentiation markers, and the cytokeratins	[28]
Naringin and naringenin	Grapefruits, oranges, and tomatoes (skin)	Reduction of tumor	[75]
Curcumin and piperine	–	Lipid peroxidation and antioxi-dant activity modulation	[68]
Withaferin-A	Winter cherry (*Withania somnifera*)	Induction of apoptosis	[77]
Curcumin and ferulic acid	–	Modulation on p53 and bcl-2 expression	[6]
Carnosic acid	Rosemary (*Rosmarinus officinalis*)	Anti-lipid peroxidative potential and modulating effect on carcin-ogen detoxification enzymes	[72]

TABLE 8.1 *(Continued)*

Constituent	Source	Probable mode of action	References
Thymoquinone	*Nigella sativa*	Protection against abnormalities of cell surface glycoconjugates	[86]
(6)-Paradol	Rhizome of ginger, *Zingiber officinale*	Anti-lipid peroxidative and antioxidant potentials, modulation of phase II detoxification enzyme and reduced glutathione (GSH), apoptosis-associated gene expression enhancement	[101,116]
Ganoderma triterpenes	–	Modulation of VEGF and caspase-3 expression	[27]
Red mold dioscorea	–	Induction of apoptosis and reduction in oxidative damage	[37]
18 β-Glycyrrhetinic acid	Liquorice, *Glycyrrhiza glabra*	Restoration of detoxification enzymes	[51]
Berberine	Berberis, e.g., *Berberis aquifolium, Berberis vulgaris, Berberis aristata*, etc.	Increased cell proliferation and induction of apoptosis	[70]
Coumarin	Cinnamon bark oil, Cassia leaf oil and Lavender oil.	anti-lipid peroxidative potential and modulating effect on carcinogen detoxification agents	[8]
Lupeol	*Mangifera indica, Acacia visco*, and *Abronia villosa*	Induction of apoptosis	[69]
Andrographolide	*Andrographis paniculata*	Antioxidant potential as well as modulating effect on xenobiotic metabolizing enzymes	[71]
Ferulic acid	Plant cell wall component	Anti-cell proliferative potential	[80]
Chlorophyllin and ellagic acid	–	Modulation of expression of TGF-β receptors, NF-κB, cyclin D1, and matrix metalloproteinases (MMPs) expression	[82]
Quercetin	Many fruits, vegetables, leaves, and grains	Suppresses cytochrome P450 mediated ROS generation and NF-κB activation	[83]
Diosgenin	Tubers of dioscorea	Antioxidant function	[84]

TABLE 8.1 *(Continued)*

Constituent	Source	Probable mode of action	References
Chlorophyllin	Semisynthetic derivatives of chlorophyll	Angiogenesis and apoptosis inhibition	[96,108]
Geraniol	Rose oil, palma-rosa oil, and citronella oil	Modulation of cell proliferation, apoptosis, inflammation and angiogenesis	[113]
Ellagic acid	Numerous fruits and vegetables	Induction of intrinsic apoptosis	[3]
Apigenin	Chamomile tea	Anti-inflammatory, antiprolifera-tive, apoptotic properties, modu-lation of antioxidant defense mechanism and the activities of phase I and phase II detoxifica-tion enzymes	[31,85,97]
Astaxanthin	Keto-carotenoid found in trout, microalgae, yeast, and shrimp, among other sea creatures	Induction of apoptosis, inhibits cell proliferation, invasion, and angiogenesis	[45,52]
Astaxanthin, blue-berry, chlorophyllin, ellagic acid, and theaphenon-E	–	potent activators of Nrf2	[46]
Xanthorrhizol	sesquiterpenoid isolated from the rhizome of *C. xanthorrhiza*	Induces apoptosis	[47]
Saffron	Flower of *Crocus sativus*	Modulating antioxidant properties	[73]
Carnosic acid	Benzenediol abietane diterpene found in rosemary (*Rosmarinus officinalis*) and common sage (*Salvia officinalis*)	Modulating effect on cell prolif-eration, apoptosis, inflammation, and angiogenesis	[31]
Phloretin	Apple tree leaves and the Manchu-rian apricot	Modulating the antioxidant and detoxification enzyme status	[1]
Mentha piperita	–	Reduction of epithelial dysplasia	[43]

TABLE 8.2   Phytochemicals as Extract on DMBA-HBM Model with their Probable Mode of Action.

Extract	Mode of action	References
Ethanol extracts of *Terminalia catappa* leaves	Induction of apoptosis	[117]
Neem leaf sub-fractions	Modulation of xenobiotic-metabolizing enzymes, inhibition of cell proliferation, induction of mitochondrial apoptosis	[67]
ZengShengPing, a mixture of six medicinal herbs	Reduction in AgNOR count and PCNS labeling index	[100]
Lyophilized strawberries	Reduction of dysplasia	[11]

TABLE 8.3   Role of Natural Chemopreventive agents in Different Normal and Cancerous Epidermoid and Immortalized Keratinocytes.

Constituent	Source	Probable mode of action	References
Lycopene	Tomato	EMT and downregulation of cell proliferation	[62,63]
Quercetin (3,5,7,3′,4′-pentahydroxyflavone)	citrus fruits, apples, onions, parsley, sage, tea, and red wine	S-phase arrest and inhibits proliferation	[33,119]
Ferulic acid and β-sitosterol	Coffee, apple, artichoke, peanut, and orange, rice, wheat, etc.	Ferulic acid-based G2/M Phase arrest and β-sitosterol-mediated apoptosis	[34]
Punicalagin, ellagic acid, and a total pomegranate tannin	Pomegranate	Cellular proliferation inhibition	[91]
Extract of *Coptidis rhizoma*		Apoptosis induction	[53]
Extracts of six popularly consumed berries, blackberry, black raspberry, blueberry, cranberry, red raspberry, and strawberry	Berries	Antiproliferative activity	[92]
Proanthocyanidin	Apples, maritime pine bark, cinnamon	Inhibition of cellular proliferation, and induction of apoptosis by G2/M arrest	[48,61, 81,94]

TABLE 8.3   *(Continued)*

Constituent	Source	Probable mode of action	References
Aloe-emodin	Aloe vera	Apoptosis and alkaline phosphatase activation	[116]
Extracts from *Cladiella australis*, *Clavularia viridis*, and *Klyxum simplex*	Soft corals	Apoptosis	[56]
Shikonin	Zicao (purple gromwell, the dried root of *Lithospermum erythrorhizon*)	Inhibition of NF-κB	[76]
Liposomal curcumin	*Curcuma longa*	Inhibition of NF-κB	[114]
Verticinone	*Bulbus fritillaria*	Apoptosis	[120]
Aliphatic acetogenin constituents	Avocado	Apoptosis induction and inhibiting cell proliferation	[21,23]
Cepharanthine	Stephania	Inhibition of NF-κB and VEGF	[35]
Diterpenoid ovatodiolide	*Anisomeles indica*	Cell cycle arrest at G2/M phase	[36]
Esculetin	Chicory	Apoptosis induction	[50]
S-Allylcysteine (SAC)	Garlic	Modulation of cell proliferation and EMT	[106]
Cranberry and grape seed extracts	–	Apoptosis and cell proliferation inhibition	[17]
Kaempferol and Quercetin	–	Apoptosis induction	[42]
Gypenosides	*Gynostemma pentaphyllum*	Inhibition of ERK1/2 and NF-κB signaling	[64]
Curcumin	Turmeric	Inhibition of cellular motility	[125]
Epigallocatechin-3 Gallate, epicatechin-3-gallate, and theaflavin 3,3′-digallate	Tea	Epithelial mesenchymal transition (EMT)	[15,19]
Buddlejasaponin IV	*Pleurospermum kamtschatidum*	Cell cycle arrest at G2/M phase and apoptosis	[38]

TABLE 8.3   *(Continued)*

Constituent	Source	Probable mode of action	References
Sulforaphane	Brussels sprouts	Down-regulation of MMP-1 And MMP-2	[39]
Genistein and Biochanin A	–	Decreases in ERK and Akt phosphorylation	[40]
Resveratrol	Grapes	DNA Damage induction, Independent of Smad4 expression, inhibited the adhesion, migration, and invasion	[93,112]
3,3'-diindolyl-methane	Broccoli, Brussels sprouts, cabbage and kale	Inducing apoptosis and G2/M arrest	[126]
Xanthorrhizol	rhizome of *C. xanthorrhiza*	Induces apoptosis	[47]
Capsaicin	Chili peppers	Induces cell cycle arrest and apoptosis	[58]
Deguelin	*Lonchocarpus utilis*	Induces apoptosis and autophagy	[118]
Kahweol	*Coffea arabica*	Antiproliferative properties and induction of apoptosis	[12]
Pterostilbene	Blueberries	Inhibiting MMP-2 expression	[59]
Tanshinone IIA	*Salvia miltiorrhiza*	Induction of apoptosis	[111]
Kuding tea	Solitary leaf	Induction of apoptosis	[123]

8.5   SPECIAL EMPHASIS TO SOME IMPORTANT DIETARY COMPONENTS

Curcumin (diferuloylmethane) is mainly obtained from *Curcuma longa*. It has chemopreventive role against oral cancer as it can modulate NF-κB signaling as well as suppress MMP-2 and cyclooxygenase 2 (COX-2) expressions associated with the signaling pathway that affect metastasis in oral cancer. It can also arrest the cells in G1/S phase of the cell cycle, activate upstream and downstream caspases for apoptosis induction and downregulate Notch-1 and inhibits CYP1A1-mediated benzo(α)-pyrenediol bioactivation. It also shown protective activity in against oral cancer induced by 4-nitroquinoline 1-oxide in rat model.[104] Another study suggests that both curcumin and ferulic acid, a phenolic acid found in many plants, exerts it protective effect by modulating p53 and bcl-2 expression.[6] It has antimotility effect too against OSCC.[94] Interestingly in low doses it exerts antioxidant activity, but in higher concentration exert cytotoxic activity when administered locally.[60]

8.5.1 TEA AND TEA COMPONENTS AGAINST OSCC

Tea is the most commonly consumed beverage worldwide. Theaflavin-3,3'-digallate, a potent theaflavin obtained from *Camellia sinensis* from the form of black tea, that is, the fermented form of the green leaves, can induce apoptosis through ROS formation.[90] Its activity is similar like epigallo-catechin-3-gallate (ECGC), a catechin obtained from the green leaves of *Camellia sinensis*. ECGC causes downregulation of MMP-2 and MMP-9 and urokinase plasminogen secretion as well as reverse the hypermethylation of RECK, a novel tumor suppressor gene by which it inhibit tumor invasion, angiogenesis and metastasis in oral cancer cells.[44] EGCG has a potent chemopreventive activity as it has also modulating activity against NF-κB and activator protein (AP-1), activating p53 to arrest cell cycle at G0–G1 phase and inducing apoptosis.[46] It also modulates expression of μ-PA and reduces phosphorylation of FAK, by which cell migration and invasion are controlled by EGCG in dose dependent manner.[19]

The chemopreventive properties of green tea polyphenols were proved by various in vivo, in vitro models as well as pilot scale studies in dose dependent manner, which showed positive results against models or patients of oral cancer or precancerous lesions.[55] Reduction of nitric oxide production by EGCG and black tea theaflavins, reduction of decreasing vascular endothelial growth factor (VEGF) production and receptor phosphorylation, and dihydrofolate reductase activity are the other probable mechanism action tea against OSCC.[4] Synergistic efficacy of tea components as well as for dose reduction of conventional chemotherapeutic drugs against OSCC is also well documented and worth mentioning.[14,55] Black tea polyphenol has epithelial mesenchymal reversal activity in oral cancer cell line,[16] while green tea EGCG was found to modulate and downregulate HGF-induced phosphorylation of c-Met.[49]

8.6 DAILY DIET AND SOME OTHER DIETARY COMPONENTS FOR CHEMOPREVENTION

8.6.1 ROLE OF DAILY DIET

Fruits, vegetables, and fish consumption reduces the risk of oral cancer, whereas high intake of nitrate-containing meat increases the chances.[65] Raw vegetable and fruits like apple, salad, and tomato are more beneficial than cooked one. Olive oil as well as herbal tea, margarine, milk, and citrus fruit or juice may have protective effects. Antioxidants and micronutrients like vitamin A, vitamin B12, vitamin C, tocopherol (vitamin E), vitamin B6, folic acid, niacin, retinoids, carotenoids, lycopene, β-carotene, folate, glutathione, glutamine, thiamin, lutein, iron, zinc, selenium, and folate show protective activity.[41,57,102,105,115] Some other dietary compounds like phenolic acids (caffeic, chlorogenic, ellagic, ferulic, gallic, p-coumaric, p-hydroxybenzoic, protocatechuic, and vanillic acids), the flavonols (isorhamntin, kaempferol, mycetrin, and quercetin), flavan-3-ols (catechins), and complex polyphenols (lignans, ellagitannin, and gallotannin) are mentioned as chemoprotective agent against oral cancer.[10,105]

8.6.2 BERRY-BASED CHEMOPREVENTION

Berry-based chemoprevention approach is new alternative for oral carcinoma. The berries like blueberry (*Vaccinium* species), strawberry (*Fragaria* × *ananassa* Duch.), raspberry (*Rubus idaeus*), cranberry (*Vaccinium* species), black berry (*Rubus occidentals*) etc. have known chemopreventive potential. The major classes of berry phenolics are anthocyanins, flavonols, flavanols, ellagitannins, gallotannins, proanthocyanidins, and phenolic acids, examples of which includes quercetin, β-carotene, ferulic acid, protocatechuic acid, etc.[118] Freeze-dried black raspberry extract has immense potential on oral cancer treatment. The component of black raspberries responsible for cell-cycle arrest at G2/M phase is ferulic acid, whereas for G0/G1 phase, it is β-sitosterol in both premalignant and malignant cell implicates its chemoprotective nature.[34]

In 2008, topical application of freeze-dried black raspberry gel warranted chemoprevention in patients as it showed significant tumor reduction in loss of heterozygosity indices and did not exert toxic response.[95] Another study suggests anticancer activity of the gel by downregulation of 17 gene expression and reducing COX-2 expressions.[66] Further study and clinical trial is being performed in this area. Proanthocyanidins present in cranberry and grapes showed efficacy in cancer through induction of apoptosis and inhibiting cell proliferation for which it is considered as a potent preventive and treatment alternative.[18,48] Strawberry-extract-inhibited growth in oral cancer cell line.[122] Lupeol, a triterpene as well as a major constituent of blueberry, can downregulate NF-κB and can synergize the effect as well as increase the sensitivity of cisplatin, a commonly used drug used in chemotherapy which often causes chemoresistance.[54]

8.7   FOOD AND ORAL CANCER INDUCTION

Study suggests that cancer can occur from food by two ways: The first type is direct carcinogenesis and the other is indirect carcinogenesis. **Direct carcinogenesis** may occur from the carcinogens present in the food and food additives, while **indirect carcinogenesis** may occur from carcinogen formed in vivo during metabolism of the food. It is also suggested that fried food, fried meat prepared at high temperature or microwave are possible carcinogen due to production of heterocyclic amines. Monounsaturated fatty acids like n-9 oleic acid and many dietary lipids, high calorie intake are also suspected as the causes of oral cancer.[102] High sugar intake is also may be a risk factor.[78] Low dietary intake of fresh fruits, vegetables, and deficiency of some other micronutrients often promotes oral cancer risks.[102,121] Nineteen carcinogenic compounds are present in roasted coffee, though do not produce oral cancer but found to potentiate the effects of other carcinogens. It is also to be mentioned here that some component of green beans of coffee has chemopreventive efficacy.[89] Meat, maize, desserts, saturated fat, and butter also showed harmful effects.[41] These foods should be avoided from daily diet for natural prevention of oral cancer.

8.8   CONCLUSIONS

Twenty five century back, Hippocrates stated "Let food be thy medicine and medicine be thy food".[32] This notion is evident till date. With the hope of reduction in global oral cancer incidence and prevalence, this study suggested that oral carcinogenesis can be strictly avoided, modified, and even arrested by the effects of natural chemopreventive agents, by which certain disadvantages associated with surgery and chemotherapy, the most common treatment options of oral cancer can even be bypassed. Ease of accessibility, low cost, and lesser side effects are the key words for the natural chemoprevention strategies. It also suggests that, inclusion of two important well studied components of turmeric and green tea, namely, curcumin[125] and EGCG,[126] may help in reduction global oral cancer incidence. Again food-based "green" chemoprevention for developing countries is indicative toward the global urge of frugal solution of cancer prevention.[26]

The data presented in this chapter depict the strategies of prevention and treatment options of the deadly disease in natural way, but furthermore many clinical trials are needed for global inclusion of the mentioned components as functional foods in daily diet.

KEYWORDS

- alternative medicine
- chemoprevention
- complementary medicine
- functional food
- nutraceuticals
- oral cancer
- phytochemical

REFERENCES

1. Anand, M. A. V.; Suresh, K. Biochemical Profiling and Chemopreventive Activity of Phloretin on 7,12-Dimethylbenz(*a*)anthracene Induced Oral Carcinogenesis in Male Golden Syrian Hamsters. *Toxicol. Int.* **2014**, *21*, 179.

2. Anantharaman, D.; Samant, T. A.; Sen, S.; Mahimkar, M. B. Polymorphisms in Tobacco Metabolism and DNA Repair Genes Modulate Oral Precancer and Cancer Risk. *Oral Oncol.* **2011**, *47*, 866–872.

3. Anitha, P.; Priyadarsini, R. V.; Kavitha, K.; Thiyagarajan, P.; Nagini, S. Ellagic Acid Coordinately Attenuates Wnt/β-catenin and NF-κB Signaling Pathways to Induce Intrinsic Apoptosis in an Animal Model of Oral Oncogenesis. *Eur. J. Nutr.* **2013**, *52*, 75–84.

4. Ann Beltz, L.; Kay Bayer, D.; Lynn Moss, A.; Mitchell Simet, I. Mechanisms of Cancer Prevention by Green and Black Tea Polyphenols. *Anti-cancer Agents Med. Chem.* **2006**, *6*, 389–406.

5. Armstrong, W. B.; Meyskens, F. L. Chemoprevention of Head and Neck Cancer. *Otolaryngol.--Head Neck Surgery* **2000**, *122*, 728–735.

6. Balakrishnan, S.; Manoharan, S.; Alias, L. M.; Nirmal, M. R. Effect of Curcumin and Ferulic Acid on Modulation of Expression Pattern of p53 and bcl-2 proteins in 7, 12-dimethylbenz[*a*]anthracene-induced Hamster Buccal Pouch Carcinogenesis. *Indian J. Biochem. Biophys.* **2010,** *47*, 7–12.

7. Balasenthil, S.; Rajamani Ramachandran, C.; Nagini, S. S-allylcysteine, a Garlic Constituent, Inhibits 7, 12-dimethylbenz[*a*]anthracene-induced Hamster Buccal Pouch Carcinogenesis. *Nutr. Cancer* **2001,** *40*, 165–172.

8. Baskaran, N.; Manoharan, S.; Karthikeyan, S.; Prabhakar, M. M. Chemopreventive Potential of Coumarin in 7, 12-dimethylbenz[*a*]anthracene Induced Hamster Buccal Pouch Carcinogenesis. *Asian Pac. J. Cancer Prevent.* **2012,** *13*, 5273–5279.

9. Carnelio, S.; Rodrigues, G. S.; Shenoy, R.; Fernandes, D. A Brief Review of Common Oral Premalignant Lesions with Emphasis on their Management and Cancer Prevention. *Indian J. Surgery* **2011,** *73*, 256–261.

10. Casto, B. C.; Knobloch, T. J.; Weghorst, C. M. Inhibition of Oral Cancer in Animal Models by Black Raspberries and Berry Components. In *Berries and Cancer Prevention*; Casto, B. C., Knobloch, T. J., Weghorst, C. M., Eds.; Springer: New York, 2011; pp 189–207.

11. Casto, B. C.; Knobloch, T. J.; Galioto, R. L.; Yu, Z.; Accurso, B. T.; Warner, B. M. Chemoprevention of Oral Cancer by Lyophilized Strawberries. *Anticancer Res.* **2013,** *33*, 4757–4766.

12. Chae, J. I.; Jeon, Y. J.; Shim, J. H. Anti-Proliferative Properties of Kahweol in Oral Squamous Cancer through the Regulation Specificity Protein 1. *Phytother. Res.* **2014,** *28*, 1879–1886.

13. Chandra Mohan, K. V. P.; Letchoumy, P. V.; Hara, Y.; Nagini, S. Combination Chemoprevention of Hamster Buccal Pouch Carcinogenesis by Bovine Milk Lactoferrin and Black Tea Polyphenols. *Cancer Investig.* **2008,** *26*, 193–201.

14. Chang, C. M.; Chang, P. Y.; Tu, M. G.; Lu, C. C.; Kuo, S. C.; Amagaya, S.; Lee, C. Y.; Jao, H. Y.; Chen, M. Y.; Yang, J. S. Epigallocatechin gallate sensitizes CAL-27 Human Oral Squamous Cell Carcinoma Cells to the Anti-metastatic effects of Gefitinib (Iressa) via Synergistic Suppression of Epidermal Growth Factor Receptor and Matrix Metalloproteinase-2. *Oncol. Rep.* **2012,** *28*, 1799–1807.

15. Chang, Y. C.; Chen, P. N.; Chu, S. C.; Lin, C. Y.; Kuo, W. H.; Hsieh, Y. S. Black Tea Polyphenols Reverse Epithelial-to-mesenchymal Transition and Suppress Cancer Invasion and Proteases in Human Oral Cancer Cells. *J. Agric. Food Chem.* **2012,** *60*, 8395–8403.

16. Chang, Y. C.; Chen, P. N.; Chu, S. C.; Lin, C. Y.; Kuo, W. H.; Hsieh, Y. S. Black Tea Polyphenols Reverse Epithelial-to-mesenchymal Transition and Suppress Cancer Invasion and Proteases in Human Oral Cancer Cells. *J. Agric. Food Chem.* **2012,** *60*, 8395–8403.

17. Chatelain, K.; Phippen, S.; McCabe, J.; Teeters, C. A.; O'Malley, S.; Kingsley, K. Cranberry and Grape Seed Extracts Inhibit the Proliferative Phenotype of Oral Squamous Cell Carcinomas. *Evidence-based Complem. Alternat. Med.* **2008,** E-pub.

18. Chatelain, K.; Phippen, S.; McCabe, J.; Teeters, C. A.; O'Malley, S.; Kingsley, K. Cranberry and Grape Seed Extracts Inhibit the Proliferative Phenotype of Oral Squamous Cell Carcinomas. *Evidence-based Complem. Alternat. Med.* **2010,** E-pub.

19. Chen, P. N.; Chu, S. C.; Kuo, W. H.; Chou, M. Y.; Lin, J. K.; Hsieh, Y. S. Epigallocatechin-3 Gallate Inhibits Invasion, Epithelial—Mesenchymal Transition, and Tumor Growth in Oral Cancer Cells. *J. Agric. Food Chem.* **2011,** *59*, 3836–3844.

20. Choi, S.; Myers, J. N. Molecular Pathogenesis of Oral Squamous Cell Carcinoma: Implications for Therapy. *J. Dental Res.* **2008,** *87,* 14–32.

21. D'Ambrosio, S. M.; Han, C.; Pan, L.; Kinghorn, A. D.; Ding, H. Aliphatic Acetogenin Constituents of Avocado Fruits Inhibit Human Oral Cancer Cell Proliferation by Targeting the EGFR/RAS/RAF/MEK/ERK1/2 Pathway. *Biochem. Biophys. Res. Commun.* **2011,** *409,* 465–469.

22. de Moura, C. F. G.; Noguti, J.; de Jesus, G. P. P.; Ribeiro, F. A.; Garcia, F. A.; Gollucke, A. P.; Ribeiro, D. A. Polyphenols as a Chemopreventive Agent in Oral Carcinogenesis: Putative Mechanisms of Action Using In-Vitro and In-Vivo Test Systems. *Eur. J. Cancer Preven.* **2013,** *22,* 467–472.

23. Ding, H.; Han, C.; Guo, D.; Chin, Y. W.; Ding, Y.; Kinghorn, A. D.; D'Ambrosio, S. M. Selective Induction of Apoptosis of Human Oral Cancer Cell Lines by Avocado Extracts Via a ROS-mediated Mechanism. *Nutr. Cancer* **2009,** *61,* 348–356.

24. Ding, Y.; Yao, H.; Yao, Y.; Fai, L. Y.; Zhang, Z. Protection of Dietary Polyphenols against Oral Cancer. *Nutrients* **2013,** *5,* 2173–2191.

25. Dutta, K. R.; Banerjee, S.; Mitra, A. Medicinal Plants of West Midnapore, India: Emphasis on Phytochemical Containment having Role on Oral Cancer. *IJP* **2012,** *3,* 198–208.

26. Fahey, J. W.; Talalay, P.; Kensler, T. W. Notes from the Field:"green" Chemoprevention as Frugal Medicine. *Cancer Preven. Res.* **2012,** *5,* 179–188.

27. Gao, Y.; Zhang, R.; Zhang, J.; Gao, S.; Gao, W.; Zhang, H.; Wang, H.; Han, B. Study of the Extraction Process and In Vivo Inhibitory effect of Ganoderma Triterpenes in Oral Mucosa Cancer. *Molecules* **2011,** *16,* 5315–5332.

28. Garg, R.; Ingle, A.; Maru, G. Dietary Turmeric Modulates DMBA-induced p21 ras, MAP kinases and AP-1/NF-κB pathway to Alter Cellular Responses During Hamster Buccal Pouch Carcinogenesis. *Toxicol. Appl. Pharmacol.* **2008,** *232,* 428–439.

29. Gillison, M. L. Current Topics in the Epidemiology of Oral Cavity and Oropharyngeal Cancers. *Head Neck* **2007,** *29,* 779–792.

30. Benner, S. E.; Lippman, S. M.; Hong, W. K. Chemoprevention of Head and Neck Cancer. *Chemopreven. Cancer* **1994,** *121.*

31. Gómez-García, F. J.; López-Jornet, M. P.; Álvarez-Sánchez, N.; Castillo-Sánchez, J.; Benavente-García, O.; Vicente Ortega, V. Effect of the Phenolic Compounds Apigenin and Carnosic Acid on Oral Carcinogenesis in Hamster Induced by DMBA. *Oral Dis.* **2013,** *19,* 279–286.

32. Gupta, S. C.; Kim, J. H.; Prasad, S.; Aggarwal, B. B. Regulation of Survival, Proliferation, Invasion, Angiogenesis, and Metastasis of Tumor Cells through Modulation of Inflammatory Pathways by Nutraceuticals. *Cancer Metastasis Rev.* **2010,** *29,* 405–434.

33. Haghiac, M.; Walle, T. Quercetin Induces Necrosis and Apoptosis in SCC-9 Oral Cancer Cells. *Nutr. Cancer* **2005,** *53,* 220–231.

34. Han, C.; Ding, H.; Casto, B.; Stoner, G. D.; D'Ambrosio, S. M. Inhibition of the Growth of Premalignant and Malignant Human Oral Cell Lines by Extracts and Components of Black Raspberries. *Nutr. Cancer* **2005,** *51,* 207–217.

35. Harada, K.; Ferdous, T.; Itashiki, Y.; Takii, M.; Mano, T.; Mori, Y.; Ueyama, Y. Cepharanthine Inhibits Angiogenesis and Tumorigenicity of Human Oral Squamous Cell Carcinoma Cells by Suppressing Expression of Vascular Endothelial Growth Factor and Interleukin-8. *Int. J. Oncol.* **2009,** *35,* 1025–1035.

36. Hou, Y.; Wu, M, L.; Hwang, Y. C.; Chang, F. R.; Wu, Y. C.; Wu, C. C. The Natural Diterpenoid Ovatodiolide Induces Cell Cycle Arrest and Apoptosis in Human Oral Squamous Cell Carcinoma Ca9–22 Cells. *Life Sci.* **2008,** *85,* 26–32.

37. Hsu, W. H.; Lee, B. H.; Pan, T. M. Effects of Red Mold Dioscorea on Oral Carcinogenesis in DMBA-induced Hamster Animal Model. *Food Chem. Toxicol.* **2011,** *49,* 1292–1297.

38. Hwang, Y. S.; Chung, W. Y.; Kim, J.; Park, H. J.; Kim, E. C.; Park, K. K. Buddlejasaponin IV Induces Cell Cycle Arrest at G2/M Phase and Apoptosis in Immortalized Human Oral Keratinocytes. *Phytother. Res.* **2011,** *25,* 1503–1510.

39. Jee, H. G.; Lee, K. E.; Kim, J. B.; Shin, H. K.; Youn, Y. K. Sulforaphane Inhibits Oral Carcinoma Cell Migration and Invasion In Vitro. *Phytother. Res.* **2011,** *25,* 1623–1628.

40. Johnson, T. L.; Lai, M. B.; Lai, J. C.; Bhushan, A. Inhibition of Cell Proliferation and MAP kinase and Akt Pathways in Oral Squamous Cell Carcinoma by Genistein and Biochanin A. *Evidence-based Complement. Alternat. Med.* **2010,** *7,* 351–358.

41. Joshipura, K.; Dietrich, T. Nutrition and Oral Health: A Two-way Relationship. In *Handbook of Clinical Nutrition and Aging*; Joshipura, K., Dietrich, T., Eds.; Humana Press, 2009; pp 247–262.

42. Kang, J. W.; Kim, J. H.; Song, K.; Kim, S. H.; Yoon, J. H.; Kim, K. S. Kaempferol and Quercetin, Components of *Ginkgo biloba* Extract (EGb 761), Induce Caspase-3-dependent Apoptosis in Oral Cavity Cancer Cells. *Phytother. Res.* **2010,** *24,* S77–S82.

43. Kasem, R. F.; Hegazy, R. H.; Arafa, M. A.; AbdelMohsen, M. M. Chemopreventive Effect of Mentha piperita on dimethylbenz[*a*]anthracene and formaldehyde-Induced Tongue Carcinogenesis in Mice (Histological and Immunohistochemical study). *J. Oral Pathol. Med.* **2014,** *43,* 484–491.

44. Kato, K.; Long, N. K.; Makita, H.; Toida, M.; Yamashita, T.; Hatakeyama, D.; Shibata, T. Effects of Green Tea Polyphenol on Methylation Status of RECK Gene and Cancer Cell Invasion in Oral Squamous Cell Carcinoma Cells. *Br. J. Cancer* **2008,** *99,* 647–654.

45. Kavitha, K.; Kowshik, J.; Kishore, T. K. K.; Baba, A. B.; Nagini, S. Astaxanthin Inhibits NF-κB and Wnt/β-catenin Signaling Pathways Via Inactivation of Erk/MAPK and PI3K/Akt to Induce Intrinsic Apoptosis in a Hamster Model of Oral Cancer. *Biochim. Biophys. Acta (BBA)-Gen. Subj.* **2013,** *1830,* 4433–4444.

46. Kim, J. W.; Amin, A. R.; Shin, D. M. Chemoprevention of Head and Neck Cancer with Green Tea Polyphenols. *Cancer Prevent. Res.* **2010,** *3,* 900–909.

47. Kim, J. Y.; An, J. M.; Chung, W. Y.; Park, K. K.; Hwang, J. K.; Kim, D. S.; Seo, J. T. Xanthorrhizol Induces Apoptosis Through ROS-mediated MAPK Activation in Human Oral Squamous Cell Carcinoma Cells and Inhibits DMBA-Induced Oral Carcinogenesis in Hamsters. *Phytother. Res.* **2013,** *27,* 493–498.

48. King, M.; Chatelain, K.; Farris, D.; Jensen, D.; Pickup, J.; Swapp, A.; Kingsley, K. Oral Squamous Cell Carcinoma Proliferative Phenotype is Modulated by Proanthocyanidins: A Potential Prevention and Treatment Alternative for Oral Cancer. *BMC Complement. Alternat. Med.* **2007,** *7,* 22.

49. Koh, Y. W.; Choi, E. C.; Kang, S. U.; Hwang, H. S.; Lee, M. H.; Pyun, J.; Park, R.; Lee, Y.; Kim, C. H. Green Tea Epigallocatechin-3-gallate Inhibits HGF-induced Progression in Oral Cavity Cancer Through Suppression of HGF/c-Met. *J. Nutr. Biochem.* **2011,** *22,* 1074–1083.

50. Kok, S. H.; Yeh, C. C.; Chen, M. L.; Kuo, M. Y. P. Esculetin Enhances Trail-induced Apoptosis Through DR5 Upregulation in Human Oral Cancer SAS Cells. *Oral Oncol.* **2009,** *45,* 1067–1072.

51. Kowsalya, R.; Vishwanathan, P.; Manoharan, S. Chemopreventive Potential of 18 Beta-glycyrrhetinic acid: An Active Constituent of Liquorice, in 7,12-Dimethylbenz(*a*) anthracene Induced Hamster Buccal Pouch Carcinogenesis. *Pak. J. Biol. Sci.: PJBS* **2011,** *14*, 619–626.

52. Kowshik, J.; Baba, A. B.; Giri, H.; Reddy, G. D.; Dixit, M.; Nagini, S. Astaxanthin Inhibits JAK/STAT-3 Signaling to Abrogate Cell Proliferation, Invasion and Angiogenesis in a Hamster Model of Oral Cancer. *PLoS ONE* **2014,** *9*, e109114.

53. Lee, H. J.; Son, D. H.; Lee, S. K.; Lee, J.; Jun, C. D.; Jeon, B. H.; Kim, E. C. Extract of Coptidis Rhizoma Induces Cytochrome-*c* Dependent Apoptosis in Immortalized and Malignant Human Oral Keratinocytes. *Phytother. Res.* **2006,** *20*, 773–779.

54. Lee, T. K.; Poon, R. T.; Wo, J. Y.; Ma, S.; Guan, X. Y.; Myers, J. N.; Yuen, A. P. Lupeol Suppresses Cisplatin-induced Nuclear Factor-κB Activation in Head and Neck Squamous Cell Carcinoma and Inhibits Local Invasion and Nodal Metastasis in an Orthotopic Nude Mouse Model. *Cancer Res.* **2007,** *67*, 8800–8809.

55. Lee, U. L.; Choi, S. W. The Chemopreventive Properties and Therapeutic Modulation of Green Tea Polyphenols in Oral Squamous Cell Carcinoma. *ISRN Oncol.* **2011,** *2011*, 7 pp. Article ID 403707, E-pub; http://dx.doi.org/10.5402/2011/403707.

56. Liang, C. H.; Wang, G. H.; Liaw, C. C.; Lee, M. F.; Wang, S. H.; Cheng, D. L.; Chou, T. H. Extracts from *Cladiella australis, Clavularia viridis* and *Klyxum simplex* (soft corals) are Capable of Inhibiting the Growth of Human Oral Squamous Cell Carcinoma Cells. *Mar. Drugs* **2008,** *6*, 595–606.

57. Lim, V.; Korourian, S.; Todorova, V. K.; Kaufmann, Y.; Klimberg, V. S. Glutamine Prevents DMBA-induced Squamous Cell Cancer. *Oral Oncol.* **2009,** *45*, 148–155.

58. Lin, C. H.; Lu, W. C.; Wang, C. W.; Chan, Y. C.; Chen, M. K. Capsaicin Induces Cell Cycle Arrest and Apoptosis in Human KB Cancer Cells. *BMC Complement. Alternat. Med.* **2013,** *13*, 46.

59. Lin, C. W.; Chou, Y. E.; Chiou, H. L.; Chen, M. K.; Yang, W. E.; Hsieh, M. J.; Yang, S. F. Pterostilbene Suppresses Oral Cancer Cell Invasion by Inhibiting MMP-2 Expression. *Expert Opin. Ther. Targets* **2014,** *18*, 1109–1120.

60. Lin, H. Y.; Thomas, J. L.; Chen, H. W.; Shen, C. M.; Yang, W. J.; Lee, M. H. In Vitro Suppression of Oral Squamous Cell Carcinoma Growth by Ultrasound-mediated Delivery of Curcumin Microemulsions. *Int. J. Nanomed.* **2012,** *7*, 941.

61. Lin, Y. S.; Chen, S. F.; Liu, C. L.; Nieh, S. The Chemoadjuvant Potential of Grape Seed Procyanidins on p53-related Cell Death in Oral Cancer Cells. *J. Oral Pathol. Med.* **2012,** *41*, 322–331.

62. Livny, O.; Kaplan, I.; Reifen, R.; Polak-Charcon, S.; Madar, Z.; Schwartz, B. Lycopene Inhibits Proliferation and Enhances Gap-junction Communication of KB-1 Human Oral Tumor Cells. *J. Nutr.* **2002,** *132*, 3754–3759.

63. Livny, O.; Kaplan, I.; Reifen, R.; Polak-Charcon, S.; Madar, Z.; Schwartz, B. Oral Cancer Cells Differ from Normal Oral Epithelial Cells in Tissue Like Organization and in Response to Lycopene Treatment: An Organotypic Cell Culture Study. *Nutr. Cancer* **2003,** *47*(2), 195–209.

64. Lu, K. W.; Chen, J. C.; Lai, T. Y.; Yang, J. S.; Weng, S. W.; Ma, Y. S.; Lu. P. J.; Weng.; J. R.; Chueh.; F. S.; Wood.; W. J.; Chung, J. G. Gypenosides Inhibits Migration and Invasion of Human Oral Cancer SAS Cells Through the Inhibition of Matrix Metalloproteinase-2/-9 and Urokinase–Plasminogen by ERK1/2 and NF-kappa B Signaling Pathways. *Hum. Exp. Toxicol.* **2010,** *30*, 406–415.

65. Lucenteforte, E.; Garavello, W.; Bosetti, C.; La Vecchia, C. Dietary Factors and Oral and Pharyngeal Cancer Risk. *Oral Oncol.* **2009**, *45*, 461–467.
66. Mallery, S. R.; Tong, M.; Shumway, B. S.; Curran, A. E.; Larsen, P. E.; Ness, G. M.; et. al. Topical Application of a Mucoadhesive Freeze-dried Black Raspberry Gel Induces Clinical and Histologic Regression and Reduces Loss of Heterozygosity Events in Premalignant Oral Intraepithelial Lesions: Results from a Multicentered, Placebo-controlled Clinical Trial. *Clin. Cancer Res.* **2014**, *20*, 1910–1924.
67. Manikandan, P.; Ramalingam, S. M.; Vinothini, G.; Ramamurthi, V. P.; Singh, I. P.; AnR. Investigation of the Chemopreventive Potential of Neem Leaf Subfractions in the Hamster Buccal Pouch Model and Phytochemical Characterization. *Eur. J. Med. Chem.* **2012**, *56*, 271–281.
68. Manoharan, S.; Balakrishnan, S.; Menon, V. P.; Alias, L. M.; Reena, A. R. Chemopreventive Efficacy of Curcumin and Piperine During 7,12-Dimethylbenz(*a*)anthracene-induced Hamster Buccal Pouch Carcinogenesis. *Singapore Med. J.* **2009**, *50*, 139.
69. Manoharan, S.; Palanimuthu, D.; Baskaran, N.; Silvan, S. Modulating Effect of Lupeol on the Expression Pattern of Apoptotic Markers in 7,12-dimethylbenz(*a*)anthracene Induced Oral Carcinogenesis. *Asian Pac. J. Cancer Prevent.* **2012**, *13*, 5753–5757.
70. Manoharan, S.; Sindhu, G.; Nirmal, M. R.; Vetrichelvi, V.; Balakrishnan, S. Protective Effect of Berberine on Expression Pattern of Apoptotic, Cell Proliferative, Inflammatory and Angiogenic Markers During 7,12-Dimethylbenz(*a*)anthracene Induced Hamster Buccal Pouch Carcinogenesis. *Pak. J. Biol. Sci.: PJBS* **2011**, *14*, 918–932.
71. Manoharan, S.; Singh, A. K.; Suresh, K.; Vasudevan, K.; Subhasini, R.; Baskaran, N. Anti-tumor Initiating Potential of Andrographolide in 7,12-dimethylbenz[*a*]anthracene Induced Hamster Buccal Pouch Carcinogenesis. *Asian Pac. J. Cancer Prevent.* **2012**, *13*, 5701–5708.
72. Manoharan, S.; VasanthaSelvan, M.; Silvan, S.; Baskaran, N.; Singh, A. K.; Kumar, V. V. Carnosic Acid: A Potent Chemopreventive Agent Against Oral Carcinogenesis. *Chemico-biol. Interact.* **2010**, *188*, 616–622.
73. Manoharan, S.; Wani, S. A.; Vasudevan, K.; Manimaran, A.; Prabhakar, M. M.; Karthikeyan, S.; Rajasekaran, D. Saffron Reduction of 7,12-dimethylbenz[*a*]anthracene-induced Hamster Buccal Pouch Carcinogenesis. *Asian Pac. J. Cancer Prevent.* **2013**, *14*, 951–957.
74. Mariadoss, A. V.; Kathiresan, S.; Muthusamy, R. Protective Effects of [6]-Paradol on Histological Lesions and Immunohistochemical Gene Expression in DMBA Induced Hamster Buccal Pouch Carcinogenesis. *Asian Pac. J. Cancer Prevent.* **2012**, *14*, 3123–3129.
75. Miller, E. G.; Peacock, J. J.; Bourland, T. C.; Taylor, S. E.; Wright, J. M.; Patil, B. S.; Miller, E. G. Inhibition of Oral Carcinogenesis by Citrus Flavonoids. *Nutr. Cancer* **2007**, *60*, 69–74.
76. Ruan, M.; Ji, T.; Yang, W.; Duan, W.; Zhou, X.; He, J.; et al. Growth Inhibition and Induction of Apoptosis in Human Oral Squamous Cell Carcinoma Tca-8113 Cell Lines by Shikonin was Partly Through the Inactivation of NF-κB pathway. *Phytother. Res.* **2008**, *22*, 407–415.
77. Panjamurthy, K.; Manoharan, S.; Nirmal, M. R.; Vellaichamy, L. Protective Role of Withaferin-A on Immunoexpression of p53 and bcl-2 in 7,12-Dimethylbenz(*a*) anthracene-Induced Experimental Oral Carcinogenesis. *Investig. New Drugs* **2009**, *27*, 447–452.

78. Petti, S. Lifestyle Risk Factors for Oral Cancer. *Oral Oncol.* **2009,** *45*, 340–350.

79. Pitiphat, W.; Diehl, S. R.; Laskaris, G.; Cartsos, V.; Douglass, C. W.; Zavras, A. I. Factors Associated with Delay in the Diagnosis of Oral Cancer. *J. Dent. Res.* **2002,** *81*, 192–197.

80. Prabhakar, M. M.; Vasudevan, K.; Karthikeyan, S.; Baskaran, N.; Silvan, S.; Manoharan, S. Anti-cell Proliferative Efficacy of Ferulic Acid against 7,12-Dimethylbenz(*a*) anthracene Induced Hamster Buccal Pouch Carcinogenesis. *Asian Pac. J. Cancer Prevent.* **2011,** *13*, 5207–5211.

81. Prasad, R.; Katiyar, S. K. Bioactive Phytochemical Proanthocyanidins Inhibit Growth of Head and Neck Squamous Cell Carcinoma Cells by Targeting Multiple Signaling Molecules. *PLoS ONE* **2012,** *7*, e46404.

82. Priyadarsini, R. V.; Kumar, N.; Khan, I.; Thiyagarajan, P.; Kondaiah, P.; Nagini, S. Gene Expression Signature of DMBA-induced Hamster Buccal Pouch Carcinomas: Modulation by Chlorophyllin and Ellagic Acid. *PLoS ONE* **2012,** *7*, e34628.

83. Priyadarsini, R. V.; Nagini, S. Quercetin Suppresses Cytochrome P450 Mediated ROS Generation and NFκB Activation to Inhibit the Development of 7,12-Dimethylbenz[*a*] anthracene (DMBA) Induced Hamster Buccal Pouch Carcinomas. *Free Radic. Res.* **2012,** *46*, 41–49.

84. Rajalingam, K.; Sugunadevi, G.; Vijayaanand, M. A.; Kalaimathi, J.; Suresh, K. Anti-tumour and Anti-oxidative Potential of Diosgenin against 7,12-dimethylbenz(*a*) anthracene Induced Experimental Oral Carcinogenesis. *Pathol. Oncol. Res.* **2012,** *18*, 405–412.

85. Rajasekaran, D.; Manoharan, S.; Silvan, S.; Vasudevana, K.; Baskaran, N.; Palanimuthu, D. Proapoptotic, Anti-cell Proliferative, Anti-Inflammatory and Antiangiogenic Potential of Carnosic Acid During 7,12-Dimethylbenz[*a*]anthracene-induced Hamster Buccal Pouch Carcinogenesis. *Afr. J. Trad., Complement. Alternat. Med.* **2013,** *10*, 102–112.

86. Rajkamal, G.; Suresh, K.; Sugunadevi, G.; Vijayaanand, M. A.; Rajalingam, K. Evaluation of Chemopreventive effects of Thymoquinone on Cell Surface Glycoconjugates and Cytokeratin Expression during DMBA Induced Hamster Buccal Pouch Carcinogenesis. *BMB Rep.* **2010,** *43*, 664–669.

87. Saba, N. F.; Haigentz, M.; Vermorken, J. B.; Strojan, P.; Bossi, P.; Rinaldo, A. Prevention of Head and Neck Squamous Cell Carcinoma: Removing the "chemo" from "chemoprevention". *Oral Oncol.* **2015,** *51*, 112–118.

88. Saba, N. F.; Khuri, F. R. Chemoprevention in Head and Neck Cancer. In *Squamous Cell Head and Neck Cancer*; Saba, N. F., Khuri, F. R. Eds.; Humana Press, 2005; pp 279–303.

89. Saroja, M.; Balasenthil, S.; Ramachandran, C. R.; Nagini, S. Coffee Enhances the Development of 7,12-dimethylbenz[*a*]anthracene (DMBA)-induced Hamster Buccal Pouch Carcinomas. *Oral Oncol.* **2001,** *37*, 172–176.

90. Schuck, A. G.; Ausubel, M. B.; Zuckerbraun, H. L.; Babich, H. Theaflavin-3,3′-digallate, a Component of Black Tea: An Inducer of Oxidative Stress and Apoptosis. *Toxicol. In Vitro* **2008,** *22*, 598–609.

91. Seeram, N. P.; Adams, L. S.; Henning, S. M.; Niu, Y.; Zhang, Y.; Nair, M. G.; Heber, D. In Vitro Antiproliferative, Apoptotic and Antioxidant Activities of Punicalagin, Ellagic Acid and a Total Pomegranate Tannin Extract are Enhanced in Combination with other Polyphenols as Found in Pomegranate Juice. *J. Nutr. Biochem.* **2005,** *16*, 360–367.

92. Seeram, N. P.; Adams, L. S.; Zhang, Y.; Lee, R.; Sand, D.; Scheuller, H. S.; Heber, D. Blackberry, Black Raspberry, Blueberry, Cranberry, Red Raspberry, and Strawberry

Extracts Inhibit Growth and Stimulate Apoptosis of Human Cancer Cells In Vitro. *J. Agric. Food Chem.* **2006,** *54*, 9329–9339.

93. Shan, Z.; Yang, G.; Xiang, W.; Pei-jun, W.; Bin, Z. Effects of Resveratrol on Oral Squamous Cell Carcinoma (OSCC) Cells In Vitro. *J. Cancer Res. Clin. Oncol.* **2014,** *140*, 371–374.

94. Shrotriya, S.; Agarwal, R.; Sclafani, R. A. A Perspective on Chemoprevention by Resveratrol in Head and Neck Squamous Cell Carcinoma. In *Biological Basis of Alcohol-Induced Cancer*; Shrotriya, S., Agarwal, R., Sclafani, R. A., Eds.; Springer International Publishing, 2015; pp 333–348.

95. Shumway, B. S.; Kresty, L. A.; Larsen, P. E.; Zwick, J. C.; Lu, B.; Fields, H. W. Effects of a Topically Applied Bioadhesive Berry Gel on Loss of Heterozygosity Indices in Premalignant Oral Lesions. *Clin. Cancer Res.* **2008,** *14*, 2421–2430.

96. Siddavaram, N.; Ramamurthi, V. P.; Veeran, V.; Mishra, R. Chlorophyllin Abrogates Canonical Wnt/β-catenin Signaling and Angiogenesis to Inhibit the Development of DMBA-induced Hamster Cheek Pouch Carcinomas. *Cell. Oncol.* **2012,** *35*, 385–395.

97. Silvan, S.; Manoharan, S.; Baskaran, N.; Anusuya, C.; Karthikeyan, S.; Prabhakar, M. M. Chemopreventive Potential of Apigenin in 7,12-Dimethylbenz(*a*)anthracene Induced Experimental Oral Carcinogenesis. *Eur. J. Pharmacol.* **2011,** *670*, 571–577.

98. Silvan, S.; Manoharan, S. Apigenin Prevents Deregulation in the Expression Pattern of Cell-proliferative, Apoptotic, Inflammatory and Angiogenic Markers During 7,12-Dimethylbenz[*a*]anthracene-induced Hamster Buccal Pouch Carcinogenesis. *Arch. Oral Biol.* **2013,** *58*, 94–101.

99. Steward, W. P.; Brown, K. Cancer Chemoprevention: A Rapidly Evolving Field. *Br. J. Cancer* **2013,** *109*, 1–7.

100. Sun, Z.; Guan, X.; Li, N.; Liu, X.; Chen, X. Chemoprevention of Oral Cancer in Animal Models, and Effect on Leukoplakias in Human Patients with ZengShengPing, a Mixture of Medicinal Herbs. *Oral Oncol.* **2010,** *46*, 105–110.

101. Suresh, K.; Manoharan, S.; Vijayaanand, M. A.; Sugunadevi, G. Chemopreventive and Antioxidant Efficacy of (6)-paradol in 7,12-Dimethylbenz(*a*)anthracene Induced Hamster Buccal Pouch Carcinogenesis. *Pharmacol. Rep.* **2010,** *62*, 1178–1185.

102. Taghavi, N.; Yazdi, I. Type of Food and Risk of Oral Cancer. *Arch. Iran Med.* **2007,** *10*, 227–32.

103. Tan, A. C.; Konczak, I.; Sze, D. M. Y.; Ramzan, I. Molecular Pathways for Cancer Chemoprevention by Dietary Phytochemicals. *Nutr. Cancer* **2011,** *63*, 495–505.

104. Tanaka, T.; Makita, H.; Ohnishi, M.; Mori, H.; Satoh, K.; Hara, A. Chemoprevention of Rat Oral Carcinogenesis by Naturally Occurring Xanthophylls, Astaxanthin and Canthaxanthin. *Cancer Res.* **1995,** *55*, 4059–4064.

105. Tanaka, T.; Tanaka, M.; Tanaka, T. Oral Carcinogenesis and Oral Cancer Chemoprevention: A Review. *Pathol. Res. Int.* **2011,** *2011*, 10 pp. Article ID 431246, E-pub.; http://dx.doi.org/10.4061/2011/431246.

106. Tang, F. Y.; Chiang, E. P. I.; Chung, J. G.; Lee, H. Z.; Hsu, C. Y. S-allylcysteine Modulates the Expression of E-cadherin and Inhibits the Malignant Progression of Human Oral Cancer. *J. Nutr. Biochem.* **2009,** *20*, 1013–1020.

107. Thanusu, J.; Kanagarajan, V.; Nagini, S.; Gopalakrishnan, M. Chemopreventive Potential of 3-[2, 6-bis(4-fluorophenyl)-3-methylpiperidin-4-ylideneamino]-2-thioximidazolidin-4-one on 7,12-dimethylbenz[*a*]anthracene (DMBA) induced hamster buccal Pouch Carcinogenesis. *J. Enzyme Inhibit. Med. Chem.* **2010,** *25*, 836–843.

108. Thiyagarajan, P.; Murugan, R. S.; Kavitha, K.; Anitha, P.; Prathiba, D.; Nagini, S. Dietary Chlorophyllin Inhibits the Canonical NF-κB Signaling Pathway and Induces Intrinsic Apoptosis in a Hamster Model of Oral Oncogenesis. *Food Chem. Toxicol.* **2012**, *50*, 867–876.

109. Toledo, A. L. A. D.; Koifman, R. J.; Koifman, S.; Marchioni, D. M. L. Dietary Patterns and Risk of Oral and Pharyngeal Cancer: A Case-control Study in Rio de Janeiro, Brazil. *Cadernos Saúde Públ.* **2010**, *26*, 135–142.

110. Tsantoulis, P. K.; Kastrinakis, N. G.; Tourvas, A. D.; Laskaris, G.; Gorgoulis, V. G. Advances in the Biology of Oral Cancer. *Oral Oncol.* **2007**, *43*, 523–534.

111. Tseng, P. Y.; Lu, W. C.; Hsieh, M. J.; Chien, S. Y.; Chen, M. K. Tanshinone IIA Induces Apoptosis in Human Oral Cancer KB Cells through a Mitochondria-dependent Pathway. *Biomed. Res. Int.* **2014**, *2014*. Article ID 540516; E-pub; DOI:10.1155/*2014*/540516.

112. Tyagi, A.; Gu, M.; Takahata, T.; Frederick, B.; Agarwal, C.; Siriwardana, S.; et al. Resveratrol Selectively Induces DNA Damage, Independent of Smad4 Expression, in its Efficacy against Human Head and Neck Squamous Cell Carcinoma. *Clin. Cancer Res.* **2011**, *17*, 5402–5411.

113. Vinothkumar, V.; Manoharan, S.; Sindhu, G.; Nirmal, M. R.; Vetrichelvi, V. Geraniol Modulates Cell Proliferation, Apoptosis, Inflammation, and Angiogenesis during 7,12-Dimethylbenz[*a*]anthracene-induced Hamster Buccal Pouch Carcinogenesis. *Mol. Cell. Biochem.* **2012**, *369*, 17–25.

114. Wang, D.; Veena, M. S.; Stevenson, K.; Tang, C.; Ho, B.; Suh, J. D. Liposome-encapsulated Curcumin Suppresses Growth of Head and Neck Squamous Cell Carcinoma In Vitro and in Xenografts Through the Inhibition of Nuclear Factor κB by an AKT-Independent Pathway. *Clin. Cancer Res.* **2008**, *14*, 6228–6236.

115. Weng, J. R.; Bai, L. Y.; Chiu, C. F.; Wang, Y. C.; Tsai, M. H. The Dietary Phytochemical 3,3′-diindolylmethane induces G2/M Arrest and Apoptosis in Oral Squamous Cell Carcinoma by Modulating Akt-NF-κB, MAPK, and p53 Signaling. *Chemico-Biol. Int.* **2012**, *195*, 224–230.

116. Xiao, B.; Guo, J.; Liu, D.; Zhang, S. Aloe-emodin Induces In Vitro G2/M Arrest and Alkaline Phosphatase Activation in Human Oral Cancer KB Cells. *Oral Oncol.* **2007**, *43*, 905–910.

117. Yang, S. F.; Chen, M. K.; Hsieh, Y. S.; Yang, J. S.; Zavras, A. I.; Hsieh, Y. H.; Chu, S. C. Antimetastatic Effects of *Terminalia catappa* L. on Oral Cancer Via a Down-regulation of Metastasis-associated Proteases. *Food Chem. Toxicol.* **2010**, *48*, 1052–1058.

118. Yl, Y.; Ji, C.; Zg, B.; Cc, L.; Wang, R. Deguelin Induces both Apoptosis and Autophagy in Cultured Head and Neck Squamous Cell Carcinoma Cells. *PLoS ONE* **2013**, *8*, e54736.

119. Yuan, Z.; Wang, H.; Hu, Z.; Huang, Y.; Yao, F.; Sun, S.; Wu, B. Quercetin Inhibits Proliferation and Drug Resistance in KB/VCR Oral Cancer Cells and Enhances Its Sensitivity to Vincristine. *Nutr. Cancer* **2015**, *67*, 126–136.

120. Yun, Y. G.; Jeon, B. H.; Lee, J. H.; Lee, S. K.; Lee, H. J.; Jung, K. H.; et al. Verticinone Induces Cell Cycle Arrest and Apoptosis in Immortalized and Malignant Human Oral Keratinocytes. *Phytother. Res.* **2008**, *22*, 416–423.

121. Zain, R. B. Cultural and Dietary Risk Factors of Oral Cancer and Precancer—A Brief Overview. *Oral Oncol.* **2001**, *37*, 205–210.

122. Zhang, Y.; Seeram, N. P.; Lee, R.; Feng, L.; Heber, D. Isolation and Identification of Strawberry Phenolics with Antioxidant and Human Cancer Cell Antiproliferative Properties. *J. Agric. Food Chem.* **2008**, *56*(3), 670–675.

123. Zhao, X.; Pang, L.; Li, J.; Song, J. L.; Qiu, L. H. Apoptosis Inducing Effects of Kuding Tea Polyphenols in Human Buccal Squamous Cell Carcinoma Cell Line BcaCD885. *Nutrients* **2014,** *6,* 3084–3100.
124. Ziech, D.; Anestopoulos, I.; Hanafi, R.; Voulgaridou, G. P.; Franco, R.; Georgakilas, A. G.; Pappa, A.; Panayiotidis, M. I. Pleiotrophic effects of Natural Products in ROS-Induced Carcinogenesis: The Role of Plant-derived Natural Products in Oral Cancer Chemoprevention. *Cancer Lett.* **2012,** *327,* 16–25.
125. Zlotogorski, A.; Dayan, A.; Dayan, D.; Chaushu, G.; Salo, T.; Vered, M. Nutraceuticals as New Treatment Approaches for Oral Cancer—I: Curcumin. *Oral Oncol.* **2013,** *49,* 187–191.
126. Zlotogorski, A.; Dayan, A.; Dayan, D.; Chaushu, G.; Salo, T.; Vered, M. Nutraceuticals as New Treatment Approaches for Oral Cancer: II. Green Tea Extracts and Resveratrol. *Oral Oncol.* **2013,** *49,* 502–506.

CHAPTER 9

FOOD ENGINEERING: AROMATIC AND MEDICINAL PLANTS

DAWN C. P. AMBROSE

ICAR—Central Institute of Agricultural Engineering—Regional Centre, Coimbatore 641 003, Tamil Nadu, India. E-mail: dawncp@yahoo.com

CONTENTS

ABSTRACT

Aromatic and medicinal plants have been in use as natural medicines since time immemorial and are gaining demand in the global market in recent times as an alternative to artificial medicines. The difference between aromatic and medicinal plants has been demarcated. The various postharvest operations on aromatic and medicinal plants have been discussed in this chapter. Also, the extraction technologies for aromatic and medicinal plants have been dealt separately.

9.1 INTRODUCTION

Since the dawn of civilization, man is dependent on aromatic and medicinal plants for medicinal purpose. Plant derived medicines coexists with the modern medicines in all the counties. Aromatic and medicinal plants are receiving considerable attention all over the world because of their vast untapped economic potential, especially in the use of herbal medicines. Aromatic plants are plants having aroma and flavor. Many of them are exclusively used also for medicinal purposes in aromatherapy as well as in various systems of medicine. Similarly, a number of medicinal plants also produce essential oils as well as being used for perfumery. Medicinal plants are those plants, cultivated or feral and used for medicinal purpose. According to the World Health Organization, "a medicinal plant is any plant which, in one or more of its organs, contains substances that can be used for therapeutic purposes, or which are precursors for chemo-pharmaceutical semi synthesis"

According to World Health Organization, there are about 21,000 plants having medicinal uses. India and Brazil are the largest exporters of medicinal plants in the world. Aromatic and medicinal plants have a high market potential with the world demand of herbal products growing of the rate of 7% per annum. The major medicinal aromatic plants are listed in Table 9.1.

Aromatic and medicinal plants contain active ingredients like alkaloids, glycosides, essential oils, etc. which is widely used in therapeutic applications. Aromatic and medicinal plants provide sustainable means of natural source of high-value industrial raw material for pharmaceutical, agrichemical, food, and cosmetic industries. The aromatic and medicinal plants are emerging as the industrial crops that require comparatively less inputs but gives better returns to the growers. Aromatic and medicinal plants have a high market potential with the world demand for herbal products growing of the rate of 7% per annum. The biologically active components present in the aromatic and medicinal plants, namely, alkaloids, glycosides, essential oil, etc. contribute to the medicinal properties. Aromatic plants provide products with are extensively used as spices, flavoring agents and in perfumes and medicine. In addition, they also provide raw materials for the production of many important industrial chemicals.

TABLE 9.1 Medicinal and Aromatic Plants.

Common name	Botanical name
Medicinal plants	
Aloe vera	*Aloe barbadensis Mill*
Amla	*Emblica officinalis*
Aswagandha	*Withania somnifera*
Brahmi	*Bacopa monnieri*
Opium poppy	*Papaver somniferum*
Pepper mint	*Mentha pipertia*
Periwinkle	*Catharanthus roseus*
Quinine	*Cinchona officinalis*
Sacred Basil	*Ocimum sanctum*
Sage	*Salvia officinalis*
Senna	*Casia augustifolia*
Aromatic plants	
Citronella	*Cymbopogon winterianus*
Geranium	*P elargonium graveolens*
Jasmine	*Jasminum sambac* L.
Lavender	*Lavandula angustifolia*
Lemongrass	*Cymbopogon flexuosus*
Mint	*Mentha arvensis* L.
Palmarosa	*Cymbopogon martini*
Patchouli	*Pogostemon patchouli*
Rose	*Rosa damascena*
Vetiver	*Vetiveria zizanioides*

The global market for herbal remedies can be classified into five strategic areas: **phyto-pharmaceuticals** the plant-based drugs containing isolated pure active compounds used to treat diseases; **medicinal botanicals/dietary supplements**—the whole plant or plant-part extracts used for maintenance of health by affecting a body structure and its function; **nutraceuticals**—the food containing supplements from natural (botanical) sources, that deliver a specific health benefit, including prevention and treatment of disease; **cosmeceuticals**—the cosmetic products, which contain biologically active ingredients having an effect on the user and herbal raw material.[7] Aromatic and medicinal plants are traded both in raw and also in processed form. The products obtainable from aromatic and medicinal plants are shown in Figure 9.1.

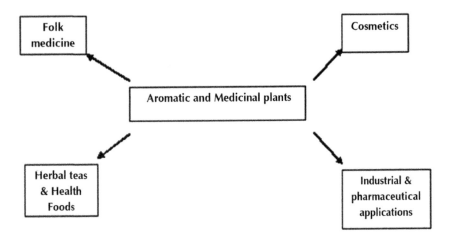

FIGURE 9.1   Uses of aromatic and medicinal plants.

Active chemical constituents of aromatic and medicinal plants are present in different parts of the plant like root, stem, bark, leaf, flower, fruit, or plant exudates. These medicinal principles are separated and preserved by different processes, namely, drying, extraction, packaging, irradiation, etc.

In this chapter, the application of food engineering to process aromatic and medicinal plants is discussed in detail.

9.2   ROLE OF FOOD ENGINEERING IN AROMATIC AND MEDICINAL PLANTS

9.2.1   DRYING

Drying is the most common and fundamental method for postharvest preservation of biological materials. The aim of drying is to lower the water activity, thereby inhibiting the microbial growth, and to achieve longer shelf life, ease of transportation, etc. The physical and chemical properties of aromatic and medicinal plants are determined by their moisture content. During drying, due to the supply of heat water, is evaporated from the material mixture. The parts of aromatic and medicinal plants are often dried before extraction to reduce moisture content. Drying of small quantities can be carried out by natural drying. For larger production, technical drying is the best choice. For preservation of active ingredients, namely, the volatile components of aromatic and medicinal plants, drying at low temperature is recommended. The quality of end product from aromatic and medicinal plants is influenced by the drying method and temperature for drying. In order to minimize the color changes of aromatic and medicinal plant components, shade drying is preferred over sun drying under natural method of drying. The plant materials should be intermittently stirred to achieve uniform drying thereby avoiding mould growth. Mechanical drying should ensure the different parameters, namely, the drying time, drying temperature, drying air velocity, humidity of drying air, etc. for better quality of the dried product.

Drying of aromatic and medicinal plants must meet the following requirements: (1) Moisture content has to be brought down to be at an equilibrium level that is defined for certain relative air humidity and temperature. This is defined as storage condition by standards; (2) minimum quality reduction in terms of active ingredients, color, flavor and aroma; and (3) microbial count must be below the prescribed limits. No chemical additives may be used.[14]

Figure 9.2 depicts the drying process for aromatic and medicinal plants. The harvested plants are dried in a convenient dryer based on the requirement to moisture content below 5% (w.b). The design of the dryer must concentrate on the plant parts to be dried since the active ingredients of aromatic and medicinal plants are dispersed in leaves, flower, bark or roots. The material should be dried immediately to avoid undesirable changes due to discoloration and chemical decomposition. For uniform and quicker drying, the harvested material may be separated into leaves and stalk either manually or using mechanical strippers. After drying, the leaves may be packed or may be size reduced in a pulveriser as powder or leaf particles for further use. The process flow chart of drying of aromatic and medicinal plants is depicted below:

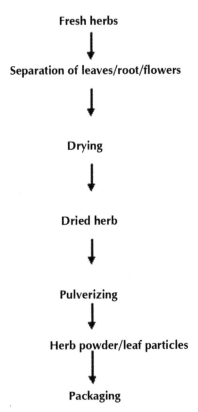

FIGURE 9.2 Drying process for aromatic and medicinal plants.

The optimum drying air temperature is the main objective in the research on aromatic and medicinal plants drying. The influence of drying on the quality of drying aromatic and medicinal plants has been carried by researchers all over the world. The effect of different drying treatments on the volatiles in bay leaf (*Laurus nobilis* L.) was studied. Simultaneous distillation extraction (SDE) and solid-phase microextraction (SPME) were compared by gas chromatography–mass spectrometry of the volatile components in bay leaves. SDE yielded better quantitative analysis results. Four drying treatments were employed: air-drying at ambient temperature, oven-drying at 45°C, freezing, and freeze-drying. Oven-drying at 45°C and air-drying at ambient temperature produced quite similar results and caused hardly any loss in volatiles as compared to the fresh herb, whereas freezing and freeze-drying brought about substantial losses in bay leaf aroma and led to increases in the concentration levels of certain components, for example, eugenol, elemicin, spathulenol, and â-eudesmol.[6]

In order to prevent extreme loss of heat sensitive properties (aromatic, medicinal, & color) when drying herbs, they are normally dried at low temperatures for longer periods of time to preserve these sensory properties. High energy consumption often results from drying herbs over a long period. Coriander (*Coriandrum sativum* L., *Umbelliferae*) was dehydrated in two different drying units (thin-layer convection and microwave dryers) in order to compare the drying and final product quality (color) characteristics. Microwave drying of the coriander foliage was faster than convective drying. The entire drying process took place in the falling rate period for both microwave and convective dried samples. The drying rate for the microwave dried samples ranged from 42.3% to 48.2% db/min and that of the convective dried samples ranged from 7.1% to 12.5% db/min. The fresh sample color had the lowest L value at 26.83 with higher L values for all dried samples. The results show that convective thin layer dried coriander samples exhibited a significantly greater color change than the microwave dried coriander samples.[15] Drying studies on lemongrass, was conducted at three temperatures (35, 45, and 55°C) and three humidities (30%, 40%, and 50%) at fixed air velocity of 1 m s^{-1}. The results revealed that the increase in the drying air temperature increased the drying process and decreased the equilibrium moisture content (EMC) of lemon grass. The drying process decreased as the air humidity increases. The effect was less than that of the temperature.[13]

Mint leaves were dried by three different types of dryers, namely, tray, freeze, and distributed (indirect)-type solar dryer. Sorption isotherms of fresh, solar, tray, and freeze-dried mint were determined at temperatures of 15, 25, and 35°C over a range of relative humidities (10–90%). The effect of drying method on the water sorption isotherms of dried mint samples was evaluated. The EMC increased decreasing temperature at constant water activity.[4] Sage plants (*Salvia officinalis* L.) were dried in the passive dryers in different times of the year. Four drying methods were used in this investigation to dry sage in two seasons before flowering stage. Plants were dried in an unglazed transpired passive solar dryer with 100% exposure to direct sunrays, in a greenhouse dryer covered with shading cloth with 50% exposure to direct sunrays, and with 0% sunrays while the medicinal plants were protected from sun, that is, in shaded barn. The study revealed that sage can be dried at different times of the year even before the flowering stage of the plants.[10]

An attempt was made to dry senna (*Cassia aungustifolia*) in a forced flow type dryer at different temperatures. Sennoside concentrate which is an important quality aspect of senna revealed that drying at 45°C was superior.[1]

Conventionally, Patchouli herbage (*Pogostemon cablin*) is shade dried for extraction of aromatic oil. However, improper drying results in poor yield and quality of the oil. A study was undertaken to find the effect of drying on the yield of volatile oil of Patchouli.[5] Patchouli herbage was dried under forced flow system of drying in a mechanical drier at 40°C for 5 h & 45°C for 4 h and also shade dried for 45 h. The essential oil was obtained by steam distillation from each treatment. Statistical analysis showed significant differences in the essential oil content of leaves dried by different drying methods. The volatile oil content of sample dried at 40°C was found to be 2.46%. In the case of 45°C drying air temperature, the oil content was 2.60%. The volatile oil content of shade-dried sample was 2.40%.

The effect of drying temperature and air velocity on the quality of *Murraya koenigii* leaves was studied. Freshly harvested, washed and stripped *M. koenigii* leaf (curry leaf) was dried at different air temperature of 40, 45, and 50°C temperature and at 2, 3, and 4 m/s air velocity in a fluidised bed dryer from an initial moisture content of 184.5% (dry basis) to a final moisture of around 5% (dry basis).[2] The drying rate decreased with the decrease in the moisture content at all drying temperatures. Drying studies revealed that fluidized bed drying at 45°C and 4 m/s air velocity was found to maintain the quality of dried curry leaf in terms of rehydration ratio and volatile oil content. The dried leaves packed in 38-µm thickness and stored under ambient condition (30.2°C) for a period of 1 month resulted in better product as seen from the volatile oil content and overall acceptability for 4 m/s fluidized bed dried sample at 45°C.

9.2.2 GRINDING AND PULVERIZING

The dried aromatic and medicinal plant can be further subjected to size reduction by milling according to standard sieve size based on the requirement of the processor. By size reduction, easier mixing is achieved in the final food, and it also favors the uniform distribution of flavor. Commercial pulverizers are used to produce powders from the dried aromatic and medicinal plants.

The powders are mainly used in the production of capsules and tablets as a health supplement. Grinding is also done prior to extraction so that its medicinal ingredients are exposed to the extraction solvent because size reduction maximizes the surface area, which in turn enhances the mass transfer of active principle from plant material to the solvent. Hammer mill is commonly used for size reduction process.

9.2.3 EXTRACTION

The bioactive components present in aromatic and medicinal plants are separated by standard methods using solvents in an extraction process. The active ingredients mainly the volatiles are extracted from the fresh or dried aromatic and medicinal plants' components. The various extract, namely, essential oil, oleoresin, gums & resins, isolates, concrete are obtained as a result of extraction process. The extract thus obtained may be ready for use as a medicinal agent in the form of tinctures and fluid extracts, it may be further processed to be incorporated in any dosage form such as tablets or capsules, or it may be fractionated to isolate individual chemical entities. There are different techniques followed separately for the extraction of these bioactive compounds for aromatic and medicinal plants.

9.2.4 EXTRACTION PROCESS FOR AROMATIC PLANTS

Aromatic plants are mainly processed for their aroma and flavor present in the form of essential oils. Essential oils are hydrophobic liquids containing volatile aromatic compounds. They are present in the specialized cell glands in plants which when ruptured release the aroma. They are mainly used in perfumery and food industries.

Extraction techniques for aromatic plants involve: Hydrodistillation techniques (water distillation, steam distillation, water and steam distillation), Hydrolytic maceration followed by distillation, expression and enfleurage (cold fat extraction) headspace trapping, SPME, protoplast extraction, microdistillation, thermo-microdistillation, molecular distillation. Basically, there are three methods of extraction: distillation, expression, and solvent extraction.

Distillation accounts for the major share of essential oils being produced today. The choice of a particular process for the extraction of essential oil is generally dictated by the following considerations (Handbook of Medicinal & Aromatic plants, NEDFI):

a. Sensitivity of the essential oils to the action of heat and water.
b. Volatility of the essential oil.
c. Water solubility of the essential oil.

Hydrodistillation is employed for the extraction of essential oil from flowers, steam distillation from woods and hydrosteam distillation from herbs. As most of the essential oils of commerce are steam volatile, reasonably stable to action of heat and practically insoluble in water hence are suitable for processing by the following distillation methods:

a. Water or hydrodistillation.
b. Steam cum water distillation.
c. Steam distillation.

9.2.4.1   WATER/HYDRODISTILLATION

Water or hydrodistillation is one of the oldest and easiest methods being used for the extraction of essential oils. In this method the plant material is fully dipped in the water. The primitive "Bhapka" method is based on this principle. In the Bhapka method, the distillation still is made of copper. The still is fitted on brick furnace and the plant material is filled in the still and entirely covered with water to the top. Another copper vessel with a long neck is placed in a water tank or natural pond to serve as condenser. A bamboo pipe is used to connect the vapor line, and mud is used to seal the various joints. The oil vapor along with steam is condensed in the copper vessel and is separated.

The influence of different methods of drying, extraction time, and type of organ on the essential oil percentage of Rosemary (*Rosmarinus officinalis* L.) was investigated.[11] Three drying methods investigated were oven-drying (45°C), shade drying and sun drying. Four extraction times were 1, 2, 3, and 4 h and three organ type were leaf, stem, mixed leaf, and stem. Essential oil was obtained by water distillation method. Results showed that effect of drying methods, extraction time, and organ type on the essential oil percentage was significant. The maximum essential oil percentage (1.8%) was obtained to leaf sample, 3 h of extraction, and shade drying. While the minimum essential oil percentage (0.12%) was obtained to stem sample, 1 h of extraction, and oven-drying.

An improved method for essential oil extraction was developed with the application of ohmic heated hydrodistillation. In this study, the parameters affecting citronella oil extraction by ohmic heated hydrodistillation such as power input, extraction time, solvent-to-solid ratio, and chopping frequency were investigated to achieve maximum oil yield.[12] The kinetics of extraction was assumed and verified based on a second-order mechanism. The initial extraction rate, the saturated extraction capacity and the rate constant of extraction were calculated using the second-order model. The optimum parameters were found at voltage input of 77 V up to boiling point and 50 V until the end of extraction, 120-min extraction time, solvent-to-solid ratio of 3:1 and once chopping frequency.

The initial extraction rate (h) was 0.134 g L^{-1} min^{-1}; the extraction capacity (C_s) was 5.787 g L^{-1}; the second order extraction constant (k) was 0.004 L g^{-1} min^{-1}, and the coefficient of determination (R^2) was 0.976.

9.2.4.2   STEAM DISTILLATION

The usual method of extraction is through steam distillation. After extraction, the properties of a good quality essential oil should be as close as possible to the "essence" of the original plant. The key to a "good" essential oil is through low-pressure and low-temperature processing. High temperatures, rapid processing, and the use of solvents alter the molecular structure, will destroy the therapeutic value, and alter the fragrance.

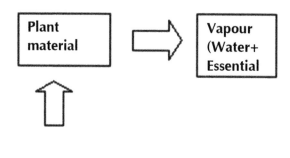

FIGURE 9.3   Steam distillation process.

Steam distillation is a of distillation process employed for heat sensitive materials like oils, resins, hydrocarbons, etc. which are insoluble in water and may decompose at their boiling point. In steam distillation, a compound or mixture of compounds is distilled at a temperature below that of the boiling point(s) of the individual constituent(s). In the presence of steam or boiling water, these substances are volatilized at a temperature close to 100°C, at atmospheric pressure (Fig. 9.3). Steam distillation exploits the twin action of heat and moisture from steam to break down the cell walls of the plant tissues to liberate the essential oil. The main components of steam distillation unit are

a. distillation tank with steam coil,
b. condenser (usually multitube tubular),
c. oil separator or receiver, and
d. boiler.

Steam is generated separately in a steam boiler and is passed through the distillation tank through a steam coil. The plant material is tightly packed above the perforated grid. Steam, along with oil vapors, is condensed in the condenser and is separated in the oil receiver.

The essential oil from Vetiver can be extracted from the roots by steam distillation. Freshly harvested roots on distillation give higher yield of oil than stored roots; the yield decreases progressively with the period of storage. The roots are soaked for 18–20 h in water prior to distillation to render the root material soft and thereby further facilitate release of oil. Fresh roots when cut to lengths 2.5–5 cm increases recovery. As the most valuable quality constituents are contained in the high-boiling fractions, the roots must be distilled for a prolonged period ranging from 20 to 24 h.[3]

9.2.4.3   SOLVENT EXTRACTION

Solvent extraction is another common method of extraction where organic solvents are used to extract the essential oils. Common solvents used are petroleum ether, methanol, ethanol, or hexane and is often used on fragile material such as jasmine flowers which would not be able to handle the heat of steam distillation. Solvent extraction is particularly suitable for botanical material that has a very low yield of essential oil, or where it is made up of mostly resinous components and as such delivers a far finer fragrance than that of distillation. A solvent extracted essential oil is very concentrated and is very close to the natural fragrance of the material used. In the solvent-extraction method of essential oils recovery, an extracting unit is loaded with perforated trays of essential oil plant material and repeatedly washed with the solvent.

A hydrocarbon solvent is used for extraction. All the extractable material from the plant is dissolved in the solvent. This includes highly volatile aroma molecules as well as nonaroma waxes and pigments. The extract is distilled to recover the solvent for future use. The waxy mass that remains is known as the concrete. The concentrated concretes are further processed to remove the waxy materials which dilute the pure essential oil. To prepare the absolute from the concrete, the waxy concrete is warmed and stirred with alcohol (ethanol). During the heating and stirring process, the concrete breaks up into minute globules. Since the aroma molecules are more soluble in alcohol than the waxes, an efficient separation of the two results. This is not considered the best method for extraction as the solvents can leave a small amount of residue behind which could cause allergies and effect the immune system.

9.2.4.4 SUPER CRITICAL CO$_2$ EXTRACTION

Supercritical fluid extraction is another common method of extraction where organic solvents are used to extract the essential oils. Supercritical carbon dioxide is used as a solvent in supercritical fluid extraction. Carbon dioxide is heated to 30°C and pumped through the plant material at around 8000 psi; under these conditions, the carbon dioxide is likened to a "dense fog" or vapor. With release of the pressure in either process, the carbon dioxide escapes in its gaseous form, leaving the essential oil behind. This method has many benefits, including avoiding petrochemical residues in the product.

The supercritical carbon dioxide will extract both the waxes and the essential oils that make up the concrete. Subsequent processing with liquid carbon dioxide, achieved in the same extractor by merely lowering the extraction temperature, will separate the waxes from the essential oils. This lowering temperature process prevents the decomposition and denaturing of compounds and provides a superior product. When the extraction is complete, the pressure is reduced to surrounding and the carbon dioxide reverts back to a gas, leaving no residue. Usually, the solid material forms a fixed bed, through which the CO$_2$ passes, extracting and transporting the solutes that, at the extractor exit, are precipitated by simple solvent expansion.

Dried seeds of coriander were subjected to extraction after grinding to particle size of 300 μm. The extraction was carried out at three different pressure levels (30, 35, and 40 MPa), three temperature levels (308, 313, 318 K), and three levels of supercritical CO_2 flow rates (10, 15, 20 g/min). The highest essential oil was obtained at 40 MPa, 313 K and 15 g/min combination of parameters and the highest yield was equal to 3.20 g/100 g. The study showed that the temperature has more significant effect than the pressure while the flow rate was having no significant effect on the yield of coriander seed oil.[9]

9.2.5 EXTRACTION PROCESSES FOR MEDICINAL PLANTS

The general techniques of extraction followed in medicinal plants are maceration, infusion, percolation, digestion, decoction, hot continuous extraction, counter-current extraction, microwave-assisted extraction, ultrasound extraction (sonication), supercritical fluid extraction, and phytonic extraction.

9.2.5.1 MACERATION

It is a method of extraction whereby the phenolic and favor compounds are leached from the plant material.

9.2.5.2 INFUSION

It is a method of extracting phenolic and favor compounds from the medicinal plants by steeping the raw material in water, oil, or alcohol.

9.2.5.3 PERCOLATION

In this method, the liquid or macerate is allowed to pass slowly through porous substance.

9.2.5.4 DECOCTION

Medicinal plants after mashing are boiled in water to extract oils and other volatile compounds.

9.2.5.5   HOT CONTINUOUS EXTRACTION

The finely ground plant material is placed in a porous filter paper, which is then placed in chamber of the Soxhlet apparatus. The extracting solvent in flask is heated, and its vapors condense in condenser. The condensed vapor drips into the thimble and is extracted by contact with the plant material.

The following flow chart depicts the steps involved in the extraction process followed for medicinal plants (Fig. 9.4):

FIGURE 9.4   The steps involved in the extraction process for medicinal plants.

9.2.5.6   RAW MATERIAL

The bioactive components in a medicinal plant may be derived from any plant part, namely, leaves, flowers, bark, roots, etc.

9.2.5.7   DRYING

The plants are dried prior to the extraction process by shade drying or drying in a convection type dryer.

9.2.5.8 SIZE REDUCTION

The dried plant material is then pulverized in a hammer mill.

9.2.5.9 EXTRACTION

Based on the quantity and quality required, the size-reduced plant material is subjected to extraction techniques accordingly.

9.2.5.10 FILTRATION

The extract obtained is separated out from the exhausted plant material by allowing it to trickle into a holding tank through filter cloth at the bottom of the extractor.

9.2.5.11 CONCENTRATION AND DRYING

The extract after filtration is fed into a film evaporator where it is concentrated under vacuum to produce a thick concentrated extract, which is then dried in a spray dryer or vacuum dryer to produce a solid mass free from solvent. The solid mass, thus obtained, is pulverized and used directly for the desired pharmaceutical formulations or further processed for isolation of its phytoconstituents.

9.2.6 IRRADIATION

Herbs are often contaminated with high levels of bacteria, molds, and yeasts; if untreated, these will result in rapid spoilage of the foods and may cause health risks, especially when contaminated with pathogenic bacteria. Microbial decontamination of herbs could be achieved by physical method involving ionizing radiation, which is approved by *Codex Alimentarius Commission 1983.*

Irradiation results in some changes in the sensory quality. Most countries require irradiated spices to be labeled with the international symbol for irradiation, and sometimes words to describe the process and/or the effect (such as treated to destroy harmful bacteria).

Irradiation, at a minimum dose of 5 kGy effectively kills bacteria, molds, and yeasts. A dose of 5–10 kGy results in an immediate minimum 2–3 log cycle reduction of bacteria. Most countries set maximum dose regulatory limits, and these higher limits allow for higher levels of microbial kill. Spore forming bacteria require higher dose levels to kill because spores are less sensitive to radiation. Surviving bacteria and spores are, however, much more susceptible to heat. The sensory properties of most spices are well maintained between 7.5 and 15 kGy. Generally, the sensory properties of spices are more resistant to irradiation than are some herbs. Irradiation can be done in bulk packages, retail packages, in gas impervious packages, and heat sealed plastics. Not all plastics are compatible with irradiation processing, and the suitability of the packaging material needs to be determined before treatment. Irradiation allows the spice package to remain closed and sealed at all times.[8]

9.2.7 PACKAGING

Packaging of aromatic and medicinal plants plays an important role in safe guarding the plant materials from contamination. The packaging material should be designed in such a way to minimize water loss and senescence. Care should be taken that constant temperature is maintained within the packaging material to avoid moisture condensation and the consequent effect of microbial growth. The packaging bags may be partially ventilated or may be partially permeable to water. Since the plant materials are soft in nature, rigid clear plastic containers are recommended.

Most fresh herbs are kept well when packed in cartons lined with folded perforated polyethylene (PE), in which water loss, leaf abscission, and decay are minimal. Perforation of the PE liner reduces the undesirable accumulation of ethylene and CO_2. However, to delay the maturity process, packaging in perforated film is not effective. Packing of yellowing—susceptible herbs is achieved by packaging in cartons with nonperforated PE liners which results in the creation of a moderate modified atmosphere capable of retarding yellowing and decay. High levels of CO_2 arrest the senescence—inducive effect of ethylene accumulation in the package, especially when combined with a decreased level of O_2. Extreme temperature fluctuations during shipment may result in anaerobic respiration in sealed film, which can be eliminated by using microperforated (MP) film.[8] The use of regular perforated films for the packaging of these plants can be effective in reducing water loss, but not for delaying the senescence of yellowing-susceptible plants and tissues. In MP packages, CO_2 concentrations would not exceed 10% and O_2 concentrations would not be dropped below 5%.

9.3 CONCLUSIONS

Plant parts containing the active constituents are processed prior to their utilisation. There is a great demand for herbal products worldwide as safe health supplements alternative to the synthetic medicines. Food engineering application in aromatic and medicinal plants processing plays a vital role. A modern processing plant with the state of the art technologies, standardization, and quality control functions would enable processing of raw herbs into value added products and formulations and would assist in realizing high value margins in the domestic and international market.

KEYWORDS

- aromatic plants
- drying
- extraction
- irradiation
- medicinal plants
- packaging
- pharmaceuticals
- size reduction
- solvent extraction
- steam distillation

REFERENCES

1. Ambrose, D. C. P.; Naik, R. Mechanical Drying of Senna Leaves (*Cassia angustifolia*). *Curr. Agric. Res.* **2013**, *1*(1), 65–68.
2. Ambrose, Dawn, C. P.; Annamalai, S. J. K.; Naik, R. Influence of Fluidised Bed Drying on the Quality and Storage of *Murraya koenigii* Leaves. *J. Appl. Hortic.* 2014, *16*(3), 222–224.

3. Balasankar, D.; Vanilarasu, K.; Selva Preetha, P.; Rajeswari, S.; Umadevi, M.; Bhowmik, D. Traditional and Medicinal Uses of Vetiver. *J. Med. Plants Stud.* 2013, *1*(3), 191–200.

4. Dalgıç, A. C.; Pekmez, H.; Belibağlı, K. B. Effect of Drying Methods on the Moisture Sorption Isotherms and Thermodynamic Properties of Mint Leaves. *J. Food Sci. Technol.* **2012**, *49*(4), 439–449.

5. Dawn, C. P.; Ambrose, Annamalai, S. J. K.; Naik, R. Effect of Drying on the Volatile Oil Yield of Patchouli. *Indian J. Sci. Technol.* **2013**, *6*(12), 5559–5562.

6. Díaz-Maroto, M. C.; Pérez-Coello, M. S.; Cabezudo, M. D. Effect of Drying Method on the Volatiles in Bay Leaf (*Laurus nobilis* L.). *J. Agric. Food Chem.* **2002**, *50*(16), 4520–4524.

7. Mishra, D.; Singh, R. K.; Srivastava, R. K. Agribusiness and Entrepreneurship Development through Medicinal and Aromatic Plants: An Indian State of Affairs. *Int. J. Food, Agric. Vet. Sci.* **2013**, *3*(1), 238–246.

8. Elhadi, M. Y. Postharvest Handling of Aromatic and Medicinal Plants. Herbal, aromatic and medicinal plants. Symposium, At Djerba, Tunisia, 2–4 November, 2006.

9. Geed, S. R.; Said, P. P.; Pradhan, R. C.; Rai, B. N. Extraction of Essential Oil from Coriander Seed. *Int. J. Food Nutr. Sci.* **2014**, *3*(3), 7–9.

10. Hassanain, A. A. Drying Sage (*Salvia officinalis* L.) in Passive Solar Dryers. *Res. Agric. Eng.* **2011**, *57*(1), 19–29.

11. Khorshidi, J.; Mohammadi, R.; Fakhr, T. M.; Nourbakhsh, H. Influence of Drying Methods, Extraction Time, and Organ Type on Essential Oil Content of Rosemary (*Rosmarinus officinalis* L.). *Nat. Sci.* **2009**, *7*(11), 42–44.

12. Muhammad, H.; Hasfalina, C. M.; Hishamuddin, J.; Zurina, Z. A. Optimization and Kinetics of Essential Oil Extraction from Citronela Grass by Ohmic Heated Hydrodistillation. *Int. J. Chem. Eng. Appl.* **2012**, *3*(3), 173–177.

13. Mustafa, I.; Sopian, K.; Daud, W. R. W. Study of the Drying Kinetics of Lemon Grass. *Am. J. Appl. Sci.* **2009**, *6*(6), 1070–1075.

14. Serdar, Ö.; Milan, M. *Medicinal and Aromatic Crops. Harvesting, Drying and Processing.* Haworth Food and Agricultural Products Press: Binghamton, NY, 2007.

15. Shaw, M.; Meda, V.; Tabil, Jr., L.; Opoku, A. Drying and Color Characteristics of Coriander Foliage using Convective Thin Layer and Microwave Drying. *J. Microwave Power Electromagn. Energy* **2007**, *41*(2), 56–65.

PART IV

Emerging Issues and Applications in Food Engineering

CHAPTER 10

THREE-DIMENSIONAL PRINTING OF FOOD

BHUPENDRA M. GHODKI[1*], BRAJESH KUMAR PANDA[2],
DEVENDRA M. GHODKI[1], and TRIDIB KUMAR GOSWAMI[2]

[1]*Department of Mechanical Engineering, N.I.T., Rourkela 769008, Odisha, India. E-mail: devendra.ch2@gmail.com, devendra.ch2@gmail.com;*

[2]*Agricultural and Food Engineering Department, IIT Kharagpur, Khargpur 721302, India. E-mail: brajeshkumarpnd2@gmail.com, tkg@agfe.iitkgp.ernet.in*

Corresponding author.

CONTENTS

ABSTRACT

Whenever a new and unconventional technology reaches to the people, it takes some time to penetrate the market, because of the little understanding of the technology among the masses. A similar trend may be seen with the 3D printing in the food sector. However, it will not be surprising for such technology to be well accepted in this industry as food is not just a mean to serve hunger, but also a source of nutrition and rejoice in the day-to-day life. When it comes to the application part, 3D printer is not just producing the single component but can fabricate the complete structure. Even at a point where 3D printing will be well accepted it may not even be called as a 3D printer rather it will be considered as a food processing unit. The food industry plays a big role in the development of the human civilization. The 3D printing technology has a broad scope in coming days includes printing in metal, plastic and human organs to printing in food. Back in the mid-2000s, this technology was mostly used to print things like peanut butter, pizza, burger, chocolate, frostings, cookie dough, and those sorts of necessary materials. But, later on the fascinating possibilities of this technology in cooking and embedding nutrients for nutrient control in food were realized. Accordingly, the idea of integrated systems came into existence, to print 3D structure with components instead of printing with raw materials.

The 3D printing technology should be analyzed with the advantages, limitations, and future prospects that may change the society. The extraordinary essence of 3D printing is creating a distinct part layer by layer; as opposed to subtractive methods of manufacturing that allow themselves to lower costs in the raw material. The 3D printing only prints according to the consumers wants and needs. The 3D printing is the ultimate just-in-time method of manufacturing. The biological material is non-homogeneous at the core of their construction. Hence, during application of AM process in 3D printing of food, it is more important to increase the number of materials to be used simultaneously, to synthesize the end product.

Further in the future 3D printing technology will fulfill unique challenges for food quality, safety, variety, and nutrient stability. Hence, 3D printing technology will soon move beyond an emerging technology into a purely transformative technology. Based on the applications, advantages, limitations, and future scope, we can conclude that the 3D printing technology will boost the food processing sector.

10.1 INTRODUCTION

Food is one of the primary elements of life and is becoming more customized as people want healthy and safe food with good taste and appearance. Hence, a method of distributing food in a personalized manner is one way to satisfy this demand. Three-dimensional (3D) printing of food is an emerging technology in the area of personalized foods and is the logical advancement of the modeling and simulation activity, which produces a tangible representation of virtual (digital) objects. The 3D printing finds the application in different fields as shown in Figure 10.1. The 3D printing technology has capabilities of solving the real problems in the physical world, which could be used efficiently in educating the students or users in the near future for better learning and understanding of the complex food process.[2]

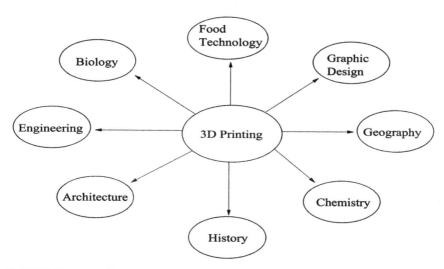

FIGURE 10.1   Application of 3D printing in different focus areas.

The 3D printing of food can be defined as a process of creating 3D food through repetitive deposition and processing of food material(s) layers like molten chocolate (printers ink) using computer controlled equipment (3D printer) that takes virtual models of food to be printed from the computer (digital file). From the preceding, a 3D food printer may be defined as a machine that prints a complete food at just one command. Generally, for a food matrix of 1-in. thickness, it may contain from 100 to 200 layers of the semiliquid food material(s). However, this process is slow, but the 3D printing technology is improving with time with an aim of reducing the cost and processing time. Various study shows the successful printing of chocolate, processed cheese, cake frosting, scallops, celery, turkey, pizza, burger, peanut butter, etc.[19] The 3D printing technology in food sector may seem to be unconventional. However, with the ever expanding food industry this technology may find its existence in the future depending on the consumer acceptance.

10.2   HISTORY OF 3D PRINTING

The concept of 3D printing emerged from the technique called as additive manufacturing (AM) which was developed in the early 1980s. By the beginning of the 2010s, the terms 3D printing and AM developed senses in which they were synonymous parasol terms for all AM technologies. It was during this decade that the term subtractive manufacturing appeared for the big family of machining operations with metal removal as their common subject. However, at present, the 3D printing still referred mostly to the polymer technologies, whereas AM was used in metalworking contexts. The technique called stereolithography (STL) was invented by Chuck Hull of 3D Systems Corporation, in which layers of material are added by curing photopolymers with UV lasers.[4] According to Hull, the STL is defined as a "system for producing 3D objects by creating a cross-sectional pattern of the object to be formed."[14] Chuck Hull also developed the STL file format broadly accepted by 3D printing. The term 3D printing at first concerned about a technique utilizing standard and custom inkjet print heads.[13] The concept of a head moving through a 3D work sleeve changing a bulk of raw material into a required shape (layer by layer) was associated uniquely with methods that takeoff metal (instead of adding it), such as CNC milling, etc. The technology used by most 3D printers to date especially by hobbyist and consumer-oriented models are fused deposition modeling (FDM), a unique application of plastic extrusion. Other terms that have appeared, which are usually used as AM synonyms have been desktop manufacturing, rapid manufacturing (as the logical production level heir to rapid prototyping), and on-demand manufacturing (which imitate on-demand printing in the 2D sense of printing).

The idea of fashioning food into fun and aesthetically pleasing shapes takes many forms in the world today. For the production of food items, various techniques are used. However, most of them are optimized for mass production. One such method involves printing of a model by the material filled in a syringe. The primary logic of the 3D printing technologies is joining the sequential layer of materials through a 3D work covered under automated control.

10.3 ADDITIVE MANUFACTURING

There are various different 3D printing techniques have been invented since the late 1980s. AM is one of the most advanced and standard methods of the 3D printing to convert engineering design more precisely a computer-aided design (CAD) files into working and durable objects created from materials like chocolate, glass, sand, and metal using a specialized 3D printer. The technology produces a product by the repeated laying of fine powdered material and binder (chemical or heat) to gradually build the finished product, which is like repeated sandwiching the binder between the layers of powdered material. Hence, vast ranges of creative geometries are possible to be manufactured for a variety of industrial, commercial, educational, and art applications. The AM works in an almost identical style to an office printer laying ink on paper that works in two-dimensional planes; hence, this process is often referred to as 3D printing.

American Society for Testing and Materials International defines AM as the "process of joining materials to make objects from 3D model data, usually layer upon layer, as opposed to subtractive manufacturing methodologies."[1] AM is an innovative technology changing the manufacturing scenario as the leading companies of manufacturing sector are transforming from analog to digital. The AM technique is used by several industries to produce a broad range of products, including: engine and engine components (automobile industry), food like chocolate, pizza, etc. (food industry), impellers and blades (aerospace industries), DNA models (education and pharmaceutical industry). AM produces almost no waste and is an energy efficient and eco-friendly process. Hence, there are enormous cost benefits as compared to the traditional manufacturing technique.

Various additive techniques are available to 3D print an object. The techniques only vary in the manner layers are deposited to create an object and the materials (printer's ink) that can be used. Few techniques melt or soften material like molten chocolate to make the layers, for example, selective laser melting (SLM) or direct metal laser sintering, selective laser sintering (SLS), FDM, whereas others fix liquid materials using different advanced technologies, for example, STL. Laminated object manufacturing (LOM) involved cutting of thin layers of a definite shape and joined together (e.g., paper, polymer, and metal). Each technique has its benefits and drawbacks.

10.4   THREE-DIMENSIONAL PRINTER FOR FOOD

The idea of the 3D printer may be coined from the fact that if an object like an apple can be sliced then it may also be possible to glue them (slices) back together. In general, a 3D food printer may be defined as a machine that creates entire food (entrees or desserts) at the one press of a button. The food printer generally works by storing and refrigerating the powdered food ingredients in the storage chamber and then mixing them together in the mixing chamber and finally printing cooked layers onto a serving tray through extruder by additive process.

The synthetic food can be produced by using progressive 3D printing and inkjet technology to meet the nutritional need of an individual. The 3D printing component will provide macronutrients (carbohydrate, protein, and fat), food structure, and texture, at the same time the inkjet will deliver micronutrients, flavor, and aroma. The macronutrients will be stored in moisture-resistant and sterile containers and supply directly to the printer. This mixture will combine with water or oil at the print head as per a digital recipe. The flavors, micronutrients, and texture modifiers can also be added to the above combination to ensure high quality and acceptability of the product. The micronutrients, as well as, flavors are also stored in dry sterile packs as an aqueous solutions or dispersions. The 3D food printers built solid foods using this blended mixture and extruded into the desired shape. The ink in the 3D printers is the foods themselves in pasty or fluid form, for example, cookies dough, thick molten chocolate, etc. However, certain foods like vegetables and meats are difficult to extrude from the printers syringe, but a novel food ink can be created by mixing the grounded powder of such food with any other edible liquid. The 3D printer usually built food through layer by layer deposition of the semiliquid (pasty) food material (mixed ingredients) until desired food is created. The 3D printers may utilize handmade ingredients and purees as well as hydrocolloid liquid ingredients permitting full control over the food and its nutritional and rheological properties.

The 3D printers may use various different methods, but they have one fundamental thing in common: they generate a 3D object by building successive layers one over the other until the entire object is complete. Figure 10.2 shows the components of Systems and Materials Research Corporation's

(SMRC's) 3D printer. SMRC (Austin, USA) has developed and demonstrated a 3D printed food system that is capable of dispensing viscous (pasty) food made from the powder and liquid ingredients.[7] The first demonstrated meal from SMRC's 3D food printer was cheese pizza the process involved dispensing of pizza dough on a hot plate where it is cooked, followed by tomato layer and cheese topping.

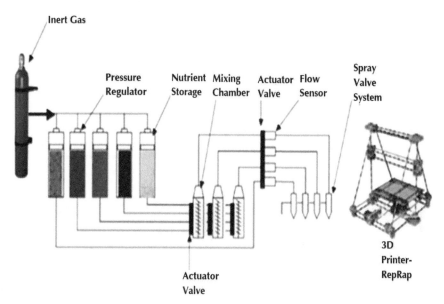

FIGURE 10.2   Schematic of a 3D printer for food (image credit by SMRC).[7]

10.5   GENERAL PRINCIPLES

10.5.1   MODELING

The virtual design is usually created in a CAD file with the help of the 3D modeling software or a 3D scanner to generate a new object. The modeling software can create a virtual design of an imaginary or existing object. However, the 3D scanner will make a digital copy of an existing object. The digital copy (design) can be imported into the 3D modeling software. The customers find difficulty in designing of 3D printable models due to lack of expertise in operating the software and time. Accordingly, there are many printing service bureaus like Shape ways, Thing verse, etc.

10.5.2 PRINTING

Once the prepared digital file is uploaded in the 3D printer, there is a touch screen interface between the user and the printer which allows to print (create) the required 3D object layer by layer as per the nutritional and design details fed to the 3D printer. The digital file created in a 3D modeling program is sliced into *n* number of horizontal layers (cross sections) by the slicing software (slicer) for printing.[16] The slicer software like slice3r, cura, etc. produces a specific G-code file containing commands tailored to a particular printer. The 3D printer follows the G-code commands to construct the specified model from the layers of printable material. The layers that represent the virtual cross sections from the CAD model are united or fused to construct the final shape of the model with invisible layering.[1] Figure 10.3 shows the making of a 3D object from a CAD file. The models created by this technique have the capabilities of producing the most challenging geometric feature.

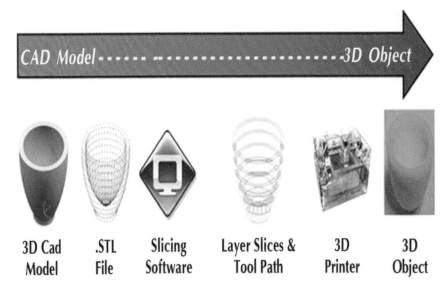

FIGURE 10.3   The general process of making a 3D object from a 3D CAD model (image credit by Foresight Investor).[6]

The resolution of a printer is described by the thickness of the layer and *X–Y* resolution in dots per inch (DPI) or micrometers (μm). The average layer thickness is around 100 μm (250 DPI); whereas some machines can print layers very close to 16 μm (1600 DPI) thickness. The particles (3D dots) diameter is about 50–100 μm (510–250 DPI). Hence, manufacturing a model with new methods can take much time even in days, relying on the method used and the shape, size, and complexity of the model. However, AM can be speedy, more pliable, and cost-effective when constructing the relatively small number of parts. The present day 3D printers offer users to produce the complex model quickly using a desktop size printer.

10.5.3 FINISHING

The advancement in the 3D printing technology permits the 3D printers to print in multiple colors and color combinations as well as different materials simultaneously while constructing a model. However, some of the AM techniques utilize supports when building. The supports are detachable or dissolvable at the time of completion of the print, and the process to perform this task is known as finishing. The supports in the model are provided to aid the overhanging features during construction.

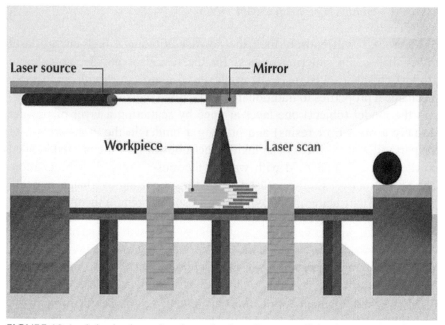

FIGURE 10.4   Selective laser sintering technology (image credit by CustomMade).[8]

10.6   METHODS AND TECHNOLOGIES OF 3D PRINTING

10.6.1   SELECTIVE LASER SINTERING

SLS technology was developed and patented by Dr. Carl Deckard at the University of Texas in the 1980s.[10] The SLS is a 3D printing approach in which selective fusing of materials in a granular bed is performed. A high-power laser is used to fuse small particles (powder form) of metal, plastic, ceramic, food components, etc. into a specified shaped 3D mass. The laser selectively fuses the powdered material by scanning the layers (cross sections) created by the 3D modeling program on the surface of a powder bed. The powder bed is lowered by one layer thickness after each layer is scanned. Then a new layer of material is enforced on the top, and the process continues until the desired shape is achieved as shown in Figure 10.4. All the untouched powder becomes a support structure for the 3D object. Hence, no support structure is required which is an advantage of SLS. The unused powder can be used for the next printing.

10.6.1.1   SELECTIVE LASER MELTING

SLM process will entirely melt the powder utilizing a high-energy laser rather than using sintering process for the fusion of powder granules to produce closely packed materials in a layer-wise procedure with comparable mechanical properties to traditional manufactured metals. The printer generates the model (object) one layer at once by scattering a layer of powder (usually a plaster, or resins) and printing a binder in the cross-section of the part utilizing a process like inkjet. The robustness of the printed bonded powder can be intensified with wax or thermoset polymer impregnation. The SLM process uses the unfused media for supporting overhangs and thin walls in the part being produced, which lessen the requirement for temporary auxiliary supports for the piece.

10.6.2   FUSED DEPOSITION MODELING

FDM or Fused Filament Fabrication was invented by S. Scott Crump in the late 1980s; however, commercialized by Stratasys in 1990.[9] In this technique, the cross sections of the model or part are printed through heated nozzles by extruding material (mostly thermoplastic) which hardens immediately to form layers. The build material is usually fed in filament form (sometime pellet form is also used) to an extrusion nozzle head that can be moved in both horizontal and vertical directions to adjust the flow. The nozzle includes resistive heaters that keep the plastic at a temperature just above its melting point so that it flows smoothly through the nozzle and forms the layer. When the heating roller achieves the melting point of the feedstock, the filament feedstock get fused into a liquefier, which enable free flowing of material through the nozzle. As the material reaches the substrate, it cools and hardens. Once the first layer is built, it acts as a platform for the second layer and deposition of the next layer begins. The thickness and vertical dimensional accuracy and precision of the layer are determined by the diameter of the extruder die; that may range from 0.013 to 0.005 in.[18] The whole mechanism is controlled by a computer-aided manufacturing software package running on a microcontroller. The machine dispenses two materials, one for the model, and one form a disposable support structure. For the construction of overhanging geometry (not in continuation with the previous layer) the disposable secondary material is deposited in between the primary layers (made with primary material) for supporting purpose, which is removed later on from the completed structure when desired hardness would be achieved as shown in Figure 10.5. The FDM finds extensive application with regards to 3D printed food.

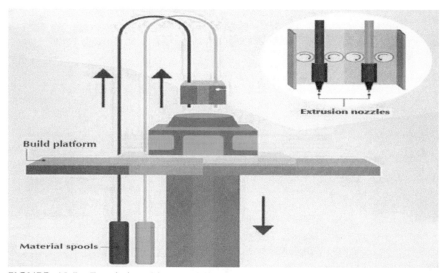

FIGURE 10.5 Fused deposition modeling technology [image credit by CustomMade (http://www.custommade.com/)].

10.6.3 STEREOLITHOGRAPHY

STL was patented in 1986 by Chuck Hull.[17] Photopolymerization is primarily used in STL to produce a solid part from a liquid. The technology employed in this process is known as continuous liquid interface production (CLIP). It uses light and oxygen to grow continuously objects from a pool of resin instead of printing them layer-by-layer. The well-known 3D printing technique works on a layer-by-layer approach whereas in STL technique objects produce from a pool of resin due to chemical reaction inhibited by controlled interaction of UV light (which inhibits the photo polymerization) and oxygen as shown in Figure 10.6. It's a tuneable photochemical process that can work at speed 25–100 times faster than traditional 3D printing. The setup required for STL primarily consists of a unique window that is both transparent to light and permeable to oxygen. The CLIP produces a dead zone in the resin pool of very fine thickness (tens of microns) by stabilizing the quantity of oxygen that comes in and out of the window, which restricts the photopolymerization to occur. In the photopolymerization, a vat of liquid polymer is exposed to control lighting under safelight conditions. The liquid polymer exposed to the ultraviolet laser light cured and solidified and joins the layer below. The build plate then proceeds down in small increments,

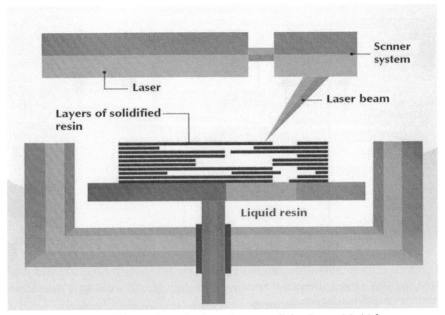

FIGURE 10.6   Stereolithography technology (image credit by CustomMade).[9]

and the liquid polymer is again exposed to the ultraviolet laser light. The process repeats till the model has been built. The liquid polymer is then drained from the vat, leaving the solid model. A gel-like support material, which is designed to support complicated geometries, is removed by hand and water jetting. Feature sizes of fewer than 100 nm, as well as complex structures with moving and interlocked parts are readily produced by this method. This technique assures a smooth outer and robust inner structure with homogeneity that gives this technology superiority over conventional 3D method.

10.6.4 LAMINATED OBJECT MANUFACTURING

Michael Feygin invented the LMO in the year 2000 which initially used ordinary sheets of office paper.[20] LOM works by layering sheets of material on top of one-another, gluing them together using a binder as shown in Figure 10.7. The printer then slices an outline of the object into that cross section to be eliminated from the adjacent excess material. Iterating this process makes up the object one layer at a moment. LOM printed objects were reliable, accurate, and durable as well as show no distortion over time that makes them well suited for each stage of the design cycle. The model can

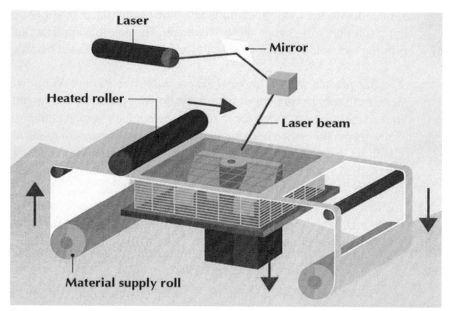

FIGURE 10.7 Laminated object manufacturing technology (image credit by CustomMade).[8]

be modified even after printing by the machining or drilling operation. The layer resolution of the LMO can be defined by the material feedstock and normally lies in the thickness range from one to a few sheets of copy paper. LOM is a low-cost 3D printing technique that is dimensionally slightly less accurate than that of STL and SLS, but no finishing step is necessary for LOM. It doesn't have any application in 3D food printing but can be applied usefully in food packaging studies.

10.7 APPLICATIONS OF 3D PRINTING IN FOOD PROCESSING

The price of the 3D printers was very high ($100,000 to $1M), when introduced in 1980. However, with the progress in the technology the price has been dropped down to less than $2500 in the current market.[9] Although the 3D printing technology has been in use for decades, still most of the peoples are not familiar with the term 3D printing. However, from the year 2011 onward the 3D printing technology has more widely known to the world, due to continuous improvement in the technology. Accordingly, the application of 3D printing has been continuously expanding.

The traditional cooking is impossible to be replaced with 3D printed food to prepare the dinner or a desert, due to the cultural importance, practice and taste. What differentiate it from the former is its ability to aid nutrition and sustainability to the food. In coming days, the health issues possibly be a strong reason that will change the way we eat. The possible applications of 3D printing technology in the sector of food processing are listed below.

- *The 3D printed food looks and tastes good, yet easy to swallow*: Senior citizens having chewing and swallowing problems are forced to take their food in puree form, which doesn't look very appetizing. They get malnourished in particular cases and leads to sort of medical condition. In such case vegetables and fruits can be 3D printed taking the mashed vegetable and gelling agent so that the desired softness is achieved without disturbing the integrity of the shape.
- *Customization of the food nutrition*: The nutritional requirement of the human body mostly depends on age, physical activity, lifestyle, and health condition of the person concern. Based on the dietary requirement of an individual a 3D printer can provide a specialized meal that can fulfill the exact need of the human body. This requirement may provide an opportunity to use the software in customize cooking through 3D printing.

- *Production of sustainable foods*: With the increasing world population, it is hard to provide nutrients like proteins, etc. from conventional sources of food. To fight such a situation alternative protein sources from algae and insects could be transformed into attractive foods with a desired texture that people may like.
- *Food for the extreme condition*: In the extreme environmental condition, it is hard to carry and food for the long period. Most prominently in defense bases there is the demand for such a meal that can last for long duration without refrigeration and survive the most extreme climatic conditions, like high-altitude airdrops. The 3D printing of food could be a solution to this problem; the fresh food can be printed according to need and environmental extremities.
- *Zero down the losses during food processing*: In food processing or even cooking handling losses is bound to happen. It can be completely eradicated in 3D printing of food as the exact amount is get used for any preparation, and any leftover amount can be used for further processing.
- *Food gardening with 3D printing*: It is a concept (by Chloé Rutzerveld) that will provide the consumer a delicious, fresh, and nutrient-rich food that grows in front of their eyes. The concept involves a specially printed outer casing made from dough that contains "edible soil" and various seeds. Multiple layers those containing seeds, spores, and yeast are printed according to a personalized 3D file. The plants and fungi will be matured in a period of 5 days whereas the yeast will ferment the solid inside into a liquid. The product's magnifying structure, aroma, and taste will be indicated in its changing appearance and can be harvested according to the need.
- Food coated with edible packaging film: The 3D printer can be used to print the food coated with an edible film that may have the long storage life.

10.8 POTENTIAL BENEFITS OF 3D PRINTING OF FOOD

Food can be produced keeping individual's interest and preference such as flavor, color, texture, size, and shape of a food item. Hence, it shows the ability to produce variety of products with individual customer requirements.

- Food can be personalized to meet the nutritional requirement of a person based on its age, sex, body weight, and profession by providing daily dietary energy through 3D designed food. Hence, sometimes 3D printed food is also called as personalized food.

- The 3D printed food can incorporate alternate health-enhancing ingredients that are uncommon in conventional food such as proteins from insects, algae, beet leaves, etc.
- Food preparation can be performed on time keeping the freshness and taste intact so that a healthy meal can be ensured every time.
- Innovative and unique design can be generated and could be replicated to increase the aesthetic aspect of food which can add value to the printed food.
- The 3D printing being an AM technique ensures proper use of raw material. Hence, resulting in almost nonsignificant waste of the food, which eliminates the waste management issues.
- Products with a great level of complexities can be produced with desired strength that is impossible to produce physically in any other manner.
- Production of any customized food can be possible at a place convenient for the end-user or consumer.

10.9 LIMITATIONS OF 3D PRINTING OF FOOD

The 3D printing technology offers following limitations, as the technology is in the initial stage of its development. The following hurdles can be eliminated in the future with the advancement in the technology:

- The 3D printed food is restricted to a limited size because of moving part involved during the printing process.
- Food with multicomponent is hard to produce through this technique as in the presently used method there is no provision for malfunctioning of a digital recipe.
- The 3D printing technique is slow as compared to the commercially available food processing method for food production.
- The technique is quite expensive and requires high power at present.
- The process of creating layers during AM, the previous layer has to be hardened to provide a platform for the next layer, failing to which deform the structure of food.
- Since it's a new technology, the availability of raw material is very limited as the printing process needs a definite structure or form of the raw materials.

10.10 FUTURE SCOPE

The present world is changing at a rapid pace; the interminable prospects of the 3D printing technology seem beyond our vision. The potential of 3D printing has been under technical and philosophical discussion, which is helping to reshape food research. The breakthroughs in the area of the 3D printers of food are fast and incredible. The 3D printers have the capabilities of producing food with different color, flavor, shape, and geometry. Accordingly, numerous materials products are already in the market and will continue to reform until the prices are further lowered with the continuous improvement in the technology. With the advancement in the present 3D printing technology, this technology will undoubtedly help to provide cost-effective, energy-efficient food in the near future in order to meet quality, nutritional stability, safety, variety (nutritional, design, etc.), and acceptability requirements for an individual consumption. NASA and SMRC are exploring this technology for providing food to the astronauts during long space missions with the minimum investment of the spacecraft resources, time, and the money.[11]

The present scenario of food production, preservation, and supply chain will not fulfill the demand of continuously growing world population. By the end of the 21st century, the anticipated world population will be around 12 billion.[3] Hence, the grant is to strive to attain the noble goal of combating the world hunger and poverty. Our traditional technologies alone are not enough for providing food at an affordable price to the growing population. Accordingly, there is a need for emerging technology like 3D printing that is limited to small scale due to sophisticated, high expense, and processing time. The novel embodiment of the 3D food printing concept with reduced cost is yet to be commercialized. With the further refinement, this technology can undoubtedly resolve future issues like food wars, malnutrition, inflation, poverty, starvation, food shortage, and famine. The 3D printed food (powdered food) is rarely as delectable as fresh food; however, the shortage in the food production will inevitably change the circumstances that people will be willing to tolerate. "3D printing, could well rewrite the rules of manufacturing in much the same way as the PC trashed the traditional world of computing," according to the Economist.[9] According to Eisenberg (2013), the 3D printing technology is referred to as "Industrial Revolution 2.0" and he notes that it "will change the nature of manufacturing."[2]

The 3D printing technology has capabilities of solving the real problems in the physical world, which could be used efficiently in educating the students and users in the near future for better learning and understanding of the complex process[2] that also includes food processing. The students could manipulate the virtual 3D model with the application of 3D printers, which will inspire and devote the student's determination toward design, increasing their self-confidence and imagination. In the future, the structure of food protein, microorganisms, DNA, etc. could be developed by the 3D printers, which would help in new ways of visualizing food structures scientifically and the gained concepts can intensify discovery and learning.[5]

Further, the efforts can be made by the educational and research institutes to develop a cost and time efficient 3D printing technology for a particular dietary requirement of an average Individual. The developed 3D printing service can be made available to the student and staff of the Institute (who belongs to different age group) on the payment basis which will definitely depends on the size, design complexity, and nutritional contents of the food. This study will also help in understanding the acceptability/ likely of 3D printed food product by the consumers of a different age group. In the future, one can make his food while traveling by sending his digital recipe to the 3D printer via, web connectivity (email), which will save the time and also prevent stress of an individual for making food after a long working day.

A study in Japan is taken as an example to compare the potential of 3D printing technology with the existing household electronics like the refrigerator, air conditioner, microwave oven, etc.[6] Figure 10.8 shows that the time to achieve 5% penetration of households was 3–5 years for most of the electronics, while for penetrating 10% of households an additional 2–4 years was required and full penetration requiring about 20–25 years;[6] the 3D printing is at initial stage and with further refinement in the technology it will be very soon in every household.

Further, an estimate showing the market value in billion US dollars of the 3D printing market, for the year 2015 (4 billion US dollars) and 2021 was provided along with the comparison with the year 2011 (1.71 billion US dollars) and 2012 (2.2 billion US dollars) as shown in Figure 10.9.[12]

The estimates show that the 3D printing industry will be growing at a rapid rate, and the market value will be around 10.8 billion US dollars by 2021.[12] It could be predicted that with the improvement in the 3D printing technology, the nature of commercial market will change, since the end users will be producing their food like pizza, burger, etc., rather than procuring from the retail shop.

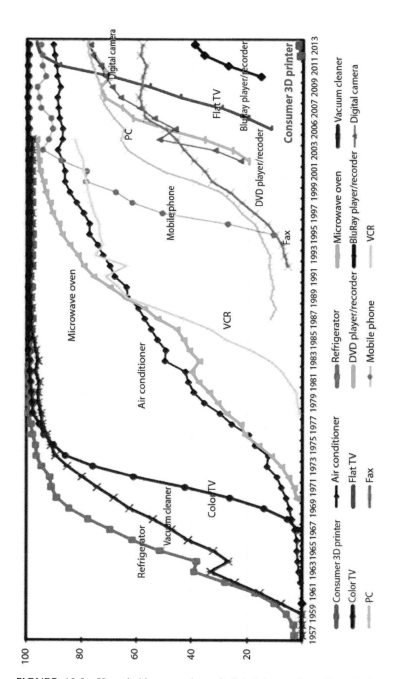

FIGURE 10.8 Household penetration of digital innovations (Japan) (image credit by Foresightinvestor).[6]

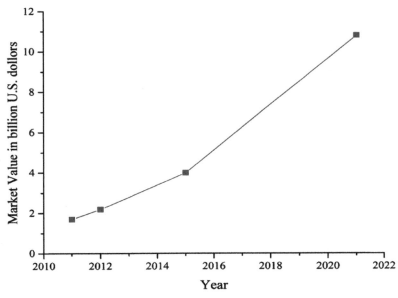

FIGURE 10.9   Graph showing the estimated value of the 3D printing market.[12]

KEYWORDS

- 3D printer
- 3D printing
- additive manufacturing
- food
- food printer
- fused deposition modeling
- inkjet technology
- laminated object manufacturing
- modeling
- selective laser melting
- selective laser sintering
- stereolithography
- synthetic food
- virtual models

REFERENCES

1. Campbell, T.; Williams, C.; Ivanova, O.; Garrett, B. Could 3D Printing Change the World? *Technologies, Potential, and Implications of Additive Manufacturing, Atlantic Council, Washington, DC*, 2011.
2. Eisenberg, M. 3D Printing for Children: What to Build Next? *Int. J. Child-Comp. Interact.* **2013,** *1*(1), 7–13.
3. Friedrichs, G.; Schaff, A. Eds. *Microelectronics and Society: For Better or for Worse*; Friedrichs, G., Schaff, A., Eds.; Elsevier: London, UK, 2013.
4. Hilton, P. *Rapid Tooling: Technologies and Industrial Applications*. CRC Press: New York, 2000.
5. http://3dprint.nih.gov/ [accessed April 21 2015].
6. http://foresightinvestor.com/articles/32393-could-3d-printing-change-the-world [accessed April 21 2015].
7. http://systemsandmaterials.com/technologies/3d-printed-food/ [accessed April 24 2015].
8. http://www.custommade.com [accessed July 15 2015].
9. http://www.economist.com/blogs/babbage/2012/09/3d-printing [accessed May 8 2015].
10. http://www.livescience.com/38862-selective-laser-sintering.html [accessed July 15 2015].
11. http://www.nasa.gov/directorates/spacetech/home/feature_3d_food_prt.htm [accessed July 15 2015].
12. http://www.statista.com/statistics/261693/3d-printing-market-value-forecast/ [accessed April 20 2015].
13. https://abdulmoidyawer.wordpress.com/ [accessed July 15 2015].
14. Hull, C. W. *U.S. Patent No. 4,929,402*. U.S. Patent and Trademark Office: Washington, DC, 1990.
15. Jain, M. C. S.; Vibhandik, M. D. V.; Gade, S. A. 3D Printing. *Int. J. Emerg. Technol. Adv. Eng.* **2013,** *3*(4), 18–23.
16. Lam, C. X. F.; Mo, X. M.; Teoh, S. H.; Hutmacher, D. W. Scaffold Development Using 3D Printing with a Starch-based Polymer. *Mater. Sci. Eng., C* **2002,** *20*(1), 49–56.
17. Melchels, F. P. Celebrating Three Decades of Stereolithography. *Virtual Phys. Prototyping* **2012,** *7*(3), 173–175.
18. Novakova-Marcincinova, L.; Novak-Marcincin, J.; Janak, M. Precision Manufacturing Process of Parts Realized by FDM Rapid Prototyping. *Key Eng. Mater.* **2014,** *581*, 292–297.
19. Periard, D.; Schaal, N.; Schaal, M.; Malone, E.; Lipson, H. Printing Food. In Proceedings of the 18th Solid Freeform Fabrication Symposium, Austin, TX, 2007; pp 564–574.
20. Wohlers, T.; Gornet, T. History of Additive Manufacturing. *Wohlers Report*; 2011; p 24.

CHAPTER 11

STRUCTURING EDIBLE OIL USING FOOD GRADE OLEOGELATORS

ASHOK R. PATEL

Vandemoortele Center for Lipid Science & Technology, Laboratory of Food Technology and Engineering, Faculty of Bioscience Engineering, Ghent University, Coupure Links 653, 9000 Gent, Belgium. E-mail: Patel.Ashok@Ugent.be

CONTENTS

ABSTRACT

Many of the functional applications of edible oils in food formulation are accomplished by structuring them into plastic-like soft matter systems. This structuring of oil is traditionally based on the colloidal network of fat crystals formed by high-melting triacylglycerol (TAG) molecules that are rich in *trans* and/or saturated fatty acids. Currently, due to huge interest in development of *trans*- and saturated fat-free food products, the research in the area of identifying alternative routes for oil structuring has gained a lot of popularity in the field of lipid science and technology. Oleogelation (gelling of liquid oil in absence of high-melting TAGs) is one such alternative which has recently attracted attention from researchers and industrial scientists alike. The possibility of transforming ≥90 wt% of liquid oil into a "gel-like" structure opens up many possibilities of developing food products with better nutritional profiles. This chapter gives a concise overview of recent research conducted in the area of oleogelation. Specifically, categories of food-grade oleogelators have been discussed along with some typical examples. The general guidelines on the properties of oleogels, their characterization, food applications, and practicality of oleogelation approach are also summarized.

11.1 INTRODUCTION

Lipid-based food products such as margarines, table spreads, ice-creams, chocolates, etc. are usually formulated using solid fats as the main component, which provides the necessary functionality, desirable texture as well as enhanced stability. From colloidal science point of view, the crystallization and subsequent network formation of crystalline fat (high-melting triacylglycerols [TAGs]) creates a structural framework that can physically trap the low-melting TAGs (liquid oil) into a three-dimensional gel-like structure which imparts functionality to the food formulations. This conventional route of structuring liquid oil which is currently used by food manufacturing industries has certain drawbacks. First, since the structured systems are more close to "particle-filled gels," the mass fraction of crystalline phase that is required for structuring is quite high (≥0.2).[5,18]

Second, the crystalline phase is made up of high-melting TAGs that are rich in *trans* and saturated fatty acids, thus arguably it makes the food product nutritionally poor. Therefore, food industries worldwide are actively working toward reformulation of lipid-based products with lowered amount of saturated fats and complete elimination of *trans*-fats.[9,29] However, replacing solid fats in food products is quite challenging because apart from imparting excellent organoleptic properties (melt-in-mouth and cooling effects), solid fats are also responsible for providing physical characteristics such as hardness, texture, crispiness, spreading and snap. Among different alternative strategies that have been explored in the field, oleogelation has been hailed as a feasible approach that can help us to formulate food products with significantly lower amount of saturated fats.[10,19] The possibility of gelling ≥90% weight of liquid oil at a relatively lower mass fraction of gelator molecules, makes oleogelation a very efficient means for structuring oil in absence of solid fats. This efficient structuring in oleogels is typically achieved by supramolecular assemblies (building blocks) of gelator molecules that organize into a 3-D network that can trap a large amount of oil into a gel-like structure.[3,10,25] In this chapter, an overview of different categories of building blocks (oleogelators) is presented along with typical examples.

11.2 OLEOGELATORS

A number of gelator molecules have been researched for edible oil structuring in the last decade or so. The basic building blocks (supramolecular assemblies) formed by these molecules can be categorized as shown in Figure 11.1:

a. crystalline particles.
b. self-assembled structures of low molecular weight compounds (fibers, strands, tubules, reverse micelles, mesophases, etc.).
c. Self-assembled structures of polymers or polymeric strands.
d. Miscellaneous structures like colloidal particles and emulsion droplets.

The building blocks can be formed by mono-component or mixture of components (mixed systems). The formation of building blocks is usually achieved through direct method, that is, by dispersing gelator molecules in oil medium at high temperatures followed by cooling. In some cases, indirect methods are utilized where dried microstructures are created from hydrated polymer solutions or surface active protein particles.[7,11,16,17,22,27]

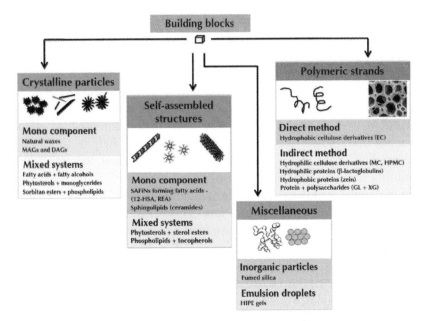

FIGURE 11.1   Categorical listing of building blocks which can be used for oil structuring.[20]

11.2.1 CRYSTALLINE PARTICLES

Oleogelators such as partial glycerides (mono and di-acyl glycerides, MAG, and DAG, respectively), natural waxes, and long-chain fatty acids are known to form well-defined crystalline particles that can organize into a 3D network and cause gelation of liquid oils at relatively lower concentrations compared to crystalline TAGs. In particular, waxes are known to be very efficient in gelling liquid oils at concentrations as low as 0.5 wt%. The linear structures of main components in waxes (wax esters, hydrocarbons and free fatty acids) promote 1D and 2D growth of the crystalline structures resulting in anisotropic particles, which favor network formation at much lower crystalline mass fraction.[24] Moreover due to the presence of polar components such as fatty alcohols, some natural waxes (such as shellac wax) are also good at stabilizing water–oil interfaces and this property has even been exploited to formulate emulsifier-free oil-continuous emulsions with water content as high as 60 wt%.[12,14]

In some cases, a combination of different gelator molecules is used in order to exploit the synergistic enhancement in the properties of structured gels. Combinations such as stearic acid + stearyl alcohol, phytosterols +

monoglycerides, and sorbitan esters + phospholipids have been explored, where synergistic interactions lead to better crystallization properties (crystal modification, network strengthening, and stabilization of complex micellar systems).[20]

11.2.2   SELF-ASSEMBLED STRUCTURES OF LOW MOLECULAR WEIGHT COMPOUNDS

Due to the structural features, some gelator molecules such as hydroxylated fatty acids, phytosterols (in combination with sterol esters), phospholipids, etc. are known to form specific self-assembled structures in liquid oil. Of these components, the gelling properties of 12-hydroxy stearic acid (12-HSA) have been studied extensively by researchers from varied fields.[1,13,26] The gelation of organic solvent by 12-HSA is attributed to the unidirectional crystal growth that results in the formation of highly anisotropic (high aspect ratio), fibrous crystalline strands (also called as self-assembled fibrillar network). As far as food-grade gelators are concerned, a synergistic combination of phytosterols (β-sitosterol) and sterol esters (γ-oryzanol) have been studied quite extensively. The building blocks in these oleogels are self-assembled tubules of nanoscale dimensions (diameter ≈ 7.2 nm, thickness ≈ 0.8 nm). These tubules have a complex helical, ribbon-like structure which is stabilized by intermolecular H-bonding between the hydroxyl group of β-sitosterol and carbonyl group of γ-oxyzanol.[2]

11.2.3   POLYMERIC OLEOGELATORS

Hydrophobic cellulose derivative (ethyl cellulose [EC]) has been studied widely as oleogelator in absence and presence of emulsifiers such as sorbitan tristearate. EC-based oleogels are usually prepared by dispersing polymer in liquid oil under shear at temperatures above the glass transition temperature of EC followed by cooling to lower temperatures. At high temperatures, the polymer chains are unfolded and on subsequent cooling, the unfolded polymer chains form a backbone for cross-linked gel network. The gel is stabilized by hydrogen bonding among polymer chains along with some hydrophobic interactions between acyl chains of oil and the side chains of EC.[6] Efforts have also been made to utilize hydrophilic polymers (hydrophilic cellulose derivatives and proteins) as oil structurants based on indirect methods, which involve hydration of polymer in aqueous phase followed by interfacial accumulation of polymers (air–water or oil–water interfaces) and removal of water to obtain dried microstructures which can then be used for oil structuring.[11,16,22] Although, the indirect methods adds complexity to the process of creating oleogels (multiple steps which are time-consuming), these do open-up newer possibilities as most food-approved polymers are hydrophilic in nature.

11.2.4 MISCELLANEOUS

Colloidal silicon dioxide or fumed silica has been recently shown to structure liquid oil at concentration of ≥10 wt%, and further this structured oil systems can be used along with water gels for fabricating relatively unexplored class of colloidal systems called bigels.[23] The oil structuring functionality of fumed silica is based on the reversible assembling of deagglomerated aggregates of silica particles in the liquid oil medium. Since, the dimensions of structuring units are well below the wavelength of visible light, the obtained oleogel is optically clear. Most interesting aspect of this oleogels is the possibility of creating novel bigels when they are combined with water gel at certain proportions. The bigels result from an arrested demixing of two structured phases leading to the formation of semicontinuous interpenetrating network of oil and water phases.[23]

It has also been demonstrated that gelled emulsion droplets can act as structuring units by providing a framework which can support "gel-like" structure, provided that the volume fraction of gelled droplets is high enough.[15] The approach is based on temperature induced gelation of dispersed water phase (structured using hydrocolloids) emulsified in an oil continuous medium, the high-volume fraction of dispersed phase results in close packing of droplets that leads to formation of droplet filled gels.[15,18]

11.3 CHARACTERIZATION OF OLEOGELS

Characterization of oleogels involves mainly the advanced microscopy studies (to understand the basic microstructure involved in creating the structural framework of gels) and detailed rheological evaluation (to study the physical properties of gels as a function of small and large deformations). Usually, a combination of different microscopic techniques including polarized light microscopy (for crystalline building blocks), confocal microscopy (for multicomponent gels), and electron microscopy (for identifying network and building blocks in de-oiled samples) are utilized to have a detailed characterization of oleogel structure. Comparative microstructures of oleogels created using wax crystals, polymer strands and MAG crystals are shown in Figure 11.2, where the continuous network formed by building blocks is clearly visible. Results from microstructure studies can be combined with rheological characterization to understand the weak link between microstructure and physical properties of the gels. Rheological characterization (oscillatory and flow measurements) can give valuable insights about the gel strength, thixotropic properties, and shear sensitivity of oleogels. For example, graphs obtained from oscillatory amplitude sweeps done on oleogels prepared from three different building blocks (wax crystals, polymeric strands, and inorganic particles) are shown in Figure 11.3. It is quite straightforward to understand that oleogel based on inorganic particles has relatively better gel strength with a prominent broad linear response region with high moduli values as compared to wax crystal-based oleogel, which shows a very brittle structure (high moduli value with a nonexistent liner region). Also, the ratio of elastic modulus/viscous modulus at lower strain/stress is significant greater for inorganic particle based gel compared to wax crystal-based gel. The polymer-based gel can be considered to be as an intermediate of these two gels.

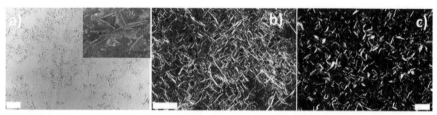

FIGURE 11.2 Microstructure of oleogels prepared using (a) wax crystals (scale bar = 100 μm, inset = cryo-SEM image of individual crystal, image width = 25 μm); (b) polymeric strands (scale bar = 200 μm); and (c) MAG crystals (scale bar = 100 μm).

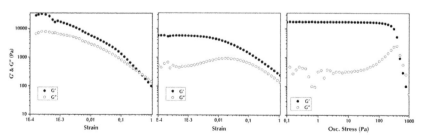

FIGURE 11.3 Rheological characterization of oleogels prepared using (from L to R) wax crystals, polymeric strands, and inorganic silica particles.

11.4 POTENTIAL FOOD APPLICATIONS

A lot of research has been done in the area of exploring newer oleogelators which could be used for food applications. Most of the work has been focused on the fundamental understanding (microstructure, structuring phenomenon, synergism, etc.). In recent years, studies exploring the functionality of oleogels for potential applications in actual food formulations have also been published. Some examples include: the use of EC oleogel as replacement of beef fat in frankfurters and heat resistant chocolates; wax-based oleogels for decreasing levels of saturated fats in ice-cream, cookies, chocolate spreads and margarines, and hydrophilic polymer-based oleogels as shortening alternatives for bakery applications,[8,14,17,28,30–32] as shown in Figure 11.4.

FIGURE 11.4   Potential food applications of oleogels. From L to R: oleogel-based spread prepared without emulsifiers, chocolate paste where oleogel was used for complete, and partial replacement of oil binder and palm oil, respectively, and 4/4 sponge cakes prepared using oleogel as shortening.[19]

Although, oleogels have definitely shown a huge potential as a possible solution for reduction of saturated fats, there are, however, certain limitations that needs to be addressed if we are to see the eventual transformation of bench scale research into commercial products. Most importantly, we need to keep in mind that a complete replacement of solid fats with gelled oil systems will most certainly lead to some loss of functionality. Thus, in order to develop end products with required qualities, a further reworking of the formulation will be required.

11.5 CONCLUSION AND FUTURE TRENDS

Oleogelation as a strategy is still in its earlier stages but it has already generated a lot of interest particularly, from the food industries due to the recent ban on *trans*-fat. It is important to clarify that the bad reputation of saturated fat has been debated based on recent reports which reveal that saturated fats might not be as bad as they are perceived to be and replacing saturated fat with carbohydrates is actually a much worse option. However, the meta-analysis studies do reveal that there is a clear benefit of replacing saturated fat with polyunsaturated fat,[4] and this is where oleogelation approach can be exploited. However, there are still some challenges that need to be addressed. The main bottleneck of this approach is identifying and selecting the right kind of structuring agent(s). Apart from the right structural features and molecular assembly properties (crystallization or supramolecular ordering), the identified structuring agents also need to satisfy most (or preferably all) of the conditions as listed below[21]:

- should be food-grade and approved for use as direct additive or formulation aid,
- should tolerate the processing conditions and be compatible with the product formulation matrix,
- should be low cost, easily available and be effective at low concentrations,
- should provide the right functionality when used in intermediate products such as cake shortening or baking margarine and be capable of mimicking the melt-in-mouth type oral sensation (provided by fat crystals) when used in final products such as spreads, chocolate products, etc.

In addition, the process used for structuring of oil should also be industrially feasible and not too energy intensive or detrimental to the oil quality. Among the structuring agents studied for their oil structuring properties, waxes and polymers can be considered to be the most promising ones. However, there are regulatory limitations on using waxes as direct additives in food product and polymer-based gels requires energy intensive process such as high-temperature dispersion, spray-drying, and/or lyophilization. Thus, for commercial application of oleogelation, we would require the food industry to step-up and take appropriate measures such as filing for regulatory approval of some promising ingredients and willingness to invest in newer process optimization.

KEYWORDS

- building blocks
- cellulose derivatives
- edible applications
- ethyl cellulose
- fumed silica
- hydrophilic polymers
- natural waxes
- oil structuring
- oleogels
- oleogelation
- shellac wax
- structurants

REFERENCES

1. Abdallah, D. J.; Weiss, R. G. Organogels and Low Molecular Mass Organic Gelators. *Adv. Mater.* **2000,** *12*, 1237–1247.
2. Bot, A.; Agterof, W. M. Structuring of Edible Oils by Mixtures of γ-oryzanol with β-Sitosterol or Related Phytosterols. *J. Am. Oil Chem. Soc.* **2006,** *83*, 513–521.
3. Bot, A.; Veldhuizen, Y. S. J.; den Adel, R.; Roijers, E. C. Non-TAG Structuring of Edible Oils and Emulsions. *Food Hydrocolloids* **2009,** *23*, 1184–1189.
4. Cassiday, L. *Big Fat Controversy: Changing Opinions about Saturated Fats*; AOCS Press: Urbana, IL, 2015; pp 342–349.
5. Co, E. D.; Marangoni, A. G. Organogels: An Alternative Edible Oil-structuring Method. *J. Am. Oil Chem. Soc.* **2012,** *89*, 749–780.
6. Davidovich-Pinhas, M.; Barbut, S.; Marangoni, A. G. The Gelation of Oil Using Ethyl Cellulose. *Carbohydr. Polym.* **2015,** *117*, 869–878.
7. Gao, Z. M.; Yang, X. Q.; Wu, N. N.; Wang, L. J.; Wang, J. M.; Guo, J. Protein-based Pickering Emulsion and Oil Gel Prepared by Complexes of Zein Colloidal Particles and Stearate. *J. Agric. Food Chem.* **2014,** *62*, 2672–2678.
8. Hwang, H. S.; Singh, M.; Bakota, E.; Winkler-Moser, J.; Kim, S.; Liu, S. Margarine from Organogels of Plant Wax and Soybean Oil. *J. Am. Oil Chem. Soc.* **2013,** *90*, 1705–1712.
9. Kodali, D. R. *Trans Fats Replacement Solutions*. AOCS Press: Urbana, IL, 2014.
10. Marangoni, A. G.; Garti, N. *Edible Oleogels: Structure and Health Implications*. AOCS Press: Urbana, IL, 2011.

11. Patel, A. R.; Schatteman, D.; Lesaffer, A.; Dewettinck, K. A Foam-templated Approach for Fabricating Organogels Using a Water-soluble Polymer. *RSC Adv.* **2013,** *3,* 2290–2293.

12. Patel, A. R.; Schatteman, D.; Vos, W. H. D.; Dewettinck, K. Shellac as a Natural Material to Structure a Liquid Oil-based Thermoreversible Soft Matter System. *RSC Adv.* **2013,** *3,* 5324–5327.

13. Patel, A. R.; Remijn, C.; Heussen, P. C. M.; den, A. R.; Velikov, K. P. Novel Low-molecular-weight-gelator-based Microcapsules with Controllable Morphology and Temperature Responsiveness. *Chem. Phys. Chem.* **2013,** *14,* 305–310.

14. Patel, A. R.; Rajarethinem, P. S.; Gredowska, A.; Turhan, O.; Lesaffer, A.; De Vos, W. H. Edible Applications of Shellac Oleogels: Spreads, Chocolate Paste and Cakes. *Food Function* **2014,** *5,* 645–652.

15. Patel, A. R.; Rodriguez, Y.; Lesaffer, A.; Dewettinck, K. High Internal Phase Emulsion Gels (HIPE-gels) Prepared Using Food-grade Components. *RSC Adv.* **2014,** *4,* 18136–18140.

16. Patel, A. R.; Cludts, N.; Bin Sintang, M. D.; Lewille, B.; Lesaffer, A.; Dewettinck, K. Polysaccharide-based Oleogels Prepared with an Emulsion-templated Approach. *Chem. Phys. Chem.* **2014,** *15,* 3435–3439.

17. Patel, A. R.; Cludts, N.; Sintang, M. D. B.; Lesaffer, A.; Dewettinck, K. Edible Oleogels based on Water Soluble Food Polymers: Preparation, Characterization and Potential Application. *Food Funct.* **2014,** *5,* 2833–2841.

18. Patel, A. R.; Dewettinck, K. Comparative Evaluation of Structured Oil Systems: Shellac Oleogel, HPMC Oleogel and HIPE Gel. *Eur. J. Lipid Sci. Technol.* **2015,** DOI:101002/ejlt201400553.

19. Patel, A. R. *Alternative Routes to Oil Structuring.* Springer International Publishing, 2015.

20. Patel, A. R. Introduction: General Considerations and Future Trends. In *Alternative Routes to Oil Structuring*; Patel, A. R., Ed.; Springer International Publishing, 2015; pp 63–70.

21. Patel, A. R.; Rajarethinem, P. S.; Cludts, N.; Lewille, B.; De Vos, W. H.; Lesaffer, A. Biopolymer-based Structuring of Liquid Oil into Soft Solids and Oleogels using Water-continuous Emulsions as Templates. *Langmuir* **2015,** *31,* 2065–2073.

22. Patel, A. R.; Mankoc, B.; Bin Sintang, M. D.; Lesaffer, A. Dewettinck, K. Fumed Silica-based Organogels and 'aqueous-organic' bigels. *RSC Adv.* **2015,** *5,* 9703–9708.

23. Patel, A. R.; Babaahmadi, M.; Lesaffer, A.; Dewettinck, K. Rheological Profiling of Organogels Prepared at Critical Gelling Concentrations of Natural Waxes in a Triacylglycerol Solvent. *J. Agric. Food Chem.* **2015,** *63,* 4862–4869.

24. Pernetti, M.; van Malssen, K. F.; Flöter, E.; Bot, A. Structuring of Edible Oils by Alternatives to Crystalline Fat. *Curr. Opin. Colloid Interface Sci.* **2007,** *12,* 221–231.

25. Rogers, M. A.; Wright, A. J.; Marangoni, A. G. Engineering the Oil Binding Capacity and Crystallinity of Self-assembled Fibrillar Networks of 12-Hydroxystearic Acid in Edible Oils. *Soft Matter* **2008,** *4,* 1483–1490.

26. Romoscanu, A. I.; Mezzenga, R. Emulsion-templated Fully Reversible Protein-in-oil Gels. *Langmuir* **2006,** *22,* 7812–7818.

27. Stortz, T. A.; Marangoni, A. G. Ethylcellulose Solvent Substitution Method of Preparing Heat Resistant Chocolate. *Food Res. Int.* **2013,** *51,* 797–803.

28. Talbot, G. Saturated Fats in Foods and Strategies for their Replacement: An Introduction. In *Reducing Saturated Fats in Foods*; Talbot, G. Ed.; Woodhead Publishing, 2011; pp 3–28.

29. Ylmaz, E.; Ogutcu, M. *The Texture, Sensory Properties and Stability of Cookies Prepared with Wax Oleogels*, 2015. food&function.com.
30. Zetzl, A. K.; Marangoni, A. G.; Barbut, S. Mechanical Properties of Ethylcellulose Oleogels and their Potential for Saturated Fat Reduction in Frankfurters. *Food Funct.* **2012,** *3*, 327–337.
31. Zulim Botega, D. C.; Marangoni, A. G.; Smith, A. K.; Goff, H. D. The Potential Application of Rice Bran Wax Oleogel to Replace Solid Fat and Enhance Unsaturated Fat Content in Ice-cream. *J. Food Sci.* **2013,** *78*, C1334–C1339.

CHAPTER 12

EXTRACTION TECHNOLOGY FOR RICE VOLATILE AROMA COMPOUNDS

DEEPAK KUMAR VERMA[1*] and PREM PRAKASH SRIVASTAV[2]

[1]Department of Agricultural and Food Engineering, Indian Institute of Technology, Kharagpur 721 302, West Bengal, India, Email: deepak. verma@agfe.iitkgp.ernet.in, rajadkv@rediffmail.com

[2]Department of Agricultural and Food Engineering, Indian Institute of Technology, Kharagpur 721 302, West Bengal, India, Email: pps@ agfe.iitkgp.ernet.in

*Corresponding author.

CONTENTS

ABSTRACT

Aroma is considered by rice consumers in India and the world as the third highest desired trait followed by taste and elongation after cooking. This desired trait is one of the critical aspects of rice quality that can determine acceptance or rejection of rice before it is tasted. It is also considered as an important property of rice that indicates the high quality and price in the market. From an assessment of all known data shows that there were around/ about 500 aroma chemical compounds which have been documented so far in various aromatic and nonaromatic rice cultivars through the world. This chapter address to importance of rice aroma, exploration of its modern and novel extraction technology and research opportunities in the development of innovative strategies that can be moved from the hope to reality in the field of rice aroma industries in the future.

12.1 INTRODUCTION

Aroma is one of the diagnostic aspects of rice quality which can determine acceptance or rejection of rice before it is tested.[98–100] It is also considered as an important property of rice that indicates its preferable high quality and price in the market.[41,68] Early research on rice aroma can be traced 30 years back,[13,14] but more scientific progress in modern analytical techniques has been achieved with discovery, development, and application of gas chromatography (GC). GC received tremendous advances and innovations when combined with mass spectrometry known as gas chromatography–mass spectrometry (GC–MS). It made identification and quantification of volatile compounds much easier in mixtures of sample matrix and greatly enhanced interest toward chemistry of rice aroma.

An assessment of all known data reveals that more than 450 compounds have been documented in various aromatic and nonaromatic rice cultivars (Table 12.1). The primary goals were to identify the compounds responsible for the characteristic rice aroma.[14] Many attempts are made till date to search the key compounds for rice aroma,[13,14] but any single compound or group of compounds could not reported which are fully responsible for rice aroma. The first volatile aroma compound, 2-Acetyl-1-pyrroline (2-AP) was reported in 1982 by Buttery et al.[14] and discovered by Buttery et al.[13] in 1983 from aromatic rice due to its impact volatile character. 2-AP is the most important flavoring compound of cooked rice since its discovery.[13,14]

Rice consists of balanced complicated mixtures of volatile aroma compounds which impart its characteristic flavor. There are no single analytical techniques that can be used for investigation of volatile aroma compounds in rice sample. Although, currently many technologies are available for extraction of rice volatile aroma compounds, and these technologies have been also modified time-to-time according to need and many of them are still under process to emerge into a new form, particularly in the distillation and extraction concept. This chapter reviews the extraction technology for rice volatile aroma compounds.

TABLE 12.1 List of Rice Volatile Aroma Compounds reported in Various Aromatic and Nonaromatic Rice Cultivars.

Aroma compounds		
1. (1*S,Z*)-calamenene	24. Heptylcyclohexane	48. Glycerin
2. 2-Methyl-decane	25. (*E*)-2,(*E*)-C heptadienal	49. (*E*)-2-octen-1-ol
3. Heptadecane	26. 2-Methyl-undecanol	50. 2-Pentylfuran
4. (2-Aziridinylethyl) aminutese	27. Hexacosane	51. Hexane
	28. (*E*)-2,(*E*)-C nonadienal	52. (*E*)-2-octenal
5. 2-Methyl-dodecane	29. 2-*n*-Butyl-furan	53. 2-Phenyl-2-methyl-aziridine
6. Heptadecyl-cyclohexane	30. Hexadecanal	
7. (2*E*)-dodecenal	31. (*E*)-2-decenal	54. Hexanoic acid
8. 2-Methyl-furan	32. 2-Nonanone	55. (*E*)-2-undecenal
9. Heptanal	33. Hexadecane	56. 2-Phenylethanol
10. (*E*)-3-Octene-2-one	34. (*E*)-2-heptanal	57. Hexanoic acid dodecate
11. 2-Methyl-hexadecanal	35. 2-Nonenal	58. (*E*)-3-nonen-2-one
12. Heptane	36. Hexadecane 2,6-bis-(*t*-butyl)-2,5	59. 2-Propanol
13. (*E*)6,10-dimethyl-5,9 undecadien-2-one	37. (*E*)-2-heptenal	60. Hexanol
14. 2-Methyl-I-propenyl benzene	38. 2-Ocetenal (*E*)	61. (*E*)-4-nonenale
	39. Hexadecanoic acid	62. 2-Tetradecanone
15. Heptanol	40. (*E*)-2-hexenal	63. Hexatriacontane
16. (*E*)-14-hexadecenal	41. 2-Octanone	64. (*E*)-hept-2-enal
17. 2-Methyl-naphthalene	42. Hexadecanol	65. 2-Tridecanone
18. Hepten-2-ol<6-methyl-5->	43. (*E*)-2-nonen-1-ol	66. Hexyl furan
	44. 2-Octenal	67. δ-Cadinol
19. (*E*)-2,(*E*)-4-decadienal	45. Hexadecyl ester, 2,6-difluro-3-methyl benzoic acid	68. 2-Undecanone
20. 2-Methyl-tridecane		69. Hexyl hexanoate
21. Heptene-dione		70. (*E,E*)-2,4-nonadienal
22. (*E*)-2,(*E*)-4-heptadienal	46. (*E*)-2-nonenal	71. Geranyl acetone
23. 2-Methyl-undecanal	47. 2-Pentadecanone	72. Hexylpentadecyl ester-sulphurous acid

TABLE 12.1 *(Continued)*

Aroma compounds

73. (*E,E*)-2,4-octadienal
74. 3-(*t*-Butyl)-phenol
75. Indane
76. (*E,E*)-farnesyl acetone
77. 3,5,23-Trimethyl-tetra-cosane
78. Indene,2,3-dihydro-1,1,3 trimethyl-3-phenyl
79. (*E,Z*)-2,4-decadienal
80. 3,5,24-Trimethyl-tetra-contane
81. Indole
82. (*S*)-2-methyl-1-dodecanol
83. 3,5-Dimethyl-1-hexene
84. I-S,*cis*-calamene lonol 2
85. (*Z*)-2-octen-1-ol
86. 3,5-Dimethyl-octane
87. Isobutyl hexadecyl ester oxalic acid
88. (*Z*)-3-Hexenal
89. 3,5-Di-*tert*-butyl-4-hydroxybenzaldehyde
90. Isobutyl nonyl ester oxalic acid
91. [(1-Methylethyl)thio] cyclohexane
92. 3,5-Heptadiene-2-one
93. Isobutyl salicylate
94. <6-methyl->Heptan-2-ol
95. 3,5-Octadien-2-ol
96. Isolongifolene
97. <*cis*-2-*tert*-Butyl->cyclohexanol acetate
98. 3,7,11,15-Tetramethyl-2-hexadecimalen-1-ol
99. Isopropylmyristate
100. 1-(1*H*-pyrrol-2-yl)-ethanone

101. 3,7,11-Trimethyl-1-dodecanol
102. Limonene
103. 1,2-Bis,cyclobutane
104. 3,7-Dimethyl-hexanoic acid
105. Lirnonene
106. 1,1-Dimethyl-2-octyl-cyclobutane
107. 3-Dimethyl-2-(4-chlorphenyl)-thio-acrylamide
108. 1-octen-3-ol
109. 1,2,3,4-Tetramethyl benzene
110. 3-Ethyl-2-methyl-heptane
111. 2-Methyl-butanal
112. 1,2,3-trimethyl-benzene
113. 3-Hexanone
114. Methoxy-phenyl-oxime
115. 1,2,4-Trimethyl-benzene
116. 3-Hydroxy-2,4,4-trimethylpentyl *iso*-butanoate
117. Methyl (*E*)-9-octadecenoate
118. 1,2-Dichloro-benzene
119. 3-Methyl-1-butanol
120. Methyl (*Z,Z*)-11,14-eicosadienoate
121. 1,2-Dimethyl-benzene
122. 3-Methyl-2-butyralde-hyde
123. Methyl (*Z,Z*)-9,12-octa-decadienoate
124. 1,3,5-Trimethyl-benzene
125. 3-Methyl-2-heptyl acetate
126. Methyl butanoate

127. 1,3-Diethyl-benzene
128. 3-Methyl-butanal
129. Methyl decanoate
130. 1,3-Dimethoxy-benzene
131. 3-Methyl-hexadecane
132. Methyl dodecanoate
133. 1,3-Dimethyl-benzene
134. 3-Methyl-pentadecane
135. Methyl ester tridecanoic acid
136. 1,3-Dimethyl-naphtha-lene
137. 3-Methyl-pentane
138. Methyl heptanoate
139. 1,4-Dimethyl-benzene
140. 3-Methyl-tetradecane
141. Methyl hexadecanoate
142. 1,4-Dimethyl-naphtha-lene
143. 3-Nonene-2-one
144. Methyl isocyanide
145. 10-Heptadecene-1-ol
146. 3-Octdiene-2-one
147. Methyl isoeugenol
148. 10-Pentadecene-1-ol
149. 3-Octene-2-one
150. Methyl linolate
151. 17-Hexadecyl-tetratria-contane
152. 3-Tridecene
153. Methyl naphthalene
154. 17-Pentatricontene
155. 4,6-Dimethyl-dodecane
156. Methyl octanoate
157. 1-Chloro-3,5-bis(1,1-dimethylethyl)2-(2-pro-penyloxy) benzene
158. 4-[2-(Methylaminuteso) ethyl]-phenol

TABLE 12.1 *(Continued)*

Aroma compounds

159. Methyl oleate
160. 1-Chloro-3-methyl butane
161. 4-Acetoxypentadecane
162. Methyl pentanoate
163. 1-Chloro-nonadecane
164. 4-Cyclohexyl-dodecane
165. Methyl tetradecanoate
166. 1-Decanol
167. 4-Ethyl-3,4-dimethyl-cyclohexanone
168. Myrcene
169. 1-Docosene
170. 4-Ethylbenzaldehyde
171. *N,N*-dimethyl chloestan-7-aminutese
172. 1-Dodecanol
173. 4-Hydroxy-4-methyl-2-pentanone
174. *N,N*-dinonyl-2-phenyl-thio ethylaminutese
175. 1-Ethyl-2-methyl-benzene
176. 4-Methyl benzene formaldehyde
177. Naphthalene
178. 1-Ethyl-4-methyl-benzene
179. 4-Methyl-2-pentyne
180. *n*-Decanal
181. 1-Ethyl-naphthalene
182. 4-Vinyl-guaiacol
183. *n*-Decane
184. 1-Heptanol
185. 4-Vinylphenol
186. *n*-Dodecanol
187. 1-Hexacosene
188. 5-Aminuteso-3,6-di-hydro-3-iminuteso-1(2*H*) pyrazine acetonitrile

189. *n*-Eicosanol
190. 1-Hexadecanol
191. 5-Butyldihydro-2(3*H*)-furanone
192. *n*-Heneicosane
193. 1-Hexadecene
194. 5-Ethyl-4-methyl-2-phenyl-1,3-dioxane
195. *n*-Heptadecanol
196. 1-Hexanol
197. 5-Ethyl-6-methyl-(*E*)3-hepten-2-one
198. *n*-Heptadecylcyclo-hexane
199. 1-Methoxy-naphthalene
200. 5-Ethyl-6-methyl-2-hep-tanone
201. *n*-Heptanal
202. 1-Methyl-2-(1-methylethyl)-benzene
203. 5-Ethyldihydro-2(3*H*)-furanone
204. *n*-Heptanol
205. 1-Methyl-4-(1-methylethylidene)-cyclohexene
206. 5-*iso*-Propyl-5*H*-furan-2-one
207. *n*-Hexadecane
208. 1-Methylene-1*H*-indene
209. 5-Methy-3-hepten-2-one
210. *n*-Hexadecanoic acid
211. 1-Nitro-hexane
212. 5-Methyl-2-hexanone
213. *n*-Hexanal
214. 1-Nonanol
215. 5-Methyl-pentadecane
216. *n*-Hexanol
217. 1-Octadecene
218. 5-Methyl-tridecane

219. Nitro-ethane
220. 1-Octanol
221. 5-Pentyldihydro-2(3*H*)-furanone
222. *n*-Nonadecane
223. 2-Methyl-5-isopropyl-furan
224. 6,10,13-Trimethyl-tetradecanol
225. *n*-Nonadecanol
226. 1-Pentanol
227. 6,10,14-Trimethyl-2-pentadecanone
228. *n*-Nonanal
229. 1-Tetradecyne
230. 6,10,14-Trimethyl-pentadecanone
231. *n*-Nonanol
232. 1-Tridecene
233. 6,10-Dimethyl-2-undecanone
234. 2-Methyl-5-decanone
235. 1-Undecanol
236. 6,10-Dimethyl-5,9 undecadien-2-one
237. *n*-Octanal
238. 2,2,4-Trimethyl-3-car-boxyisopropyl, isobutyl ester pentanoic acid
239. 6,10-Dimethyl-5,9-un-decandione
240. *n*-Octane
241. 2,2,4-Trimethyl-heptane
242. 6-Dodecanone
243. n-Octanol
244. 2,2,4-Trimethyl-pentane
245. 6-Methyl-2-heptanone
246. Heptacosane
247. 2,2-Dihydroxy-1-phenyl-ethanone

TABLE 12.1 *(Continued)*

Aroma compounds

248. 6-Methyl-3,5-hepta-diene-2-one
249. *n*-Octyl-cyclohexane
250. 2,3,5-Trimethyl-naph-thalene
251. 6-Methyl-5-ene-2-hep-tanone
252. Nonadecane
253. 2,3,6-Trimethyl-naph-thalene
254. 6-Methyl-5-heptanone
255. Nonanal
256. 2,3,6-Trimethyl-pyridine
257. 6-Methyl-5-hepten-2-one
258. Nonane
259. 2,3,7-Trimethyl-octanal
260. 6-Methyl-octadecane
261. Nonanoic acid
262. 2,3-Butandiol
263. 7,9-Di-*tert*-butyl-1-oxaspiro-(4,5) deca-6,9-diene-2,8-dione
264. Nonene
265. 2,3-Dihydrobenzofuran
266. 7-Chloro-4-hydroxy-quinoline
267. Nonnenal
268. 2,3-Dihydroxy-succinic acid
269. 7-Methyl-2-decene
270. Nonyl-cyclohexane
271. 2,3-Octanedione
272. 7-Tetradecene
273. *n*-Pentanol
274. 2,4,4-Trimethylpentan-1,3-diol di-*iso*-butanoate
275. 9-Methyl-nonadecane
276. *n*-Tetradecanol
277. 2,4,6-Trimethyl-decane

278. Acetic acid
279. *n*-Tricosane
280. 2,4,6-Trimethylpyridine (TMP) or collidine
281. Acetic acid tetradecate
282. *n*-Undecanal
283. 2,4,7,9-Tetramethyl-5-decyn-4,7-diol
284. Acetone
285. *n*-Undecanol
286. 2,4-Bis(1,1-dimethylethyl)-phenol
287. Acetonitrile
288. Oct-1-en-3-ol
289. 2,4-Di(*tert*-butyl) phenol
290. Acetophenone
291. Octacosane
292. 2,4-Diene dodecanal
293. Alk-2-enals
294. Octadecane
295. 2,4-Hexadienal
296. Alka-2,4-dienals
297. Octadecyne
298. 2,4-Hexadiene aldehyde
299. Alkanals
300. Octanal
301. 2,4-Nonadienal
302. Alkyl-cyclopentane
303. Octanoic acid
304. 2,4-Pentadiene aldehyde
305. Anizole
306. Octanol
307. 2,4-Pentandione
308. Azulene
309. Octyl formate
310. 2,5,10,14-Tetramethyl-pentadecane
311. Benzaldehyde

312. *O*-decyl hydroxaminutese
313. 2,5,10-Trimethyl-pentadecane
314. Benzene
315. Oxalic acid-cyclohexyl methyl nonate
316. 2,5-Dimethyl-undecane
317. Benzene acetaldehyde
318. Palmitate
319. 2,6,10,14-Tetramethyl-heptadecane
320. Benzene formaldehyde
321. Pentacontanal
322. 2,6,10,14-Tetramethyl-hexadecane
323. Benzothiazole
324. Pentacosane
325. 2,6,10,14-Tetramethyl-pentadecane
326. Benzyl alcohol
327. Pentadecanal
328. 2,6,10-Trimethyl-dodecane
329. Bezaldehyde
330. Pentadecane
331. 2,6,10-Trimethyl-pentadecane
332. Bicyclo[4.2.0] octa-1,3,5-triene
333. Pentadecanoic
334. 2,6,10-Trimethyl-tetra-decane
335. Bis-(I-methylethyl) hexadecanoate
336. Pentadecyl-cyclohexane
337. 2,6-Bis(*t*-butyl)-2,5-cy-clohexadien-I-one
338. Butan-2-one-3-Me
339. Pentanal

TABLE 12.1 *(Continued)*

Aroma compounds

340. 2,6-Bis(*t*-butyl)-2,5-cyclohexadien-1, 4-dione
341. Butanal
342. Pentanoic acid
343. 2,6-Di(*tert*-butyl)-4-methylphenole
344. Butandiol
345. Pentyl hexanoate
346. 2,6-Diisopropyl-naphthalene
347. Butanoic acid
348. Phenol
349. 2,6-Dimethoxy-phenol
350. Butylated hydroxy toluene
351. Phenyl acetic acid-4-tridecate
352. 2,6-Dimethyl-aniline
353. Citral
354. Phenylacetaldehyde
355. 2,6-Dimethyl-decane
356. Cubenol
357. Phthalic acid
358. 2,6-Dimethyl-heptadecane
359. Cyclodecanol
360. Phytol
361. 2,6-Dimethyl-naphthalene
362. Cyclosativene
363. Propane
364. 2,7-Dimethyl-octanol
365. *d*-Dimonene
366. Propiolonitrile
367. 2-Acetyl-1-pyrroline (2-AP)
368. Decanal
369. Propyl acid
370. 2-Acetyl-naphthalene

371. Decane
372. *p*-Xylene
373. 2-Butyl-1,2-azaborolidine
374. Decanoic acid
375. Pyridine
376. 2-Butyl-1-octanol
377. Decanol
378. Pyrolo[3,2-d]pyrimidin-2,4(1*H*,3*H*)-dione
379. 2-Butyl-2-octenal
380. Decyl aldehyde
381. Styrene
382. 2-Butylfuran
383. Decyl benzene
384. Tetracosane
385. 2-Butyl-octanol
386. Diacetyl
387. Tetradecanal
388. 2-Chloro-3-methyl-1-phenyl-1-butanone
389. Dichloro benzene
390. Tetradecane
391. 2-Chloroethyl hexyl ester isophthalic
392. Diethyl phthalate
393. Tetradecanoic acid
394. 2-Decanone
395. Diethyl phthalate
396. Tetradecanol
397. 2-Decen-1-ol
398. Di-*iso*-butyl adipate
399. Tetradec-I-ene
400. 2-Decenal
401. Dimethyl disulphide
402. Tetrahydro-2,2,4,4-tetramethyl furan
403. 2-Dodecanone
404. Dimethyl sulphide

405. Toluene
406. 2-Ethyl-1-decanol
407. Dimethyl trisulphide
408. *trans*-2-octenal
409. 2-Ethyl-1-dodecanol
410. *d*-Limonene
411. *trans*-Caryophylene
412. 2-Ethyl-1-hexanol
413. Docosane
414. Triacontane
415. 2-Ethyl-2-hexenal
416. Dodecanal
417. Tricosane
418. 2-Ethyl-decanol
419. Dodecane
420. Tridecanal
421. 2-Heptadecanone
422. Dotriacontane
423. Tridecane
424. 2-Heptanone
425. Eicosane
426. Trimethylheptane
427. 2-Heptenal
428. Eicosanol
429. Tritetracontane
430. 2-Heptene aldehyde
431. Ethanol
432. Turmerone
433. 2-Heptylfurane
434. Ethenyl cyclohexane
435. Undecanal
436. 2-Hexadecanol
437. Ethyl (*E*)-9-octadecenoate
438. Undecane
439. 2-Hexanone
440. Ethyl benzene
441. Undecanol

TABLE 12.1 *(Continued)*

Aroma compounds		
442. 2-Hexyl-1-decanol	455. Ethyl linoleate	467. Ethyl tetradecanoate
443. Ethyl decanoate	456. α-Cadinol	468. γ-Nonalacton
444. Undecyl-cyclohexane	457. 2-Methoxy-phenol	469. 2-Methyl-2,4-diphenyl-pentane
445. 2-Hexyl-1-octanol	458. Ethyl nonanoate	470. Ethylbenzene
446. Ethyl dodecanoate	459. α-Terpineol	471. γ-Nonalactone
447. Vanillin	460. 2-Methyl-butanol	472. 2-Methyl-3-furanthiol (2-MF)
448. 2-Hexyl-decanol	461. Ethyl octanoate	
449. Ethyl hexadecanoate	462. β-Bisabolene	473. Farnesol
450. Z-10-pentadecen-1-ol	463. 2-Methyl-1,3-pentane-diol	474. γ-Terpineol
451. 2-Hexylfurane		475. 2-Methyl-3-octanone
452. Ethyl hexanoate	464. Ethyl oleate	476. Formic acid hexate
453. α-Cadinene	465. β-Terpineol	477. δ-Cadinene
454. 2-Methoxy-4-vinylphenol	466. 2-Methyl-1-hexadecanol	

12.2 IMPORTANCE AND CLASSIFICATION OF RICE VOLATILE AROMAS

12.2.1 IMPORTANCE OF RICE VOLATILE AROMAS

Aroma in rice grains is contributed by several volatile aroma compounds, synthesized due to distinct biochemical pathways. More than 200 rice cultivars are recognized containing numerous volatile aroma compounds.[63] These volatile aroma compounds include a series of compounds like: aldehydes, ketones, organic acids, alcohols, esters, hydrocarbons, phenols, pyrazines, pyridines, and other compounds. Apart of these, some other compounds are also present, but they are extracted by the breakdown chemicals like fatty acids which are presented in samples.[6] The volatile aroma compounds released after cooking are different from those released in field at the time of flowering.

Aroma is considered as world's third highest desired trait as well as in India followed by taste and elongation after cooking.[8,98,99,100] Several types of aroma are reported in cooked rice; for instance, bland-like, bran-like, brown rice, burned-like, burnt-like, buttery-like, cold-steam-bread-like, corn-like, corn-leaf-like, cracker-like, dusty-like, earthy-like, fermented-sour-like, floral-like, gasoline aroma-like, grainy-like, hot-steam-bread-like, musty-like, nut-like, paint-like, pandan-like, pear-barley, plastic-like, popcorn-like, potato-like, rancid-like, raw-dough-like, rice milk-like, smoky-like, spicy-like, sulfur-like, tortilla-like, vegetable-like, and white glue-like.[19,68,105,110] Popcorn aroma, which sometimes described as pandan (*Pandanus amaryllifolius*) like aroma,[48] is due to presence of 2-AP in rice. The 2-AP is most important and prominent aroma chemical since it was used to identify in rice cultivars.[73] Yajima et al.[108] identified α-pyrrolidone & indole. The former is found as key odorant during the study of volatile flavor components in cooked Kaorimai (scented rice, *O. sativa japonica*) Buttery et al.[16] identified seven compounds, namely, (*E*)-2-decenal, (*E*)-2-nonenal, (*E, E*)-2,4-decadienal, 2-AP, decanal, nonanal, and octanal with low-odor thresholds from listed 64 aroma volatile chemical compounds from rice. Later, the chemical compound 2-AP was considered as the major odorant contributor of popcorn like rice aroma.[14,73] Although this odorant contributor in rice is found at different concentration ranged from minimum 1–10 ppb to maximum of 2 ppm level.[16] Widjaja et al.[104] made comparative study of the products of lipid oxidation and aroma volatile compounds, namely, (*E*)-2-decenal, (*E,E*)-2,4-nonadienal, and (*E, E*)-2,4-decadienal from fragrant and nonfragrant rice and reported that these aroma chemical compounds are responsible for aroma in waxy rice. These three aroma compounds are also reported by Grimm et al.[35] for contributing its distinctive aroma in glutinous or waxy rice.

12.2.2 CLASSIFICATION OF RICE VOLATILE AROMAS

Majority of rice volatile aroma chemicals have popcorn-like smell due to 2-AP content. Aroma, similar like taste and elongation after cooking of rice, is considered as desirable quality for consumer acceptability.[6,8] Rice aromas were classified into five majors groups as given in Table 12.2. Although, any single chemical compound is not responsible for any specific aroma in cooked rice, but a mixture of compounds in the fix proportions are responsible specific aromas in cooked rice. It has been shown very clearly in Table 12.2. For instance, fresh green or woody smell were contributed mainly by aldehydes, ketones and some alcohol; fruity and floral smell comes due to presence of heptanone, ketones, and 6-methyl-5-hepten-2-one while benzaldehyde and 2-pentyfuran provided nutty aromas.[103]

TABLE 12.2 Classification of Rice Volatile Aromas.

Type of rice aroma	Responsible chemical compounds
Bitter	Benzaldehyde and pyridine
Fruity/Floral	(E)-2,(E)-4-hexadienal, (E)-3-octen-2-onea, 2-hexanone, 2-nonanone, 2-undecanone, 6-methyl-5-hepten-2-one, Heptanone, Ketones, Methyl heptanoate, n-heptanal, n-nonanal and n-octanol
Green	Aldehydes, some alcohol and Ketones, (E)-2,(E)-4-hexadienal, (E)-2-hexenal, (E)-2-octenal, (E)-3-octen-2-onea, 2-heptanone, 2-methyl-2-pentanol, benzaldehyde, decanal, geranyl acetone, methyl heptanoate and n-hexanal
Nutty	(E,E)-2,4-nonadienal, 2-pentyfuran, 4-vinyl guaiacol and Benzaldehyde
Roasty	2,3-Octanedione

12.3 TECHNOLOGY FOR EXTRACTION OF RICE VOLATILE AROMA COMPOUNDS

It is important to assess the aroma compounds for the extraction of volatile aroma compounds in rice by extraction technology. There are no single technologies that will prove optimal for aroma extraction in rice. Several traditional and modern methods for rice aroma chemicals extraction (Table 12.3) coupled with analytical techniques were studied and have been used for extraction and quantification of rice aroma compounds at different levels of concentration ranges from 1 to 10 ppb level to 2 ppm.

TABLE 12.3 Traditional and Modern Extraction Technology for Rice Volatile Aroma Compounds.

Traditional technology	Modern technology
Purge and trap	Solid phase microextraction
Simultaneous steam distillation	Supercritical fluid extraction with CO_2
Micro steam distillation	
Direct solvent extraction	

These extraction technologies made a cocktail of over 450 compounds—alcohols, aldehydes, esters, heterocyclic, hydrocarbons, ketones, and organic acids (Table 12.1) from various aromatic and nonaromatic cultivars of rice.[52,103] In addition of all these rice volatile aroma compounds, 2-AP is a particular chemical which has been shown to be the predominantly active volatile aroma compound in aromatic rice at several hundred parts per billion (ppb) level of concentrations that emits a popcorn-like aroma. The presence of this particular rice volatile aroma compound was reported by Yoshihashi[112] in various parts of the rice plant except roots and is also reported in very trace amounts in nonaromatic rice.[13,34,65] Zhou et al.[119] reviewed the methods, namely, direct extraction, distillation, and headspace for volatile aroma compounds analysis and their contribution to flavor in cereals in which headspace complemented by solid phase microextraction (SFME), and it has become one of the best isolation and extraction methods. The typical concentration levels (in ppb) of different volatile aroma compounds from rice were alcohols 1869, aldehydes 5952, disulphides 79, heterocyclic compounds 1220, hydrocarbons 548, ketones 234, phenols 534, and terpenes 257. The discussed methods have been employed for extraction of aroma volatiles that affect considerably on the quality and quantity of rice[84,85] Weber et al.[101] reported that one of the major limitations is the lack of a quantitative assay in improvement of aromatic rice for aroma through breeding because there is no method to quantify relative concentration of different compounds or activity of some critical enzyme. Many researchers worked on best analytical technique for extraction and isolation of rice volatile aroma compounds. In order to get higher extraction yield (referred as high efficiency) and potency of the extract (referred as efficacy), researchers have made considerable effort to find efficient methods for extraction of volatile compounds for rice aroma.

Currently, simultaneous distillation extraction (SDE), SFME, and supercritical fluid extraction (SFE) are most important and common procedures coupled with GC–MS reported in the literature on aroma compounds. The impact of chemical character in rice, to be identified by the classical techniques of the 21st century. The present chapter addressed and discussed in details to review the extraction approaches for rice aroma compounds; principle and instrumentation behind operation of extraction methods; factors influencing method performance; and case study on research progress, strength, and weakness of rice aroma research. There are number of references that exist on this topic entitled *Extraction Technology for Rice Volatile Aroma Compounds* which can provide different perspectives.

12.3.1 SIMULTANEOUS DISTILLATION EXTRACTION TECHNIQUES

SDE is one of the most important extraction methods and more popular among the all known methods for rice aroma chemical analysis. This valuable extraction method is also known as Likens–Nickerson steam distillation,[82] was first reported in 1964 by Likens and Nickerson[50] and described in 1966.[62] Later, the SDE method was modified by Maarse and Kepner[54] and Schultz et al.[80] This method simplifies the experimental procedures by merging vapor distillation and solvent extraction shown in Figure 12.1.[17] In SDE method, the final product is a solvent extract.[9,66]. SDE method has been considered as one of the most cited methods till date that is used for the extraction and isolation of volatile aroma chemical components.[17,74,75]. Previously, many researchers has successfully applied SDE method for the extraction and widely used to analyze various rice volatile aroma chemicals from different cultivars of rice depicted in Table 12.4.[7,13,15,48,65,73,89,92,103,104,106]

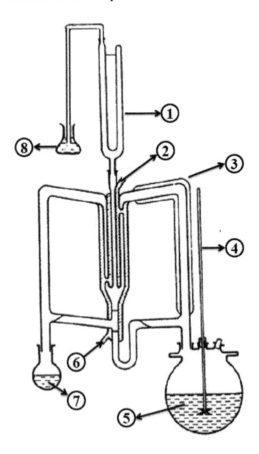

FIGURE 12.1 Simultaneous distillation extraction apparatus by Sunthonvit et al.[89]: (1) carbon dioxide (CO_2)–acetone condenser, (2) water out, (3) insulation, (4) stirrer, (5) solution of rice & water, (6) water in, (7) solvent, and (8) water bubbler.

TABLE 12.4 List of Extracted Major Rice Volatile Aroma Compounds Using SDE Techniques.

Aroma compounds		
(E)-2,(E)-4-Decadienal^{d,e}	2-Pentadecanone^{e,f}	Indole^{e,g}
(E)-2,(E)-4-heptadienal^e	2-Pentylfuran^{d,e}	1-Octen-3-ol^{d,e}
(E)-2,(E)-C heptadienal^e	2-Phenylethanol^{d,e}	Methyl heptanoate^e
(E)-2,(E)-C nonadienal^e	2-Tridecanone^e	n-Decanal^e
(E)-2-decenal^{e,j}	2-Undecanone^e	n-Heptanol^{e,j}
(E)-2-heptenal^{d,f}	4-Vinylguaiacol^{d,e,g}	n-Hexanal^{d,e,g}
(E)-2-hexenal^e	4-Vinylphenol^{d,e}	n-Nonanal^{d,e,j}
(E)-2-nonenal^{e,g}	6-Methyl-5-hepten-2-one^e	n-Nonanol^e
(E)-2-octenal^{d,e}	Acetophenone^e	n-Octanal^{e,f,j}
(E)-Hept-2-enal^f	Alk-2-enals^d	n-Octanol^{e,j}
2,4,6-Trimethylpyridine (TMP) or Collidine^e	Alka-2,4-dienals^d	n-Pentanol^{e,f}
2-Acetyl-1-pyrroline (2-AP)^{a,b,c,d,e,f,g,h,i}	Alkanals^d	n-Undecanal^e
2-Heptanone^{e,j}	Benzaldehyde^{e,f,g}	Oct-1-en-3-ol^g
2-Hexanone^e	Butan-2-one-3-me^j	Pentanal^{g,j}
2-Methyl-3-furanthiol (2-MF)^i	Decanol^j	Phenylacetaldehyde^e
2-Nonanone^e	Hexadecanol^f	Pyridine^e
2-Octanone^j	Hexanol^{f,g}	Undecane^j

Sources:

[a]Buttery et al.[13]

[b]Buttery et al.[15]

[c]Laksanalamai and Ilangantileke[48]

[d]Widjaja et al.[103]

[e]Widjaja et al.[104]

[f]Petrov et al.[73]

[g]Tava and Bocchi[92]

[h]Sunthonvit et al.[89]

[i]Park et al.[65]

[j]Bhattacharjee et al.[7]

12.3.1.1 MERITS AND DEMERITS OF SDE TECHNIQUE

12.3.1.1 Merits

1. SDE technique involves simple apparatus in one-step extraction operation that can be handled to the range of samples for extraction of the volatile compounds and concentrates the compounds, further enables obtained extracts for trace component detection that are free from nonvolatile materials (nonvolatiles compounds not present) as volatiles are separated from nonvolatiles by the process of distillation.[17]
2. This technique is quite safe because GC liners and columns cannot be decontaminated as there is absence of nonvolatile or high boiling materials during extraction.
3. It can be performed under reduced pressure at room temperature (25°C) in laboratory.[55]
4. This technique consumes less time and a little volume of solvent (due to the continuous recycling) to allow the extraction of aroma compounds.[17] Thus, extraction of volatile compounds can be performed without further concentrating solvents.
5. The extraction technique can be achieved to the higher concentration recoveries of aroma compounds under certain conditions.
6. SDE technique reduces the problems of artifact and breakdown products are reduced or generate fewer products that build-up as solvents and further concentrates.
7. SDE technique has high extraction efficiency that is always associated with the high reproducibility. Due to its efficiency, this technique has been employed for quantification of aroma compounds in various aspects matrices.[17]

12.3.1.2 Demerits

1. SDE is not best extraction technique as it concludes comparatively lower quantitative extracts than solvent extraction methods.
2. The artifact formation in SDE technique is maximum possible because of the thermal degradation of the rice sample.
3. The extraction of acids and alcohols are very poor as they are not visible in final extract as they are hydrophilic/polar in nature.

4. This technique may introduce silicone contaminants due to use of antifoaming agents for avoiding foaming.
5. SDE techniques is not suitable for fresh rice samples, as the extraction of aroma compounds is more akin to a cooked rice (previously thermally processed).
6. Major difficulties in SDE technique are to balance two boiling flasks, constant pressure, and minimize evaporation of boiling solvent.

12.3.1.2   CASE STUDY ON SDE OF RICE AROMA VOLATILE COMPOUNDS

There are considerably huge numbers of different rice cultivars are grown around the world that possess aroma. The knowledge of the volatile aroma components of rice is finding out by number of researches using SDE. In 1983, steam distillation continuous extraction method was used by Buttery et al.[13] to determine 2-AP from 10 different rice varieties. The determined levels of concentration in the steam volatile oils of rice varieties were based on the dry weight and varied from less than 0.006–0.09 ppm for Calrose and Malagkit Sungsong rice variety, respectively. Same method was again used with an acid-phase solvent extraction in 1986 by Buttery et al.[15] for extraction and quantification of a potent rice volatile compound, 2-AP from 200 g of bland rice sample. The aim of the study was development of a simple and practical method for extraction and isolation of rice volatile aroma chemicals and findings showed that bland rice sample with this developed method contain known added concentration level of 2-AP and found best suited method for the purpose of quantitative analysis of 2-AP volatile aroma chemical from rice sample. A comparative study was made by Laksanalamai and Ilangantileke[48] using a SDE method to identify the aroma chemical compound 2-AP in fresh and aged (shelf-stored) Khao Dawk Mali (KDML) 105 (a well-accepted Thai aromatic rice variety), in nonaromatic rice and in pandan leaves.

In 1996, Widjaja et al.[103] conducted a comparative study on volatile compounds of fragrant and nonfragrant rice and reported (E)-2,(E)-4-decadienal, (E)-2-octenal, (E)-Z heptenal, 2-AP, 2-pentylfuran, 2-pentylfuran, 2-phenylethanol, 2-phenylethanol, 4-vinylguaiacol, 4-vinylphenol, alk-2-enals, alka-2,4-dienals, alkanals, 1-octen-3-ol, n-hexanal, and n-nonanal were most important volatile compounds out of total 70 identified, but other identified volatile chemical components were responsible for contributing total rice aroma profile. In the same year, Widjaja et al.[104] conducted another study in storage condition to know the variation in the volatile chemical compounds and reported increase in the level of total volatile aroma chemical components in all three forms of rice namely paddy, brown, and white fragrant rice. Petrov et al.[73] isolated 100 compounds by simultaneous steam distillation-solvent extraction method from scented rice and non-scented rice in cooking water among them, 78 were identified including (*E*)-hept-2-enal, 2-AP, 6,10,14-trimethyl-pentadecan-2-one, benzaldehyde, hexadecanol, hexanol, octanal, pentadecan-2-one, and pentanol.

In commercial basmati and Italian rice, Tava and Bocchi[92] identified and quantified volatiles compounds which were collected by steam distillation as alcohols, aldehydes, disulfides, heterocyclic, hydrocarbons, ketones, phenols, and terpenes in rice samples. The compound present most abundantly was identified as hexanal in both commercial samples, followed by 2-AP, 4-vinylguaiacol, benzaldehyde, hexanol, indole, oct-1-en-3-ol, pentanal, and *trans*-2-nonenal. The presence of 2-AP was reported at 570 and 2350 ppb level of concentration in Basmati and B5-3 rice sample, respectively. Bhattacharjee et al.[7] conducted a comparative study of aroma profile from commercial brand of Basmati rice using two extraction techniques (viz., Likens–Nickerson and SFE) to find out the best method. Likens–Nickerson were reported best for smaller sample size, shorter time of extraction and negligible possibility of artifacts compare *vis-n-vis* with the SFE technique, because of its superiority for recovery of volatiles chemical constitutes from rice.

Sunthonvit et al.[89] used SDE combined with distillation–extraction apparatus (modified Likens and Nickerson apparatus) for extraction of volatiles from Thai fragrant rice varieties and reported total of 94 volatile compounds in which 17 acids, 21 alcohols, 19 aldehydes, 9 heterocyclic compounds, 8 hydrocarbons, 14 ketones, and 6 miscellaneous compounds. 2-AP was found as major volatile aroma chemical component. Park et al.[65] studied on potent aroma chemical component of cooked Korean nonaromatic rice using SDE techniques and characterized total 16 aroma volatiles chemicals with log3 flavor dilution >1. The detected volatile aroma chemical components of rice were 2-MF and 2-AP that were considered as potent active aroma chemical components. The volatile aroma chemical, 2-MF was first-time reported for its aroma potential in cooked Korean nonaromatic rice.

12.3.1.3 SUMMARY

Presently, SDE involves steam distillation followed by solvent extraction. It is one of the most popularly used extraction methods. It has always played a key role for solvent extraction. This method has two important features. First is ability to exploit the differences between volatile compounds and nonvolatile compounds present in the rice matrix. SDE technique has been selected to improve extraction of rice volatile aroma compounds as well as to limit or exclude the non-volatile components that have no importance and only causes hindrance during analysis. Second is time duration for extraction that is strong parameter for analysis. However, generally rice extraction consumes 1–4 h.[7,17] Due to solvent nature and extraction time, the extraction yield is always best in SDE techniques. This is the main reason that researchers are trying to find out the best suitable experimental conditions to recover the selected volatile aroma components of rice.

12.3.2 SOLID PHASE MICROEXTRACTION

SPME was introduced in 1990s[109] and was developed by Arthur and Pawliszyn[4,69,70] as a solvent-free extraction technique which can be used in both in laboratory and on-site. SPME has been used to analyze volatiles due to some advantages, like comparatively simple, rapid, solvent-free, less expensive, and consumes less time for extraction when compared with other techniques like SDE.[25,40,44,57,118]

SPME is ideal extraction method to characterize the unknown mixtures of volatile aroma organic chemical compounds.[1] It is a relatively and rapidly emerging as a new robust solvent free, simple, rapid low cost, easy to operate, sensitive, and reliable technique among the all.[24,25,43,119] SPME overcome the difficulties of extraction and used in detection of flavoring volatile chemical compounds[118] from various type of sample matrices by dividing them into a polymeric liquid coating (i.e., an immobilized stationary phase) from a gaseous or liquid sample matrix.[88]

SPME techniques include very simple setup, and there is no requirement of any additional instrumentation other than a GC–MS with injection port shown in Figure 12.2.[38,88] This device is a combination of two steps that deals separately. First step is related to partitioning the analytes of desired target between the matrix of sample and the fiber surface, while other step is related to the direct desorption of absorbed analytes into injection port of

GC (Fig. 12.2A).[24] Now-a-days, many researchers consider it as one of the most brilliant inventions as it is simple, efficient and eco-friendly method and widely used in sample preparation for analysis of rice volatile aroma compounds.[5,47,58,72,96] Total identified volatile aroma compounds from rice sample till date are given in Table 12.5.

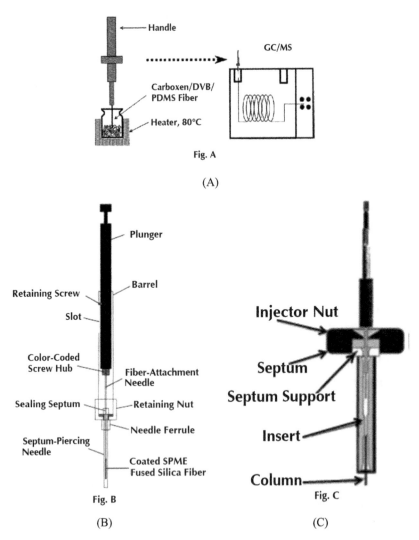

FIGURE 12.2   Solid phase microextraction from Pawliszyn[71]: (A) Block diagram of SPME analysis, (B) commercial SPME device, and (C) SPME–GC interface.

TABLE 12.5 List of Extracted Major Rice Volatile Aroma Compounds Using SPME Techniques.

Aroma compounds		
(1S,Z)-Calamenene[j]	2-Methyl-5-decanone[j]	Farnesol[f,l]
(2-Aziridinylethyl) aminutese[c]	2-Methyl-5-isopropyl-furan[g]	Formic acid hexate[g]
(2E)-dodecenal[f]	2-Methyl-butanal[e]	Geranyl acetone[f,i,j,k,l]
(E)-3-Octene-2-one[e]	2-Methyl-dodecane[g]	Glycerin[l]
(E) 6,10-dimethyl-5,9 undecadien-2-one[e]	2-Methyl-furan[g]	Heptacosane[g]
(E)-14-hexadecenal[c]	2-Methyl-hexadecanal[a]	Heptadecane[a,c,e,f,g,k,l]
(E)-2-decenal[e,f,i,j,k,l]	2-Methyl-I-propenyl benzene[e]	Heptadecyl-cyclohexane[g]
(E)-2-heptanal[b,e]	2-Methyl-naphthalene[f]	Heptanal[a,c,d,g,h,i,j,k,l]
(E)-2-heptenal[f,j,k,l]	2-Methyl-tridecane[f]	Heptane[e]
(E)-2-hexenal[d,e,k,l]	2-Methyl-undecanal[f]	Heptanol[h]
(E)-2-nonen-1-ol[a]	2-Methyl-undecanol[g]	Hepten-2-ol<6-methyl-5->[f]
(E)-2-nonenal[a,e,f,i,j,k,l]	2-n-butyl furan[c]	Heptene-dione[g]
(E)-2-octen-1-ol[k,l]	2-Nonanone[c,d,g,k,l]	Heptylcyclohexane[a]
(E)-2-octenal[f,h,j,k,l]	2-Nonenal[h]	Hexacosane[c,f,g]
(E)-2-undecenal[f,j,k,l]	2-Ocetenal (E)[a]	Hexadecanal[g,k,l]
(E)-3-nonen-2-one[k,l]	2-Octanone[g,l]	Hexadecane 2,6-bis-(t-butyl)-2,5[e]
(E)-4-nonenale[k,l]	2-Octenal[e,g]	Hexadecanoic acid[d,e,j,k,l]
(E,E)-2,4-heptadienal[f,k,l]	2-Pentadecanone[d,f,g,j,k,l]	Hexadecyl ester, 2,6-difluoro-3-methyl benzoic acid[a]
(E,E)-2,4-nonadienal[e,f,i,k,l]	2-Pentylfuran[a,b,c,e,f,g,h,i,j,k,l]	Hexane[e,h]
(E,E)-2,4-octadienal[f]	2-Phenyl-2-methyl-aziridine[f]	Hexanoic acid[b,e,k,l]
(E,E)-farnesyl acetone[k,l]	2-Propanol[d]	Hexanoic acid dodecate[g]
(E,Z)-2,4-decadienal[f,j,k,l]	2-Tetradecanone[d]	Hexanol[a,h]
(S)-2-methyl-1-dodecanol[d]	2-Tridecanone[g,j,k,l]	Hexatriacontane[a,g]
(Z)-2-octen-1-ol[k,l]	2-Undecanone[e,g,j,k,l]	Hexyl furan[f]
(Z)-3-Hexenal[e]	3-(t-Butyl)-phenol[e]	Hexyl hexanoate[k]
[(1-methylethyl)thio] cyclohexane[a]	3,5,23-Trimethyl-tetracosane[g]	Hexylpentadecyl ester-sulphurous acid[a]
<6-methyl->Heptan-2-ol[f]	3,5,24-Trimethyl-tetracontane[a]	Indane[e]
<cis-2-tert-butyl->Cyclohexanol acetate[f]	3,5-Dimethyl-1-hexene[a]	Indene,2,3-dihydro-1,1,3 trimethyl-3-phenyl[e]
1-(1H-pyrrol-2-yl)-ethanone[a]	3,5-Dimethyl-octane[g]	Indole[a,e,f,g,j,k,l]
1,2-Bis,cyclobutane[h]	3,5-Di-tert-butyl-4-hydroxy-benzaldehyde[a]	I-S,cis-calamene lonol 2[e]

TABLE 12.5 *(Continued)*

Aroma compounds		
1,1-dimethyl-2-octyl-cyclobutane[g]	3,5-heptadiene-2-one[g]	Isobutyl hexadecyl ester oxalic acid[a]
1,2,3,4-Tetramethyl benzene[e]	3,5-Octadien-2-ol[c]	Isobutyl nonyl ester oxalic acid[a]
1,2,3-Trimethyl-benzene[g]	3,7,11,15-Tetramethyl-2-hexadecimalen-1-ol[g]	Isobutyl salicylate[f]
1,2,4-Trimethyl-benzene[c,e,f]	3,7,11-Trimethyl-1-dodecanol[g]	Isolongifolene[e]
1,2-Dichloro-benzene[h]	3,7-Dimethyl-hexanoic acid[g]	Isopropylmyristate[d]
1,2-Dimethyl-benzene[e]	3-Dimethyl-2-(4-chlorphenyl)-thioacrylamide[a]	Limonene[f]
1,3,5-Trimethyl-benzene[h,f]	3-Ethyl-2-methyl-heptane[g]	Lirnonene[e]
1,3-Diethyl-benzene[c]	3-Hexanone[e]	l-Pentanol[e]
1,3-Dimethoxy-benzene[f]	3-Hydroxy-2,4,4-trimethyl-pentyl *iso*-butanoate[l]	Methoxy-phenyl-oxime[a,g]
1,3-Dimethyl-benzene[c,f]	3-Methyl butanal[e,h]	Methyl (*E*)-9-octadecenoate[j,k,l]
1,3-Dimethyl-naphthalene[f]	3-Methyl pentane[h]	Methyl (*Z,Z*)-11,14-eicosadienoate[j,k,l]
1,4-Dimethyl-benzene[e]	3-Methyl-1-butanol[d,e]	Methyl (*Z,Z*)-9,12-octadecadienoate[j,k,l]
1,4-Dimethyl-naphthalene[f]	3-Methyl-2-butyraldehyde[g]	Methyl butanoate[e]
10-Heptadecene-1-ol[g]	3-Methyl-2-heptyl acetate[e]	Methyl decanoate[e]
10-Pentadecene-1-ol[g]	3-Methyl-hexadecane[g]	Methyl dodecanoate[e]
17-Hexadecyl-tetratriacontane[g]	3-Methyl-pentadecane[f]	Methyl ester tridecanoic acid[c]
17-Pentatricontene[a]	3-Methyl-tetradecane[f]	Methyl hexadecanoate[e,j,k,l]
1-Chloro-3,5-bis(1,1-dimethylethyl)2-(2-propenyloxy) benzene[a]	3-Nonene-2-one[g]	Methyl isocyanide[c]
1-Chloro-3-methyl butane[a]	3-Octdiene-2-one[g]	Methyl isoeugenol[f]
1-Chloro-nonadecane[a]	3-Octene-2-one[g]	Methyl linolate[e]
1-Decanol[l]	3-Tridecene[a]	Methyl octanoate[e]
1-Docosene[a]	4,6-Dimethyl-dodecane[c]	Methyl oleate[e]
1-Dodecanol[j,k,l]	4-[2-(Methylaminuteso) ethyl]-phenol[c]	Methyl pentanoate[e]
1-Ethyl-2-methyl-benzene[g]	4-Acetoxypentadecane[a]	Methyl tetradecanoate[e]
1-Ethyl-4-methyl-benzene[e]	4-Cyclohexyl-dodecane[a]	Methyl-naphthalene[g]
1-Ethyl-naphthalene[f]	4-Ethyl-3,4-dimethyl-cyclohexanone[f]	Myrcene[c]

TABLE 12.5 *(Continued)*

Aroma compounds		
1-Heptanol[d,e,f,k,l]	4-Ethylbenzaldehyde[d]	*N,N*-dimethyl chloestan-7-aminutese[a]
1-Hexacosene[a]	4-Hydroxy-4-methyl-2-pentanone[g]	*N,N*-dinonyl-2-phenylthio ethylaminutese[a]
1-Hexadecanol[a,j,k,l]	4-Methyl benzene formaldehyde[g]	Naphthalene[a,b,e,f]
1-Hexadecene[e]	4-Methyl-2-pentyne[a]	*n*-Decane[f]
1-Hexanol[d,e,j,k,l]	4-Vinyl-guaiacol[i]	*n*-Dodecanol[g]
1-Methoxy-naphthalene[f]	4-Vinylphenol[i,k,l]	*n*-Eicosanol[f,g]
1-Methyl-2-(1-methylethyl)-benzene[c,e]	5-Aminuteso-3,6-dihydro-3-iminuteso-1(2*H*) pyrazine acetonitrile[a]	*n*-Heneicosane[f]
1-Methyl-4-(1-methylethylidene)-cyclohexene[c]	5-Butyldihydro-2(3*H*)-furanone[e]	*n*-Heptadecanol[g]
1-Methylene-1*H*-indene[c]	5-Ethyl-4-methyl-2-phenyl-1,3-dioxane[a]	*n*-Heptadecylcyclohexane[a]
1-Nitro-hexane[c,f]	5-Ethyl-6-methyl-(*E*)3-hepten-2-one[e]	*n*-Heptanal[f]
1-Nonanol[d,f,k,l]	5-Ethyl-6-methyl-2-hep-tanone[g]	*n*-Heptanol[g]
1-Octadecene[f]	5-Ethyldihydro-2(3*H*)-fura-none[e]	*n*-Hexadecane[g]
1-Octanol[d,e,f,i,j,k,l]	5-*iso*-Propyl-5*H*-furan-2-one[k,l]	*n*-Hexadecanoic acid[c]
1-Octene-3-ol or 1-octen-3-ol[a,c,f,g,i,j,k,l]	5-Methy-3-hepten-2-one[a]	*n*-Hexanal[a,b,c,d,e,f,g,h,i,j,k,l]
1-Pentanol[d,h,k,l]	5-Methyl-2-hexanone[a]	*n*-Hexanol[g,f]
1-Tetradecyne[a]	5-Methyl-pentadecane[g]	Nitro-ethane[c]
1-Tridecene[e]	5-Methyl-tridecane[g]	*n*-Nonadecane[f]
1-Undecanol[l]	5-Pentyldihydro-2(3*H*)-furanone[e]	*n*-Nonadecanol[a]
2,2,4-Trimethyl-3-carboxyisopropyl, isobutyl ester pentanoic acid[a]	5-Propyldihydro-2(3*H*)-furanone[e]	*n*-Nonanal[b,c,d,e,f,g,h,i,k,l]
2,2,4-Trimethyl-heptane[e]	6,10,13-Trimethyl-tetradecanol[g]	*n*-Nonanol[c,e,g]
2,2,4-Trimethyl-pentane[c]	6,10,14-Trimethyl-2-pentadecanone[e,g,j,k,l]	*n*-Octanal[a,c,d,f,g,h,i,j,k,l]
2,2-Dihydroxy-1-phenyl-ethanone[a]	6,10,14-Trimethyl-pentadecanone[d]	*n*-Octane[e,f]

TABLE 12.5   *(Continued)*

Aroma compounds		
2,3,5-Trimethyl-naphthalene[f]	6,10-Dimethyl-2-undecanone[g,k,l]	*n*-Octanol[f,g]
2,3,6-Trimethyl-naphthalene[f]	6,10-Dimethyl-5,9 undecadien-2-one[a]	*n*-Octyl-cyclohexane[f]
2,3,6-Trimethyl-pyridine[e]	6,10-Dimethyl-5,9-undecandione[g]	Nonadecane[a,e,f,g]
2,3,7-Trimethyl-octanal[g]	6-Dodecanone[k,l]	Nonane[g]
2,3-Butandiol[e]	6-Methyl-2-heptanone[g]	Nonanoic acid[j,k,l]
2,3-Dihydrobenzofuran[g]	6-Methyl-3,5-heptadiene-2-one[g,j,k,l]	Nonene[a]
2,3-Dihydroxy-succinic acid[g]	6-Methyl-5-ene-2-heptanone[g]	Nonnenal[g]
2,3-octanedione[g,l]	6-Methyl-5-heptanone[g]	Nonyl-cyclohexane[g]
2,4,4-Trimethylpentan-1,3-diol di-*iso*-butanoate[l]	6-Methyl-5-hepten-2-one[a,e,j,k,l]	*n*-Tetradecanol[f]
2,4,6-trimethyl-decane[g]	6-Methyl-octadecane[a]	*n*-Tricosane[f]
2,4,6-Trimethylpyridine (TMP) or collidine[e,m]	7,9-Di-*tert*-butyl-1-oxaspiro-(4,5) deca-6,9-diene-2,8-dione[a]	*n*-Undecanol[f]
2,4,7,9-Tetramethyl-5-decyn-4,7-diol[a]	7-Chloro-4-hydroxyquinoline[a]	Octacosane[g]
2,4-Bis(1,1-dimethylethyl)-phenol[a]	7-Methyl-2-decene[l]	Octadecane[c,e,f,g]
2,4-Di(*tert*butyl) phenol[j,k,l]	7-Tetradecene[a]	Octadecyne[a]
2,4-Diene dodecanal[g]	9-Methyl-nonadecane[f]	Octanal[e]
2,4-Hexadienal[h]	Acetic acid[g]	Octanoic acid[k,l]
2,4-Hexadiene aldehyde[g]	Acetic acid tetradecate[g]	Octyl formate[f]
2,4-Nonadienal[g]	Acetone[b,e,g]	*O*-decyl hydroxaminutese[a]
2,4-Pentadiene aldehyde[g]	Acetonitrile[c]	Oxalic acid-cyclohexyl methyl nonate[g]
2,4-Pentandione[e]	Acetophenone[c]	Palmitate[g]
2,5,10,14-Tetramethyl-pentadecane[e]	Alkyl-cyclopentane[g]	Pentacontonal[a]
2,5,10-Trimethyl-pentadecane[e]	Anizole[f]	Pentacosane[f,g]
2,5-Dimethyl-undecane[g]	Azulene[c]	Pentadecanal[a]
2,6,10,14-Tetramethyl-heptadecane[g]	Benzaldehyde[b,c,d,e,f,j,k,l]	Pentadecane[c,e,f,g,k,l]
2,6,10,14-Tetramethyl-hexadecane[e,f,g]	Benzene acetaldehyde[e,g]	Pentadecanoic[c]

TABLE 12.5 *(Continued)*

Aroma compounds		
2,6,10,14-tetramethyl-pentadecane[f]	Benzene[c]	Pentadecyl-cyclohexane[g]
2,6,10-Trimethyl-dodecane[c,g]	Benzene formaldehyde[g]	Pentanal[a,e,g,h,j,k,l]
2,6,10-Trimethyl-pentadecane[c,f,g]	Benzothiazole[a,d,e]	Pentanoic acid [e]
2,6,10-Trimethyl-tetradecane[g]	Benzyl alcohol[e]	Pentyl hexanoate[k]
2,6-Bis(*t*-butyl)-2,5 cyclohexadien-I-one[e]	Bezaldehyde[h]	Phenol[l]
2,6-Bis(t-butyl)-2,5-cyclohexadien-1,4-dione[e]	Bicyclo[4.2.0] octa-1,3,5-triene[c]	Phenyl acetic acid-4-tridecate[g]
2,6-Di(*tert*-butyl)-4-methylphenole[j,k,l]	Bis-(I-methylethyl) hexadecanoate [e]	Phthalic acid[f]
2,6-Diisopropyl-naphthalene [e]	Butanal[e]	Phytol[g]
2,6-dimethoxy-phenol[f]	Butandiol[e]	Propane[g]
2,6-dimethyl-aniline[f]	Butanoic acid[e]	Propiolonitrile[a]
2,6-dimethyl-decane[g]	Butylated hydroxy toluene[a,e,f,h]	Propyl acid[g]
2,6-dimethyl-heptadecane[g]	Citral[g]	*p*-xylene[c]
2,6-dimethyl-naphthalene [c,f]	Cubenol[l]	Pyridine [e]
2,7-dimethyl-octanol[h]	Cyclodecanol[a]	Pyrolo[3,2-d] pyrimidin-2,4(1*H*,3*H*)-dione[a]
2-acetyl-naphthalene[f]	Cyclosativene[f]	Styrene[c,e]
2-AP[a,c,d,e,f,i,m]	*d*-Dimonene[d]	Tetracosane[f,g]
2-Butyl-1,2-azaborolidine[a]	Decanal[c,d,e,f,h,i,j,k,l]	Tetradecanal[g]
2-Butyl-1-octanol[a,d,g]	Decane[e,g,j,l]	Tetradecane[b,c,e,f,g,h]
2-Butyl-2-octenal[b,e,g]	Decanoic acid[l]	Tetradecanoic acid[d,j,k,l]
2-Butylfuran [e,f,k,l]	Decyl aldehyde[g]	Tetradecanol[g]
2-Butyl-octanol[h]	Decyl benzene[a]	Tetradec-I-ene[e]
2-Chloro-3-methyl-1-phenyl-1-butanone[a]	Diacetyl[c]	Tetrahydro-2,2,4, 4-tetramethyl furan[a]
2-Chloroethyl hexyl ester isophthalic[a]	Dichlorobenzene[b]	Toluene[c,d,e,g]
2-Decanone[k,l]	Diethyl phthalate[a]	*trans*-2-Octenal[c]
2-Decen-1-ol[a]	Diethyl phthalate[c,f,j,k,l]	*trans*-Caryophylene[e]
2-Decenal[g]	Di-*iso*-butyl adipate[l]	Triacontane[g]
2-Dodecanone[d,g]	Dimethyl disulphide[c,e]	Tricosane[g]
2-Ethyl-1-decanol[g]	Dimethyl sulphide [e]	Tridecanal[g,l]
2-Ethyl-1-dodecanol[d]	Dimethyl trisulphide[e]	Tridecane[c,e,f,g,k,l]

TABLE 12.5 *(Continued)*

Aroma compounds		
2-Ethyl-1-hexanol[a,e]	d-Limonene[c]	Trimethylheptane[e]
2-Ethyl-2-hexenal[g]	Docosane[f,g]	Tritetracontane[a]
2-Ethyl-decanol[h]	Dodecanal[b,g,k,l]	Turmerone[f]
2-Heptadecanone[j,k,l]	Dodecane[c,e,f,g,h]	Undecanal[e,j,k,l]
2-Heptanone[c,e,g,h,l]	Dotriacontane[a,g]	Undecane[b,c,e,f,g]
2-Heptenal[g]	Eicosane[c,e,g]	Undecanol[f]
2-Heptene aldehyde[g]	Eicosanol[a]	Undecyl-cyclohexane[g]
2-Heptylfurane[k]	Ethanol[e,f,j,k,l]	Vanillin[k,l]
2-Hexadecanol[a]	Ethenyl cyclohexane[e]	Z-10-pentadecen-1-ol[a]
2-Hexyl-1-decanol[a]	Ethyl (E)-9-octadecenoate[j,k,l]	α-Cadinene[l]
2-Hexyl-1-octanol[a,c,e]	Ethyl benzene[e]	α-Cadinol[l]
2-Hexyl-decanol[a]	Ethyl decanoate[e]	α-Terpineol[e]
2-Hexylfurane[k,l]	Ethyl dodecanoate[e]	β-Bisabolene[f]
2-Methoxy phenol[a]	Ethyl hexadecanoate[e,j,k,l]	β-Terpineol[e]
2-Methoxy-4-vinylphenol[f,j,k,l]	Ethyl hexanoate[e]	γ-Nonalacton[k]
2-MethyI-butanol[e]	Ethyl linoleate[e]	γ-Nonalactone[j,l]
2-Methyl decane[a]	Ethyl nonanoate[e]	γ-Terpineol[e]
2-Methyl-1,3-pentanediol[a]	Ethyl octanoate[e]	δ-Cadinene[l]
2-Methyl-1-hexadecanol[a]	Ethyl oleate[e]	δ-Cadinol[l]
2-Methyl-2, 4-diphenylpentane[e]	Ethyl tetradecanoate[e]	
2-Methyl-3-octanone[k,l]	Ethylbenzene[c]	

Sources:

[a]Bryant and McClung.[10]

[b]Bryant et al.[11]

[c]Ghiasvand et al.[31]

[d]Goufo
et al.[32]

[e]Grimm et al.[33]

[f]Khorheh et al.[45]

[g]Lin et al.[52]

[h]Monsoor et al.[59]

[i]Tananuwonga and Lertsirib[91]

[j]Zeng et al.[113]

[k]Zeng et al.[115]

[l]Zeng et al.[114]

12.3.2.1 FACTORS INFLUENCING SPME TECHNIQUE

There are several factors, on which the sensitivity of the SPME technique affects. However, these factors highly affect the quantitative accuracy with SPME techniques. These factors are depicted in Table 12.6 that is very important to investigate the determination of optimum sample materials and operating condition for the isolation and extraction of rice volatile aroma compounds.

TABLE 12.6 Factors' Influence to SPME Technique.

Factors influencing SPME
Analyte characteristics
Coating material of the fiber
Extraction time
Fiber exposure time
Ionic strength of the matrix
Sample matrix
Temperature of the sample
Type of fiber

12.3.2.2 SELECTION OF FIBER IN SPME

Selection of fiber is one of the most important factors that affect the result of SPME extraction technology because the extracted amounts of volatile chemical compounds depends upon selected fiber. The chemistry of rice volatile aroma compounds may decide their behaviors of adsorption and desorption on a particular type of fiber. Analytes varied in polarities may require different fiber chemistry. Presently, various type of SPME fibers are available in market with different variety of thicknesses coated with polymers ranging from the nonpolar (e.g., PDMS) to the more polar (e.g., Carbowax) (Table 12.7). These SPME fibers depend on the desired target volatile aroma compounds which are extracted from the sample matrix.[4,29,61,70] This also provides additional advantages as it allow the choice of absorption or adsorption characteristics for extraction of volatile aroma compounds.

According to available literatures initially, only three types of commercial fibers of a different composition and polarities were present, namely, (1) PDMS, (2) CAR/PDMS, and (3) DVB/CAR/PDMS in SPME technology are commonly used to extract aroma volatiles compounds from the rice samples. A fiber of 100-μm polydimethylsiloxane coating is opted for rice volatile aroma compounds. Thirty-micrometer polydimethylsiloxane-coated fiber is much suitable for rapid equilibration and 7-μm polydimethylsiloxane-coated fiber is best for high boiling component sample (e.g., polyaromatic hydrocarbons) as it is chemically bonded the fused silica support and this fiber is also best for those operations which require higher temperature to thermally desorb in the injection port of the gas chromatograph. Ghiasvand et al.[31] used three commercial fibers, namely PDMS, CAR/PDMS, and DVB/CAR/PDMS for investigating flavor profiling in the headspace of the Iranian fragrant rice samples at the beginning of the experiments by using cold-fiber SPME–GC–TOF–MS. Each fiber offered particular advantages. For instance, PDMS is well efficient to perform for wide range of nonpolar analytes as it has non polar coating. Hence, the chromatogram in GC utilizes PDMS fiber coating for analysis of rice volatile aroma compounds is relatively poor first 15 min in beginning and turns richer at peak at more than 15 min retention time period. This ability of PDMS fibers proves that they have more efficiency for extraction of nonpolar volatile aroma compounds. On the other hand, CAR–PDMS fiber reveals contrary effect on compression with PDMS coating as chromatogram is richer in peaks at the beginning only. Relative to the PDMS and CAR–PDMS fibers, the DVB–CAR–PDMS is most suitable for extraction of target analytes shown in Figure 12.3.[31]

TABLE 12.7   Available Commercial SPME Fibers in Market.

Commercial available fibers	Used in volatile aroma extraction	
	In general	In rice
Carbowax/Divinylbenzene (CW/DVB)	Used	Not used
Carbowax/Templated Resin (CW/TPR)	Used	Not used
Polydimethylsiloxane (PDMS)	Used	Used
Carboxen/Polydimethylsiloxane (CAR/PDMS)	Used	Used
Divinylbenzene/Carboxen/Polydimethylsiloxane (DVB/CAR/PDMS)	Used	Used
Polyacrylate	Used	Not used
Polydimethylsiloxane/Divinylbenzene (PDMS/DVB)	Used	Not used

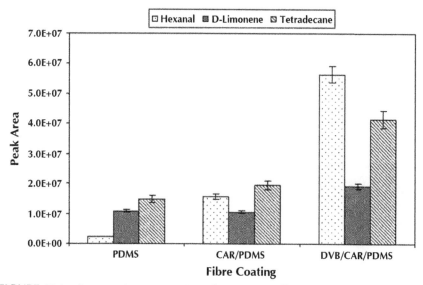

FIGURE 12.3 Comparative presentation of extraction efficiencies of fibers for volatile compounds in the headspace of rice samples from Ghiasvand et al.[31]

12.3.2.3   MODES OF SPME

The volatiles analytes of rice sample can be extracted by two modes from different kinds of media either in liquid or gas phase. These are as described in the following sections.

12.3.2.3.1   Direct Extraction (Liquid Phase)

This method (Fig. 12.4A) involves immersion of coated fiber into sample without headspace and direct transport of analytes from sample matrix to the extraction phase. Agitation is required to some extent for transportation of analytes from bulk to the adjacent fiber for the rapid extraction. This direct extraction mode of SPME aids to calculate required time period for extraction of analytes from the sample.

12.3.2.3.2   Headspace Extraction (Gas Phase)

This mode of headspace extraction of SPME involves direct contact of fiber into the gaseous and relatively clean water sample as analytes shows high affinity for the fiber and interfering contaminants don't exist. In the head-space mode (Fig. 12.4B), analytes are extracted from the gas phase equilibrated with sample. Nowadays, Headspace SPME (HS-SPME) is widely used for the analysis of rice aroma compounds.

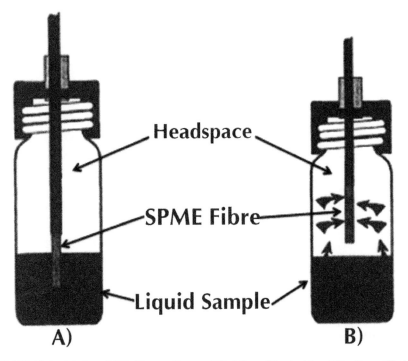

FIGURE 12.4 Modes of SPME sampling modified from Wercinski and Pawliszyn[102]: (A) direct extraction (solid or liquid phase) and (B) headspace extraction (gas phase).

12.3.2.4   MERITS AND DEMERITS OF SPME

12.3.2.4.1   Merits

1. SPME technique is suitable for the extraction of volatile aroma compounds from liquids (water), gases, or solids rice matrix.
2. This one-step extraction microtechnology is free from calibration as all analytes from rice sample are transferred to the extraction phase.
3. The most obvious advantages of SPME technique are solvent-free sampling (100%), comparatively low cost, reproducible (minimizes the use of solvents and their disposal), fast (reduces sample

preparation time by 70%), and easy to use. Volatile aroma compounds from rice are isolated and concentrated without any hindrance.

4. Equilibrium methods in SPME technique are more selective because analysis can be take full advantage of the differences in extracting phase/matrix distribution constants to separate target volatile compounds from interferences and enhanced by the selection of the stationary phase that best suits the rice aroma compounds.

5. SPME is best suited sample preparation technique to quantify the rice aroma compound coupled with gas chromatography.

12.3.2.4.2 Demerits

1. In SPME technique, the comparison of obtained data are only possible when exact uses same fiber on all the samples.

2. The aroma profiling of collected volatiles from rice matrix in SPME technique always depends upon fiber type, thickness and length, sampling time, and temperature.

3. Mixed fibers have largely solved the difficulty of differentiation against polar components in SPME technique, whereas early fibers support differentiation against such components.

12.3.2.5 CASE STUDIES ON SPME OF RICE VOLATILE AROMA COMPOUNDS

SPME is one of the most widely used techniques among 204 techniques in various fields (Fig. 12.5) and also gained popularity among scientists and researchers due to its superiority over other techniques.[4,5,18,39,47,58,72,96] Several studies have been done for the analysis of rice volatile aroma compounds using SPME, and many components have been identified. Grimm et al.[34] screened for 2-AP in rice headspace using SPME on twenty one experimental rice varieties. SPME method involves 100-µL water to 0.75 g of rice sample at 80°C temperature. A 25-min preheated rice headspace was exposed to Carboxen/DVB/PDMS SPME fiber for 15 min followed by subsequent GC–MS analysis of 35 min. Several nanograms 2-AP was recovered from 0.75-g aromatic rice sample whereas nonaromatic rice produced only trace amount of 2-AP. Single SPME headspace analysis recovered 0.3% of the total 2-AP in the sample.

In 2003, SPME was used to identify 2-AP first time by Wongpornchai et al.[106] in headspace of fresh bread flowers (*Vallaris glabra* Ktze). Monsoor et al.[59] analyzed volatile aroma component of commercially milled head and broken rice samples and observed that broken rice contains high concentration of volatile compounds than head rice. He also reported that the concentration of volatile components was increased after storage of both rice types and pentanal, pentanol, hexanal, pentyfuran, octanal, and nonanal are major volatile components in head and broken rice sample.

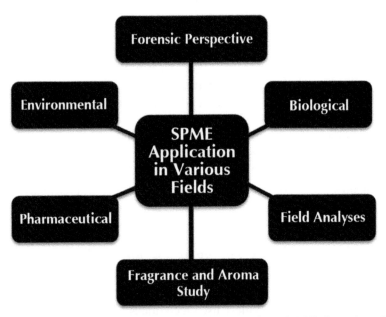

FIGURE 12.5 Application of solid phase microextraction (SPME) in various fields: biological,[58] environmental,[72] field analyses,[47] forensic perspective,[18] pharmaceutical,[96] and fragrance & aroma study.[5]

In 2006, Bryant et al.[11] experimented upon 'Koshihikari' and 'Basmati' rice for texture profile and volatile compound analysis and cooked both in different rice cookers. He reported the higher amount of dodecanal and hexanal in the sample prepared in Hitachi cooker and higher amount of acetone and nephthalene in the sample prepared in National cooker using SPME/GC/MS method. Similar method was used by Bryant and McClung[10] to identify volatile profile of aromatic and nonaromatic rice cultivars. They reported 93 volatile compounds, out of which 64 compounds were not reported previously. They differentiated aromatic rice from nonaromatic rice by the presence of 2-AP in former. Bryant et al.[11] studied the diversity of volatiles in aromatic and nonaromatic rice cultivars. He stated that further research may conclude in better understanding of flavoring compounds. Again in 2011, Bryant et al.[12] used SPME with GC/MS and differentiated between nonaromatic rice (*Oryza sativa,* L.) kernels and aromatic rice kernels. He detected the presence of 2-AP in aromatic rice germplasm by development of single kernel analysis method as aromatic rice cultivars were higher in demand. Determination of 2-AP is very important as it can be utilized in breeding methods to detect aromatic and nonaromatic rice in marketplace.

In 2007, Zeng et al.[116] designed a new apparatus using SPME for direct extraction of flavor volatiles from rice during cooking and tested in the Japanese rice cultivar Akitakomachi. Zeng et al.[113] used modified HS-SPME/GC–MS for analysis of flavor volatiles of 3 Japanese rice cultivars, namely, Nihonbare, Koshihikari, and Akitaomachi and detected 46 components like aldehydes, ketones, alcohols, heterocyclic compounds, fatty acids, esters, phenolic compounds, hydrocarbons, etc. On cooking the amount of low boiling volatiles were decreased whereas the amount key odorant compounds were increased. The three rice cultivars were compared on the basis of similarities and differences of their volatile compounds. Zeng et al.[115] again used modified HS-SPME with GC–MS for analyzing flavor volatiles in three rice cultivars during cooking with less amount of digestive protein. He detected 77 volatile compounds, out of which 13 components were new and not reported previously. These volatile compounds were indole, vanillin, (*E,E*)-2,4-decadienal, (*E*)-2-nonenal, 2-pentyfuran, and 2-methoxy-4-vinylphenol, etc. Zeng et al.[114] identified a broad range of the total 96 flavor volatile compounds from cooked glutinous rice that could be extracted and detected by single HS-SPME/GC–MS by which 27 compounds, namely, 2-methyl-3-octanone, 2-hexylfuran, 7-methyl-2-decane, 2-heptyfuran, (*E*)-4-nonenal, (*E*)-3-nonen-2-one, 2-methyl-5-decanone, (*Z*)-2-octen-1-ol, 6-dodecanone, 2,3-octanedione, 6,10-dimethyl-2-undecanone, dodecanal, alpha-cadinene, 1-decanol, delta-cadinene, 1-undecanol, tridecanal, (1*S,Z*)-calamenene, 3-hydroxy-2,4,4-trimethylpentyl isobutanoate, 2,4,4-trimethylpentan-1,3-diol di-iso-butanoate, hexadecanal, cubenol, delta cadinol, 5-iso-propyl-5*H*-furan-2-one, di-iso-butyl adipate, alpha-cadinol, and glycerin. These detected volatile components in three glutinous rice cultivars belongs to various groups like aldehyde, ketone, alcohols, and heterocyclic compounds, fatty acids, esters, phenolic compounds, and hydrocarbons etc. In 2010, Lin et al.[52] presented a paper on volatile compounds in 10th International

Working Conference on Stored Product Protection. He experimented, 20-g sample of heated rice for 30 min at 80°C prior to headspace absorption, and then extracted for 30 min using HS–SPME with GCMS. He reported alcohols, aldehydes, ketones, esters, hydrocarbons, organic acids, as well as heterocyclic compounds as volatile compounds in indica and japonica rice and also reported that aldehydes were present in abundance, around 30% of volatiles by weight.

An another new powerful device cold-fiber solid-phase microextraction (CF–SPME) was developed in 2007 by Ghiasvand et al.[31] for rapid detection of flavoring agents in sample of fragrant rice headspace (Fig. 12.6). In CF–SPME volatile compounds concentration showed that two uncooked varieties of Indian fragrant rice and nine Iranian rices were successfully analyzed as dry kernels by utilizing fully automated CF–SPME–GC–TOF–MS without addition of water. Different experimental parameters were optimized including 2-AP as a key odorant based on four target analysis.

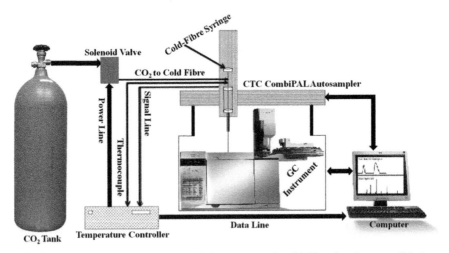

FIGURE 12.6 Schematic diagram of the automated cold-fiber headspace solid-phase microextraction system: Concept and picture modified from Ghiasvand et al.[31]

Tananuwonga and Lertsirib[91] studied the changes in volatile aroma compounds of red organic fragrant rice cv. Hom Daeng under different conditions of storage and evaluated effects of temperature at storage, time, and effects of packing by using SPME/GC–MS. The study detected total 13 key volatile compounds which includes 10 lipid oxidation products. In 2010, Goufo et al.[68] conducted a study to investigate two fragrant rice cultivars, namely, Guixiangzhan and Peizaruanxiang, grown in South China. He used HS–SPME and static headspace with GC–MS to find out the factors that affects concentration of aroma compound that is 2-AP and noted 5 times difference of 2-AP levels which is highest content (3.86 µg·g^{-1}) in Guixiangzhan with comparison of Thai KDML 105 rice. Despite 2-AP other compounds were supposed to contribute the characteristic aroma of Peizaruanxiang. The findings showed that manipulation of pre- and postharvest treatment can largely improve the specific attributes of domestically produced cultivars. He reported that China could increase the share in domestic market of fragrant rice as well as into the international market also.[32]

HS–SPME method combined with GC–MS was employed by Khorheh et al.[45] for assessment of flavor volatiles of Iranian rice, two modified Iranian rice and Hashemi rice cultivar using gelatinization process. The proposed combination provided a powerful system for easy and rapid screening of a wide range of flavors in fragrant rice samples. For optimization of different experimental parameters fiber composition effects, water content of rice sample and equilibrium time was evaluated. The result showed that the amounts of volatile compounds were increased during gelatinization. All free flavored volatiles, bound flavor components were liberated during gelatinization process were formed by thermal decomposition of nonvolatile constituents present in rice. The identified volatile components during gelatinization process were 75, 55, and 66 from Hashemi, HD5, and HD6 Iranian rice cultivars, respectively, out of which 58 unique compounds were not detected previously. The chemical groups of identified volatile components in three Iranian rice cultivars belong to Aldehydes, ketones, alcohols, heterocyclic compounds, fatty acids, esters, phenolic compounds, and hydrocarbon.

12.3.2.6   SUMMARY

SPME provide significant benefits to rice volatile aroma compounds researches as an extraction technique prior to GC analysis. The quantitative and qualitative results obtained from SPME techniques are always better than traditional/conventional extraction techniques. Since past decades, SPME technique has been matured as a tool for the analysis of rice volatile aroma compounds till date. The strategies of SPME method is straightforward applications ranging from the quantitation of specific chemicals at low ppb levels to the complete identification of the volatile aroma compounds in rice sample. A number of publications listed due to sustainability of SPME techniques in the field of rice volatile aroma compounds since more than 10 years back. The researchers of this focus area are continuing to realize the benefits from SPME. As numerous findings are examples on SPME techniques for rice aroma study are illustrated in this chapter which are potent extraction tool for the analysis of impact rice volatile aroma compounds and also superior than other traditional/conventional extraction techniques in terms of accuracy.

12.3.3   SUPERCRITICAL FLUID EXTRACTION

SFE is one of the most innovative and reliable methods for separation and extraction process of volatile aroma compounds by use of supercritical fluids (SCFs) as the extraction solvent.[3,37,46,56,83] SCFs are a kind of clean solvent in the view of 'Green Chemistry,' less toxic than organic solvents. A widely used SCF is carbon dioxide (CO_2), sometimes modified by ethanol or methanol as such co-solvents.[117] In addition of CO_2, there are a number of solvent which have been used for SFE (Table 12.8). Therefore, it has been found that SFE is an alternative to liquid extraction, and this has proved to be an efficient sample preparation technique.[53] It is eco-friendly as CO_2 is easily removed or almost leaving no trace by decreasing pressure.[86,87] Therefore, the consequent treatment is simple when SFE is used for extraction of volatile aroma compounds. Presently, SFE is well known method for separation and extraction because of its design and easy operating method depicted in Figure 12.7.[49]

TABLE 12.8   Chemical Solvents for SFE Techniques.

Solvents	T_c (K)	P_c (Mps)	Critical density (kg/m³)
Ammonia	406	11.3	235
Carbon dioxide	304	7.38	468
Ethane	305	4.88	203
Ethylene	283	5.03	218
Methane	192	4.60	162
Propane	370	4.24	217
Propylene	365	4.62	233
Water	647	22.0	322

Source: http://www.separationprocesses.com/Extraction/SE_Chp06.htm

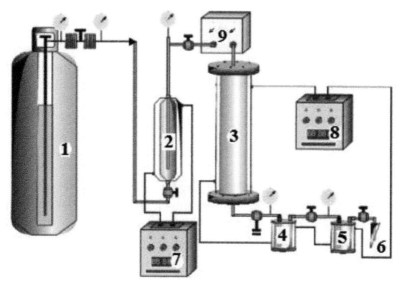

FIGURE 12.7 Schematic diagram of supercritical fluid extraction (SFE) set up from Sarmento et al.[79]: (1) CO_2 cylinder, (2) surge tank, (3) extraction, (4) and (5) separators, (6) collector and gas measuring device, (7) and (8) thermostatic baths, and (9) isocratic pump.

12.3.3.1 MERITS AND DEMERITS SUPERCRITICAL FLUID EXTRACTION

12.3.3.1.1 Merits

SFE is an environment friendly and very advantageous extraction technology in the course of analysis of volatile aroma compounds from rice sample than the other conventional extraction methods. Lili et al.[51] reviewed the recent advances and developments of SFE in the extraction from the plant materials, during 2005–2011. The merits of SFE over conventional methods can be summarized as[2,36]

1. Possibility of aroma volatiles compounds extraction at low temperature which avoids damage from heat (because of susceptible to thermal degradation due to low operating conditions).
2. There is no solvents residue (i.e., reduces the risk of storage), better diffusivity and fast extraction (since the complete extraction step is performed in about 20 min) over the other extraction technology.

3. More selective extractions are allowed by SFE technology by low viscosity of SCF and connected directly to a chromatograph immediately after analysis of extraction.
4. This technology is also best suit for extraction and purification of low-volatile aroma chemical compounds present in solid or liquid sample matrix.

12.3.3.1.2   Demerits

1. SCF penetration take little time and rapid into the interior of a solid as compare to solute diffusion from solid is not rapid and take prolonged time into the SCF.
2. Scale in SFE is not possible because of absence of fundamental and molecular-based model of solutes.
3. SFE technology is very expensive compare to others and consistency and reproducibility may vary in continuous production.

12.3.3.2   CASE STUDIES ON SFE OF RICE VOLATILE AROMA COMPOUNDS

SFE has already been used for a number of years for the extraction of food constituents. SFE has recently become an alternative to conventional extraction procedures that avoids some of the drawbacks linked to previous extraction technology. There are numerous articles that deal with the advantages, applications, and possibilities of SFE in flavor analysis.[3,22,30,42,64,76,93,97]

A comparative evaluation of the extraction of the aroma constituents of a popular commercial brand of Basmati rice using Likens–Nickerson extraction and SFE with CO_2 was carried out. SFE at 50°C and 120 bar for 2 h provided appreciable extraction of the volatile constituents of the rice as compared with Likens–Nickerson extraction. The advantages of smaller sample size, shorter time of extraction, and negligible possibility of artifacts with the SFE technique merit its use for recovery of aroma volatiles from Basmati rice.[7]

Yahya[107] experimented pandan leaf for extraction of aroma compound to use the compound that enhance rice flavor using supercritical carbon dioxide (SC-CO_2) and Soxhlet extraction and hexane as solvent to extract 2-AP from pandan leaves. The effect of different extraction pretreatments, such as particle size and drying on the extraction yield and concentration of 2-AP were investigated. The identification and quantification of 2-AP were carried out by GC–MS and gas chromatography–flame ionization detection (GC–FID), respectively. This work aims to provide an understanding of the phenomenon that occurs during cooking and storage; typically by the changes of 2-AP absorption when cooking rice grains with pandan leaves. The parameters were investigated at optimal water conditions. Although 2-AP concentration was obtained from supercritical carbon dioxide extraction in low amount and the extracts were pure without any contamination. Grinding and freeze–drying methods were proved best pretreatment for SFE. The absorption of 2-AP during the cooking of rice grains did not increase with time. This unexpected result indicated that the phenomena occurring during cooking are quite complex. This work also quantified the potential of pandan leaves to enhance the flavor of cooked rice, particularly under excess water conditions. Storage for 15 min at $24.0 \pm 1.0°C$ is considered as the optimum time for obtaining cooked rice with a high quality of flavor. In the experiment, the volatile components of rice were isolated by extraction with supercritical CO_2 under pressure followed by atmospheric steam distillation and enrichment of steam volatiles on Porapak Q.

Although SFE is able to recover the majority of the aroma compounds, it is also more selective, and therefore less effective, than solvent extraction, that is, fewer compounds are extracted with SFE than with solvent extraction.[76] At present, SFE has been applied for the extraction of volatile aroma compounds from rice (Table 12.9) and other plant samples such as fruits, vegetables, spices, etc.[23,26,60,81,94,95]

TABLE 12.9 List of Extracted Major Rice Volatile Aroma Compounds Using SFE Techniques.[7]

Aroma compounds	
1. 2-Decenal	2. Hexanal
3. 2-Heptanone	4. Hexanol
5. 2-Octanone	6. Nonanal
7. Butan-2-one-3-Me	8. Octanal
9. Decanol	10. Octanol
11. Dodecane	12. Pentanal
13. Heptadecane	14. Tetradecane
15. Heptanol	16. Undecane
17. Hexadecane	

12.3.3.3 SUMMARY

In the recent decades, use of SFE technology has received special attention that provides promising extraction and fractionation of food.[51] There are many literatures about the extraction by using SFE[20,21,27,28,67,77,78,90,111] but very few research studies have been found in literature on application of SFE in rice volatile aroma compound extraction.[7,107] Study of these modern and novel extraction technology are necessary for the further application of SFE for aroma chemical extraction of the rice in future.

12.4 FUTURE RESEARCH ON RICE VOLATILE AROMA COMPOUNDS

Does the rice industry have a future in volatile aroma compounds analysis? Authors indicate that they should have reviewed the research on rice aroma compounds up to at least four past decades. They further indicate that the rice industries have broad and prospective future in volatile aroma chemical compound analysis. On the basis of the collected literature in this chapter, it can be concluded that there are various extraction technologies for volatile aroma compounds analysis from rice sample but presently, researchers and scientists who are working in this area are continuously looking for economic and environment friendly feasible technology. There are various extraction technologies available, but constant efforts are being made in the development of new analytical methods for rice volatile aroma compounds analysis. In the traditional methods, large quantities of solvent and time are consumed, thus causing additional environmental problems as well as increase operating costs due to use of large amount of solvent. Beside from these traditional and normalized methods, several modern novel technologies for extraction of rice volatile aroma compounds such as SPME and SFE are appearing as an alternative to conventional and traditional methods of extraction coupled with modern instruments: GC–MS, gas chromatography–olfactometry, GC–FID and gas chromatography–pulsed flame photometric detection (GC–PFPD), etc., which are partially or fully automated. These modern novel technologies offer great potential and advantages with respect to solvent consumption, extraction time, yield, and reproducibility. However, most of these modern novel extraction technologies are still conducted successfully for the determination of volatile aroma compounds from rice sample at the laboratory level, although very few applications like SFE are available for rice aroma extraction. More research is needed to exploit applications of these modern novel extraction technologies, which can be characterized by low limits of detection and quantification, and also high precision and accuracy of the results obtained.

ACKNOWLEDGMENTS

Authors are indebted to Department of Science and Technology, Ministry of Science and Technology, Govt. of India for an individual research fellowship (INSPIRE Fellowship Code No.: IF120725; Sanction Order No. DST/ INSPIRE Fellowship/2012/686 & Dated: 25/02/2013).

KEYWORDS

- analytes
- analytical techniques
- aroma
- aroma chemicals
- aroma compounds
- aroma profile
- aromatic rice
- basmati rice
- chemical component
- cooked rice
- cultivars
- distillation
- extraction
- extraction method
- extraction technique
- extraction technology
- flavor
- fragrant rice
- headspace
- identification

- identification and quantification
- isolation
- isolation and extraction
- nonaromatic rice
- nonfragrant rice
- *Oryza sativa*
- pandan
- popcorn-like
- quality
- quantification
- rice
- rice grain
- rice volatile
- smell
- solvent extraction
- volatile
- volatile aroma
- volatile aroma compounds
- volatile compound
- volatile profile

REFERENCES

1. Adahchour, M.; Beens, J.; Vreuls, R. J. J.; Batenburg, A. M.; Rosing, E. A. E.; Brinkman, U. A. T. Application of Solid-phase Micro-extraction and Comprehensive Two-dimensional Gas Chromatography (GC × GC) for Flavor Analysis. *J. Chromatogr.* **2002,** *55*(5/6), 361–367.

2. Ahuja, S.; and Diehl, D.; Sampling and Sample preparation. In *Comprehensive Analytical Chemistry*; Ahuja, S., Jespersen, N., Eds.; Elsevier (Wilson & Wilson): Oxford, UK, 2006; Vol 47, pp 15–40.

3. Anklam, E.; Berg, H.; Mathiasson, L.; Sharman, M.; Ulberth, F. Supercritical Fluid Extraction (SFE) in Food Analysis: A Review. *Food Addit. Contam.* **1998,** *15*(6), 729–750.

4. Arthur, C. L.; Pawliszyn, J. Solid Phase Microextraction with Thermal Desorptionusing Fused Silica Optical Fibers. *Anal. Chem.* **1990,** *62*, 2145–2418.

5. Augusto, F.; Lopes, A. L.; Zini, C. A. Sampling and Sample Preparation for Analysis of Aromas and Fragrances. *Trends Anal. Chem.* **2003**, *22*, 160–169.

6. Bergman, C. J.; Delgado, J. T.; Bryant, R.; Grimm, C.; Cadwallader, K. R.; Webb, B. D. Rapid Gas Chromatographic Technique for Quantifying 2-Acetyl-1-pyrroline and Hexanal in Rice (*Oryza sativa* L.). *Cereal Chem.* **2000**, *77*, 454–458.

7. Bhattacharjee, P.; Ranganathan, T. V.; Singhal, R. S.; Kulkarni, P. R. Comparative Aroma Profiles Using Supercritical Carbon Dioxide and Likens–Nickerson Extraction from a Commercial Brand of Basmati Rice. *J. Sci. Food Agric.* **2003**, *83*, 880–883.

8. Bhattacharjee, P.; Singhal, R. S.; Kulkarni, P. R. Basmati Rice: A Review. *Int. J. Food Sci. Technol.* **2002**, *37*, 1–12.

9. Blanch, G. P.; Tabera, J.; Herraiz, M.; Reglero, G. Preconcentration of Volatile Components of Foods: Optimization of the Steam Distillation–Solvent Extraction at Normal Pressure. *J. Chromatogr.* **1993**, *628*, 261–268.

10. Bryant, R. J.; McClung, A. M. Volatile Profiles of Aromatic and Non-aromatic Rice Cultivars Using SPME/GC–MS. *Food Chem.* **2011**, *124*(2), 501–513.

11. Bryant, R. J.; Jones, G.; Grimm, C. Texture Profile and Volatile Compound Analyzes of 'Koshihikari' and 'Basmati' Rice Prepared in Different Rice Cookers. B. R. Wells Rice Research Studies, *AAES Res. Ser.* **2006**, *550*, 370–376.

12. Bryant, R. J.; McClung, A. M.; Grimm, C. Development of a Single Kernel Analysis Method for Detection of 2-Acetyl-1-pyrroline in Aromatic Rice Germplasm. *Sens. Instrumen. Food Qual.* **2011**, *5*(5), 147–154.

13. Buttery, R. G.; Ling L. C.; Juliano B. O.; Turnbagh J. G. Cooked Rice Aroma and 2-Acetyl-1-pyrroline. *J. Agric. Food Chem.* **1983**, *31*, 823–826.

14. Buttery, R. G.; Ling, L. C.; Juliano, B. O. 2-Acetyl-1-pyrroline—An Important Aroma Component of Cooked Rice. *Chem. Ind. (Lond.)* **1982**, *23*, 958–959.

15. Buttery, R. G.; Ling, L. C.; Mon, T. R. Quantitative Analysis of 2-Acetyl-1-pyrroline in Rice. *J. Agric. Food Chem.* **1986**, *34*(1), 112–114.

16. Buttery, R.; Turnbaugh, J.; Ling, L. Contributions of Volatiles to Rice Aroma. *J. Agric. Food Chem.* **1988**, *36*, 1006–1009.

17. Chaintreau, A. Simultaneous Distillation–Extraction: From Birth to Maturity—Review. *Flav. Fragr. J.* **2001**, *16*, 136–148.

18. Chang, K.; Abdullah, A. F. L. A Review on Solid Phase Microextraction and Its Applications in Gunshot Residue Analysis. *Malay. J. For. Sci.* **2010**, *1*(1), 42–47.

19. Chastril, J. Chemical and Physicochemical Changes of Rice During Storage at Different Temperatures. *J. Cereal Sci.* **1990**, *11*, 71–85.

20. Comim, S. R. R, Madella, K.; Oliveira, J. V.; Ferreira, S. R. S. Supercritical Fluid Extraction from Dried Banana Peel (*Musa* spp., Genomic Group AAB), Extraction Yield, Mathematical Modeling, Economical Analysis and Phase Equilibria. *J. Supercrit. Fluid* **2010**, *54*, 30–37.

21. Corso, M. P.; Fagundes-Klen, M. R.; Silva, E. A.; Cardozo Filho, L.; Santos, J. N.; Freitas, L. S.; Dariva, C. Extraction of Sesame Seed (*Sesamun indicum* L.) Oil Using Compressed Propane and Supercritical Carbon Dioxide. *J. Supercrit. Fluid* **2010**, *52*, 56–61.

22. Diaz, O.; Cobos, A.; de la Hoz, L.; Ordonez, J. A. Supercritical Carbon Dioxide in the Production of Food from Plants. *Aliment. Equipos. Technol.* **1997**, *16*(8), 55.

23. Díaz-Maroto, M. C.; Pérez-Coello, M. S.; Cabezudo, M. D. Supercritical Carbon Dioxide Extraction of Volatiles from Spices: Comparison with Simultaneous Distillation–Extraction. *J. Chromatogr.* **2002**, *947*, 23–29.

24. Djozan, D.; Ebrahimi, B. Preparation of New Solid Phase Microextraction Fiber on the Basis of Atrazine-molecular Imprinted Polymer: Application for GC/MS Screening of Triazine Herbicides in Water, Rice and Onion. *Anal. Chim. Acta* **2008**, *616*, 152–159.
25. Djozan, D.; Makham, M.; Ebrahimi, B. Preparation and Binding Study of Solid-phase Microextraction Fiber on the Basis of Ametryn-imprinted Polymer: Application to the Selective Extraction of Persistent Triazine Herbicides in Tap Water, Rice, Maize and Onion. *J. Chromatogr. A* **2009**, *1216*, 2211–2219.
26. Duarte, C.; Moldão-Martins, M.; Gouveia, A. F.; Costa, S. B.; Leitão, A. E.; Bernardo-Gil, M. G. Supercritical Fluid Extraction of Red Pepper (*Capsicum frutescens* L.) *J. Supercrit. Fluid* **2004**, *30*, 155–161.
27. Fiori, L. Grape Seed Oil Supercritical Extraction Kinetic and Solubility Data: Critical Approach and Modeling. *J. Supercrit. Fluid* **2007**, *43*, 43–54.
28. Fiori, L. Supercritical Extraction of Sunflower Seed Oil: Experimental Data and Model Validation. *J. Supercrit. Fluid* **2009**, *50*, 218–224.
29. Furton, K. G.; Almirall, J. R.; Bi, M.; Wang, J.; Wu, L. Application of Solid-phase Microextraction to the Recovery of Explosives and Ignitable Liquid Residues from Forensic Specimens. *J. Chromatogr. A* **2000**, 885, 419–432.
30. Gere, D. R.; Randall, L. G.; Callahan, D. Supercritical Fluid Extraction: Principles and Applications. In *Instrumental Methods in food analysis, Techniques and Instrumentation in Analytical Chemistry*; Paré, J. R. J., Bélanger, J. M. R., Eds.; Elsevier, 1997; Vol 18, pp 421–484.
31. Ghiasvand, A. R.; Setkova, L.; Pawliszyn, J. Determination of Flavor Profile in Iranian Fragrant Rice Samples Using Cold-fiber SPME–GC–TOF–MS. *Flav. Fragr. J.* **2007**, *22*(5), 377–391.
32. Goufo, P.; Duan, M.; Wongpornchai, S.; Tang, X. Some Factors Affecting the Concentration of the Aroma Compound 2-Acetyl-1-pyrroline in Two Fragrant Rice Cultivars Grown in South China. *Front. Agric. China* **2010**, *4*(1), 1–9.
33. Grimm, C. C, Champagne, E. T.; Lloyd, S. W.; Easson, M.; Condon, B.; McClung, A. Analysis of 2-Acetyl-l-Pyrroline in Rice by *HSSE/GCIMS*. *Cereal Chern.* **2011**, *88*(3), 271–277.
34. Grimm, C. C.; Bergman, C.; Delgado, J. T.; Bryant, R. Screening for 2-Acetyl-1-pyrroline in the Headspace of Rice Using SPME/GC–MS. *J. Agric. Food Chem.* **2001**, *49*(1), 245–249.
35. Grimm, C.; Champagne, E.; Bett-Garber, K.; Ohtsubo, K. The Volatile Composition of Waxy Rice. In Proceedings of the United States–Japan Cooperative Program in Natural Resources (UJNR) 29th Protein Resources Panel Meeting, Honolulu, Hawaii, 19–25th November, 2000.
36. Handa, S. S. An Overview of Extraction Techniques for Medicinal and Aromatic Plants. In *Extraction Technologies for Medicinal and Aromatic Plants*; Handa, S. S., Khanuja, S. P. S., Longo, G., Rakesh, D. D., Eds. International Centre for Science and High Technology ICS-UNIDO: Trieste, Italy, 2008; pp 21–52.
37. Hawthorne, S. B.; Krieger, M. S.; Miller, D. J. Analysis of Flavor and Fragrance Compounds Using Supercritical Fluid Extraction Coupled with Gas Chromatography. *Anal. Chem.* **1988**, *50*, 472–477.
38. Hawthorne, S. B.; Miller, D. J, Pawliszyn, J.; Arthur, C. L. Solventless Determinutesation of Caffeine in Beverages Using Solid-phase Microextraction with Fused-silica Fibers. *J. Chromatogr.* **1992**, *603*(1–2), 185–191.

39. Heydari, S.; Haghayegh, G. H. Extraction and Microextraction Techniques for the Deter-minutesation of Compounds from Saffron. Cana. *Chem. Trans.* **2014**, *2*(2), 221–247.
40. Holt, R. U. Mechanisms Effecting Analysis of Volatile flavor Components by Solid-phase Microextraction and Gas Chromatography. *J. Chromatogr. A* **2001**, *937*, 107–114.
41. Ishitani, K.; Fushimi, C. Influence of Pre- and Post-harvest Conditions on 2-Acetyl-1-pyrroline Concentration in Aromatic Rice. *J. Food Sci.* **1994**, *67*(2), 619–622.
42. Kallio, H.; Kerrola, K. Dense Gas Extraction as a Preparation Method in Food Anal-ysis. In Curr. Status Future Trends Anal. Food Chem.; Proc. Eur. Conf. Food Chem., 8th; Sontag, G., Pfannhauser, W., Eds.; Austrian Chemical Society: Vienna, 1995; Vol 1, p 30.
43. Kaseleht, K.; Leitner, E.; Paalme, T. Determinutesing Aroma-active Compounds in Kama flour using SPME–GC/MS and GC–olfactometry. *Flav. Fragr. J.* **2011**, *26*, 122–128.
44. Kataoka, H.; Lord, H. L.; Pawliszyn, J. Applications of Solid-phase Micro-extraction and Gas Chromatography. *J. Chromatogr. A* **2000**, *800*, 35–62.
45. Khorheh, N. A.; Givianrad, M. H.; Ardebili, M. S.; Larijani, K. Assessment of Flavor Volatiles of Iranian Rice Cultivars during Gelatinization Process. *J. Food Biosci., Technol. IAU* **2011**, *1*, 41–54.
46. King, J. W. Supercritical Fluid Extraction: Present Status and Prospects. *Gras. Aceit.* **2002**, *53*(1), 8–21.
47. Koziel, J.; Jia, M. Y.; Khaled, A.; Noah, J.; Pawliszyn, J. Field Air Analysis with SPME device. *Anal. Chim. Acta* **1999**, *400*, 153–162.
48. Laksanalamai, V.; Ilangantileke, S. Comparison of Aroma Compound (2-Acetyl-l-pyrroline) on Leaves from Pandan (*Pandanum amaryllifolius*) and Thai Fragrant Rice (Khao Dawk Mali-105). *Cereal Chem.* **1993**, *70*, 381–384.
49. Li, H.; Wu, J.; Rempel, C. B.; Thiyam, U. Effect of Operating Perameters on Oil and Phenolic Extraction Using Supercritical CO_2. *J. Am. Oil Chem. Soc.* **2010**, *87*, 1081–1089.
50. Likens, S. T.; Nickerson, G. B. Detection of Certain Hop Oil Constituents in Brewing Products. *Am. Soc. Brew. Chem. Proc.* **1964**, 5–13.
51. Lili, X.; Zhan, X.; Zeng, Z.; Chen, R.; Li, H.; Xie, T.; Wang S. Recent Advances on Supercritical Fluid Extraction of Essential Oils. *Afr. J. Pharm. Pharmacol.* **2011**, *5*(9), 1196–1211.
52. Lin, J.-Y.; Fan, W.; Gao, Y.-N.; Wu, S.-F.; Wang, S.-X. Study on Volatile Compounds in Rice by HS–SPME. 10th International Working Conference on Stored Product Protec-tion. *Julius-Kühn-Arch.* **2010**, *425*, 125–134.
53. Lou, M. T.; Chen, R. Advances of Application Techniques of Supercritical Fluid Extrac-tion. *Anal. Instrum.* **1996**, *4*, 5–9.
54. Maarse, H.; Kepner, R. Changes on Composition of Volatile Trpenes in Douglas Fir Needles During Maturation. *J. Agric. Food Chem.* **1970**, *18*, 1095–1101.
55. Maignial, L.; Pibarot, P.; Bonetti, G.; Caintreau, A.; Marion, J. P. Simultaneous Distil-lation-extraction Under Static Vacuum: Isolation of Volatile Compounds at Room Temperature. *J. Chromatogr.* **1992**, *606*, 87–94.
56. McHugh, M.; Krukonis, V. Supercritical Fluid Extraction, Principles and Practice. Butterworth: USA, 1986.
57. Mehdinia, A.; Mousavi, M. F.; Shamsipur, M. Nano-structured Lead Dioxide as a Novel Stationary Phase for Solid-phase Microextraction. *J. Chromatogr. A* **2006**, *1134*, 24–31.

58. Mills, G. A.; Walke, V. Headspace Solid-phase Microextraction Procedures for Gaschromatographic Analysis of Biological Fluids and Materials. *J. Chromatogr.* **2000**, *902*, 267–287.

59. Monsoor, M. A.; Proctor, A. Volatile Component Analysis of Commercially Milled Head and Broken Rice. *J. Food Sci.* **2004**, *69*(8), C632- C636.

60. Morales, M. T.; Berry, A. J.; McIntyre, P. S.; Aparicio, R. Tentative Analysis of Virgin Olive Oil Aroma by Supercritical Fluid Extraction–High-resolution Gas Chromatography–mass Spectrometry. *J. Chromatogr.* **1998**, *819*, 267–275.

61. Namiesnik, J.; Zygmunt, B.; Jastrzebska, A. Application of Solid-phase Microextraction for Determination of Organic Vapours in Gaseous Matrices. *J. Chromatogr. A* **2000**, *885*, 405–418.

62. Nickerson, G. B.; Likens, S. T. Gas Chromatographic Evidence for the Occurrence of Hop Oil Components in Beer. *J. Chromatogr.* **1966**, *21*, 1–3.

63. Nijssen, L. M.; Vischer, C. A.; Maarse, H.; Willemsens, L. C.; Boelens, M. H. *Volatile Compounds in Food Qualitative and Quantitative Data. Rice (No. 123)*. Central Institute for Nutrition and Food Research: Zeist, The Netherlands, 1996.

64. Obaya-Valdivia, A.; Guerrero-Barajas, C. Supercritical Fluid Extraction Applied to Food Processing Industries. *Tecnol. Aliment.* **1993**, *28*, 22.

65. Park, J. S.; Kim, K.; Baek, H. H. Potent Aroma-active Compounds of Cooked Korean Nonaromatic Rice. *Food Sci. Biotechnol.* **2010**, *19*(5), 1403–1407.

66. Parliament, T. H. 1997. Solvent Extraction and Distillation Techniques. I. In *Techniques for Analyzing Food Aroma*; Marsili, R., Ed. Marcel Dekker, Inc.: New York, pp 1–27.

67. Passos, C. P.; Silva, R. M.; Da Silva, F. A.; Coimbra, M. A.; Silva, C. M. Enhancement of the Supercritical Fluid Extraction of Grape Seed Oil by Using Enzymatically Pretreated Seed. *J. Supercrit. Fluid* **2009**, *48*, 225–229.

68. Paule, C. M.; Powers, J. J. Sensory and Chemical Examinutesation of Aromatic and Nonaromatic Rices. *J. Food Sci.* **1989**, *54*, 143–146.

69. Pawliszyn, J. New Directions in Sample Preparation for Analysis of Organic Compounds. *Trends Anal. Chem.* **1995**, *14*, 113–122.

70. Pawliszyn, J. *Solid Phase Microextraction: Theory and Practice*. Wiley-VCH: New York, 1997.

71. Pawliszyn, J. Solid-Phase Microextraction. In *Extraction*; Academic Press, 2000; pp 1416–1424.

72. Peñalve, A.; Pocurull, E.; Borrull F.; Marcé, R. M. Trends in Solid-phase Microextraction for Determining Organic Pollutants in Environmental Sample. *Trends Anal. Chem.* **1999**, *18*, 557–568.

73. Petrov, M.; Danzart, M.; Giampaoli, P.; Faure, J.; Richard, H. Rice Aroma Analysis: Discriminutesation Between a Scented and a Non-scented Rice. *Sci. Des Alim.* **1996**, *4*(16), 347–360.

74. Pino, J. A. Marbot. R.; Vazquez, C. Volatile Compounds of Psidium Salutare (H.B.K.) Berg. Fruit. Characterizarion of Volatiles in Strawberry Guava (*Psidium cattleianum* Sabine) Fruits. *J. Agric. Food Chem.* **2001**, *49*, 5883–5887.

75. Pino, J. A.; Marbot, R.; Bello, A. Volatile Compounds of Psidium Salutare (H.B.K.) Berg. Fruit. *J. Agric. Food Chem.* **2002**, *50*, 5146–5148.

76. Polesello, S.; Lovati, F.; Rizzolo, A.; Rovida, C. Supercritical Fluid Extraction as a Preparative Tool for Strawberry Aroma Analysis. *J. High Res. Chromatogr.* **1993**, *16*, 555–559.

77. Salgin, U. Extraction of Jojoba Seed Oil Using Supercritical CO_2 + Ethanol Mixture in Green and High-tech Separation Process. *J. Supercrit. Fluid* **2007**, *39*, 330–337.
78. Salgin, U.; Doker, O.; Calimli, A. Extraction of Sunflower Oil with Supercritical CO_2: Experiments and Modelling. *J. Supercrit. Fluid* **2006**, *38*, 326–331.
79. Sarmento, C. M. P.; Ferreira, S. R. S.; Hense, H. Supercritical Fluid Extraction (SFE) of Rice Bran Oil to Obtain Fractions Enriched with Tocopherols and Tocotrienols. *Braz. J. Chem. Eng.* **2006**, *23*(2), 243–249.
80. Schultz, T. H.; Flath, R. A.; Mon, T. R.; Eggling, S. B.; Teranishi, R. Isolation of Volatile Compounds from a Model System. *J. Agric. Food Chem.* **1977**, *25*, 446–449.
81. Schultz, W. G.; Randall, J. M. Liquid Carbon Dioxide for Selective Aroma Extraction. *Food Technol.* **1970**, *24*, 1282–1286.
82. Siegmund, B.; Leitner, E.; Mayer, I.; Farkas, P.; Sadecka, J.; Pfannhauser, W.; Kovac, M. Investigation of the Extraction of Aroma Compounds Using the Simultaneous Distillation Extraction According to Likens–Nickerson Method. *Dtsch. Lebensm. Rundsch. (German)* **1996**, *92*(9), 286–290.
83. Sinha, N. K.; Guyer, D. E.; Gage, D. A.; Lira, C. T. Supercritical Carbon Dioxide Extraction of Onion Flavors and their Analysis by Gas Chromatography–Mass Spectrometry. *J. Agric. Food Chem.* **1992**, *40*, 842–845.
84. Sriseadka, T.; Wongpornchai, S.; Kitsawatpaiboon, P. Rapid Method for Quantitative Analysis of Aroma Impact Compound, 2-acetyl-1 Pyrroline, in Fragrant Rice Using Automated Headspace Gas Chromatography. *J. Agric. Food Chem.* **2006**, *54*, 8183–8189.
85. Stashenko, E. E.; Martínez, J. R. Sampling Volatile Compounds from Natural Products with Headspace/Solid-phase Micro-extraction. *J. Biochem. Biophys. Methods* **2007**, *70*(2), 235–242.
86. Stashenko, E. E.; Jaramillo, B. E.; Martínez, J. R. Analysis of Volatile Secondary Metabolites from Colombian *Xylopia aromatica* (Lamarck) by Different Extraction and Headspace Methods and Gas Chromatography. *J. Chromatogr.* **2004**, *1025*, 105–113.
87. Stashenko, E. E.; Puertas, M. A.; Combariza, M. Y. Volatile Secondary Metabolites from *Spilanthes americana* obtained by Simultaneous Steam Distillation-solvent Extraction And Supercritical Fluid Extraction. *J. Chromatogr.* **1996**, *752*, 223–232.
88. Steffen, A.; Pawliszyn, J. Analysis of Flavor Volatiles Using Headspace Solid-phase Microextraction. *J Agric. Food Chem.* **1996**, *44*, 2187–2193.
89. Sunthonvit, N.; Srzednicki, G.; Craske, J. Effects of High Temperature Drying on the Flavor Components in Thai Fragrant Rice Varieties. *Dry. Technol.* **2005**, *23*, 1407–1418.
90. Talansier, E.; Braga, M.; Rosa, P.; Paoluccijeanjean, D.; Meireles, M. Supercritical Fluid Extraction of Vetiver Roots: A Study of SFE Kinetics. *J. Supercrit. Fluid* **2008**, *47*, 200–208.
91. Tananuwonga, K.; Lertsirib, S. Changes in Volatile Aroma Compounds of Organic Fragrant Rice During Storage Underdifferent Conditions. *J. Sci. Food Agric.* **2010**, *90*, 1590–1596.
92. Tava, A.; Bocchi, S. Aroma of Cooked Rice (*Oryza sativa*), Comparison Between Commercial Basmati and Italian Line B5-3. *Cer. Chem.* **1999**, *76*(4), 526–529.
93. Taylor, D. L.; Larick, D. K. Volatile Content and Sensory Attributes of Supercritical Carbon Dioxide Extracts of Cooked Chicken Fat. *J. Food Sci.* **1995**, *60*(6), 1197–1200.
94. Temelli, F.; Chen, C. S.; Braddock, R. J. Supercritical Fluid Extraction in Citrus Oil Processing. *Food Technol.* **1988**, *42*, 145–150.

95. Tuan, D. Q.; Ilangantileke, S. G. Liquid CO_2 Extraction of Essential Oil from Star Anise Fruits (*Illicium verum* H.). *J. Food Eng.* **1997**, *31*, 47–57.

96. Ulrich, S. Solid-phase Microextraction in Biomedical Analysis. *J. Chromatogr.* **2000**, *902*, 167–194.

97. Valcarcel, M.; Tena, M. T. Applications of Supercritical Fluid Extraction in Food Analysis. *Fresenius J. Anal. Chem.* **1997**, *358*(5), 561–573.

98. Verma, D. K.; Mohan, M.; Asthir, B. Physicochemical and Cooking Characteristics of Some Promising Basmati Genotypes. *Asian J. Food Agro-Ind.* **2013**, *6*(2), 94–99.

99. Verma, D. K.; Mohan, M.; Prabhakar, P. K.; Srivastav, P. P. Physico-chemical and Cooking Characteristics of Azad Basmati. *Int. Food Res. J.* **2015**, *22*(4), 1380–1389.

100. Verma, D. K.; Mohan, M.; Yadav, V. K.; Asthir, B.; Soni, S. K. Inquisition of Some Physico-Chemical Characteristics of Newly Evolved Basmati Rice. *Environ. Ecol.* **2012**, *30*(1), 114–117.

101. Weber, D. J.; Rohilla, R.; Singh, U. S. Chemistry and Biochemistry of Aroma in Scented Rice. In *Aromatic Rices*; Singh, R. K., Singh, U. S., Khush, G. S., Eds.; Oxford & IBH, 2000; pp 135–151.

102. Wercinski, S. A. S.; Pawliszyn, J. Solid Phase Microextraction Theory. In *Solid Phase Microextraction: A Practical Guide*; Wercinski, S. A. S., Ed. CRC Press, Marcel Dekker, Inc.: New York, 1999.

103. Widjaja, R.; Craske, J. D.; Wootton, M. Comparative Studies on Volatile Components of Non-Fragrant and Fragrant Rices. *J. Sci. Food Agric.* **1996a**, *70*(2), 151–161.

104. Widjaja, R.; Craske, J. D.; Wootton, M. Changes in Volatile Components of Paddy, Brown and White Fragrant Rice During Storage. *J. Sci. Food Agric.* **1996b**, *71*, 218–224.

105. Withycombe, D. A.; Lindsay, R. C.; Stuiber, D. A. Isolation and Identification of Volatile Components from Wild Rice Grain (*Zizania aquatica*). *J. Agric. Food Chem.* **1978**, *26*(4), 816–821.

106. Wongpornchai, S.; Sriseadka, T.; Choonvisase, S. Identification and Quantitation of the Rice Aroma Compound, 2-Acetyl-1-pyrroline, in Bread Flowers (*Vallaris glabra* Ktze). *J. Agric. Food Chem.* **2003**, *51*(2), 457–462.

107. Yahya, F. B. Extraction of Aroma Compound from Pandan Leaf and Use of the Compound to Enhance Rice Flavor, PhD thesis submitted to School of Chemical Engineering, The University of Birminutesgham, Birminutesgham, UK, 2011.

108. Yajima, I.; Yani, T.; Nakmura, M.; Sakakibara, H.; Hayashi, K. Volatile Flavor Components of Cooked Kaorimai (Scented Rice, *O. sativa japonica*). *Agric. Biol. Chem.* **1979**, *43*(12), 2425–2429.

109. Yang, X.; Peppard, T. Solid-phase Microextraction for Flavor Analysis. *J. Agric. Food Chem.* **1994**, *42*, 1925–1930.

110. Yau, N. J. N.; Liu, T. T. Instrumental and Sensory Analysis of Volatile Aroma of Cooked Rice. *J. Sens. Stud.* **1999**, *14*, 209–233.

111. Yilmaz, E. E.; Özvural, E. B.; Vural, H. Extraction and Identification of Proanthocyanidins from Grape Seed (*Vitis vinifera*) Using Supercritical Carbon Dioxide. *J. Supercrit. Fluid* **2011**, 55(3), 924–928.

112. Yoshihashi. T. Quantitative Analysis on 2-Acetyl-1-pyrroline of an Aromatic Rice by Stable Isotope Dilution Method and Model Studies on its Formation During Cooking. *J. Food Sci.* **2002**, *67*, 619–622.

113. Zeng, Z.; Zhang, H.; Chen, J. Y.; Zhang, T.; Matsunaga, R. Flavor Volatiles of Rice During Cooking Analyzed by Modified Headspace SPME/GC–MS. *Cereal Chem.* **2008a**, *85*(2), 140–145.

114. Zeng, Z.; Zhang, H.; Zhang, T.; Tamogami, S.; Chen, J. Y. Analysis of Flavor Volatiles of Glutinous Rice During Cooking by Combined Gas Chromatography–Mass Spectrometry with Modified Headspace Solid-phase Microextraction Method. *J. Food Compos. Anal.* **2009,** *22*(4), 347–353.
115. Zeng, Z.; Zhang, H. Zhang, T.; Chen, J. Y. Flavor Volatiles in Three Rice Cultivars with Low Levels of Digestible Protein During Cooking. *Cereal Chem.* **2008b,** *85*(5), 689–695.
116. Zeng, Z.; Zhang, H.; Chen, J. Y.; Zhang, T.; Matsunaga, R. Direct Extraction of Volatiles of Rice During Cooking Using Solid-phase Microextraction. *Cereal Chem.* **2007,** *84*(5), 423–427.
117. Zhang, Z.; Li, G. A Review of Advances and New Developments in the Analysis of Biological Volatile Organic Compounds. *Microchem. J.* **2010,** *95*, 127–139.
118. Zhang, Z.; Yang, M. J.; Pawliszyn, J. Solid-phase Microextraction. *Anal. Chem.* **1994,** *66*, 844A–853A.
119. Zhou, M.; Robards, K.; Glennie-Holmes, M.; Helliwell, S. Analysis of Volatile Compounds and their Contribution to Flavor in Cereals. *J. Agric. Food Chem.* **1999,** *47*(16), 3941–3953.

CHAPTER 13

NONTHERMAL FOOD PROCESSES: PULSED ELECTRIC FIELDS, PULSED LIGHT, HIGH HYDROSTATIC PRESSURE, AND IONIZING RADIATION

BRAJESH KUMAR PANDA*, GAYATRI MISHRA, and VIVEK KUMAR

*Department of Agricultural and Food Engineering, Indian Institute of Technology, Kharagpur 721302, West Bengal, India.
E-mail: brajeshkumarpnd2@gmail.com, gayatri.mishra21@gmail.com, vivek.btag@gmail.com*

*Corresponding author.

CONTENTS

ABSTRACT

Nonthermal technologies are the novel and most promising technologies emerging in the food processing industries with many challenges out of which pulse electric field, pulse light, high hydrostatic pressure and ionizing radiations are the most used sources of various food processing operations has been commercialized successfully in the food industries. Pulsed electric field processing, in which short electrical pulses to the disruption of cells are used are generally used to pasteurization of foods, extraction of sugars and other cellular content from plant cells, and reduction in the solid volume (sludge) of wastewater. PEF has many potential but limitation to the processing of foods with no air bubbles and with low electrical conductivity.

Pulse light is a broad spectrum light source, with a general spectral range of 200–1100 nm, which is used to purification and sterilization of food items, decontamination of surfaces of processing, and packaging materials. The main disadvantage of pulse light is the shadowing effect of light to microorganisms reducing the inactivation efficiency. High hydrostatic pressure processing (HHPP) is a novel nonthermal method of sterilization of food, in which a product is processed under very high pressure, leading to inactivation of certain microorganisms and enzymes in the food. The food gets decontaminated and enzymes are also got deactivated at room temperature or slight higher than that leading to superior quality food with high retention of nutritional components. Irradiation is the process of treatment of food using ionizing radiations such as microwave, UV, IR, X-ray, gamma ray, etc.

Irradiation used for various food processing operations such as inactivation of microorganisms, delaying the ripening process of fruits and vegetables, inhibition of sprouts, extraction of oils, volatiles, etc. For every radiation, FDA specified certain limit beyond that the exposure of ionizing radiation may be harmful to human body. In nonthermal food processing, bulk heat is not provided to food directly and the processing methods are hardly affect the covalent bonds of nutritional and flavor components of food, which helps in retention of color, flavor, and also the essential vitamins and minerals. Therefore, nonthermal processing technology is ahead in the line of food processing technologies providing safe and fresh processed food which consumer demands nowadays.

13.1   INTRODUCTION

Food processing is the process of transformation of raw ingredients by physical or chemical means by maintaining the desired properties or nature of food for as long as possible. As microbial and enzymatic activity in food are the major factors for spoilage, control of those can extent the shelf life of food ensuring safety due to pathogens. Microorganisms can be controlled by either slowing their rate of metabolic activity (e.g., freezing, reduction in a_w, acidification, and addition of preservatives) or inactivated using elevated temperatures (e.g., pasteurization, sterilization), but both of the methods having limitations of level of uncertainty regarding safety and degradation of food quality, respectively. Conventional food thermal processing methods often result in a number of undesired changes in foods, such as loss of smell, color, flavor, texture, and nutritional value or we can say, a reduction in the apparent freshness and quality of the final product.

As consumer nowadays' demands food of high quality, fresh alike, and safe, researches has been conducted to find out preservation techniques capable of inactivating microorganisms at temperatures below the critical temperature above which food quality deteriorates. These novel technologies are known as nonthermal food processing technologies, in which rather giving bulk heat, the food is processed at ambient temperatures.

Nonthermal technologies represent a novel area of food processing and are currently being explored on a global scale. These have attracted the interest of the industries and include: pulsed electric fields (PEF), pulsed light (PL), ionizing radiation (e.g., gamma-ray, X-ray, ultraviolet), high hydrostatic pressure processing (HPP).

13.2   PULSE ELECTRIC FIELD TECHNOLOGY

Low frequency alternative currents were found to be having lethal effects on microorganisms by a process called "electropure" was first introduced in Europe and the USA in the 1920s. Electropure is the process of application of external electrical field for a few microseconds induces local structural changes and a rapid breakdown of the cell membrane of most of the microorganisms. The first attempt of use of electricity is for milk pasteurization and was by the application of a (not pulsed) 220-V alternating current within a carbon electrode treatment chamber.[37] The concept of microbial inactivation by application of high-voltage electricity across two electrodes was first investigated in the 1950s. PEF can be applied to inactivate the microorganisms by an irreversible breakdown of the cell membrane, when applied with high intensity. PEF can also be applied as a mild-preservation technique for liquid food as well as a substitute for conventional cell disintegration methods, such as enzymatic treatment, as a pretreatment step for mass transfer improvement prior to dehydration, extraction, or pressing.

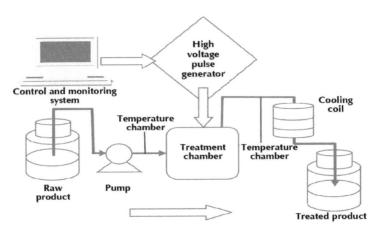

FIGURE 13.1   Basic components of a pulse electric field system.

13.2.1   WORKING PRINCIPLE AND SYSTEM COMPONENTS OF PEF

The basic principle of the PEF technology is the application of short pulses of high electric fields with duration of microseconds to milliseconds and intensity in the order of 10–80 kV/cm. A simplified diagram for generation of exponential decay pulses is shown in Figure 13.1, consisting of a charging unit, pulse generator, and treatment chamber. The main process of generating pulse electric field is the slow charging and a fast discharging of the energy, as the pulse width is short in comparison to the time between pulses. The charging voltage required to generate pulses of sufficient electric field strength is principally depends upon the distance between two electrodes, termed as gap. The process is based on pulsed electrical currents delivered to a product placed between the gaps of a set of electrodes in a treatment chamber. The electric power is most commonly stored in a bank of capacitors connected in series or parallel and discharged into the treatment chamber across a high voltage switch and protective resistors within microseconds. The electric field may be applied in various forms, namely, exponentially decaying, square wave, bipolar, or oscillatory pulses and at ambient, subambient, or slightly above-ambient temperature.

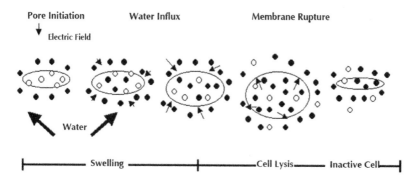

FIGURE 13.2 Electroporation of a cell membrane due to pulse electric field treatment.[47]

The PEF processing system is composed of a high-voltage repetitive pulser, a treatment chamber(s), a cooling system(s), voltage-, and current-measuring devices, a control unit, and a data-acquisition system. A pulsed power supply is used to obtain high voltage from low-utility level voltage, and the former is used to charge a capacitor bank and switch to discharge energy from the capacitor across the food in the treatment chamber. Treatment chambers are designed to hold the food during PEF processing and house the discharging electrodes. After processing the product is cooled, if necessary, packed aseptically, and then stored at refrigerated or ambient temperatures depending on the type of food.[35,50] Liquid food may be processed in a static treatment chamber or in a continuous treatment chamber through a pump. For preliminary laboratory-scale studies, the static treatment chamber is used, but a continuous treatment chamber(s) is desirable for the pilot plant or industrial-scale operations. Total power of the system is limited by the number of times a capacitor can be charged and discharged in a given time. Total power of the system is limited by the number of times a capacitor can be charged and discharged in a given time. The capacitance C_0 (F) of the energy storage capacitor is given below[27]:

$$C_0 = \frac{\tau}{R} = \frac{\tau \sigma A}{d} \qquad (13.1)$$

where τ is the pulse duration (s), R is the resistance (Ω), σ is the conductivity of the food (S/m), d is the treatment gap between electrodes (m), and A is the area of the electrode surface (mm²).

It is believed that that the primary effect of PEF on biological cells is related to local structural changes and permeabilization of microbial membranes resulting synthesis of RNA and DNA, protein, and cell wall components is hampered and metabolic activities are stopped. The permeabilization of a cell membrane requires two key steps: first, the formation of a pore has to be induced by the electric field applied, and second, this pore has to be stable enough to allow interaction of the intra- and extracellular media (Fig. 13.2). But, until now there has been no clear evidence on time sequence and the dynamics of the electroporation process as well as on reversible-irreversible structural changes of cells during and after PEF treatments. Factors that affect the efficiency of inactivation of microorganisms by PEF treatment are presented in Table 13.1.

TABLE 13.1 Factors that Affecting Inactivation of Microorganisms by PEF Treatment.[14]

Process	Media	Micro organisms
Pulse wave width	pH	Type
Electric field intensity	Antimicrobials	Concentration
Temperature	Ionic compounds	Growth stage
Time	Medium ionic strength	
	Electrical conductivity	

13.2.2 DESCRIPTION OF PULSED WAVEFORMS

PEF can be applied to food materials in various wave forms, namely exponentially decaying, square wave, bipolar, or oscillatory pulses. An exponential decay voltage wave is a unidirectional voltage that rises rapidly to a maximum value and decays slowly to zero. A DC power supply charges a capacitor bank connected in series with a charging resistor. When a trigger signal is applied, the charge stored in the capacitor flows though the food in the treatment chamber. Square pulse waveforms are more lethal and more energy efficient than exponential decaying pulses. A square waveform can be obtained by using a pulse-forming network (PFN) consisting of an array of capacitors. The instant-charge-reversal pulses or similar to bipolar pulses are characterized by a positive part and a negative part of pulse with various widths and peak field strengths. Instant-charge-reversal pulses can drastically reduce energy requirements to as low as 1.3 J/ml.[12] Oscillatory decay pulses are less preferred because they prevent the cell from being continuously exposed to a high-intensity electric field for an extended period of time, thus preventing the cell membrane from irreversible breakdown over a large area.

13.2.2.1   TYPES OF TREATMENT CHAMBER

The heart of the pulse electric field system is the treatment chamber which is the most complicated component in the processing system. The treatment chamber is generally designed to keep the product during pulsing, although the uniformity of the process is highly dependent on the characteristic design of the treatment chamber. The chamber is composed of two electrodes held in position by insulating material, which forms an enclosure containing food material. Treatment chambers are mainly grouped together to operate in either a batch or continuous manner; batch systems are generally found in early designs for handling of static volumes of solid or semi-solid foods. The chambers are generally classified in to two categories: (a) parallel plate type, which is usually used for batch process and (b) Co-axial type which has been used for continuous fluid flow processing, where semisolid or liquid foods are generally pumped in to the chamber for pulsing. For batch operation, there are several designs of treatment chambers, namely (Fig. 13.3):

- U-shaped static treatment chamber.
- Disk-shaped static treatment chamber.
- Wire–cylinder static treatment chamber.
- Rod–rod static treatment chamber.

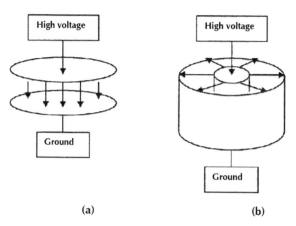

(a)   (b)

FIGURE 13.3   Types of treatment chamber according to electrode configuration: (a) parallel plate and (b) co-axial.[27]

13.2.3 APPLICATION OF PULSE ELECTRIC FIELD TECHNOLOGY

Pulse electric field technology has been successfully applied to many food pasteurization operations such as pasteurization of foods such as juices, milk, yogurt, soups, and liquid eggs. The limitation of application PEF processing is for foods with no air bubbles and with low electrical conductivity. The maximum particle size in the liquid must be smaller than the gap of the treatment region in the chamber in order to ensure proper treatment. PEF is only suitable for liquid or semi liquid foods which can be pumpable. PEF is also applied to enhance extraction of sugars and other cellular content from plant cells, such as sugar beets. PEF also found application in reducing the solid volume (sludge) of wastewater.

PEF has been proven to be most promising in inactivation of pathogenic food borne microorganisms such as *Escherichia coli, Clostridium botulinum, Bacillus cereus, Salmonella, Campylobacter, Listeria monocytogenes*, etc. with the potential of being an alternative for pasteurization of liquid foods. Studies related to inactivation of microorganisms by PEF are given in Table 13.2. Other microorganisms which must be taken into consideration for adequate effectiveness of PEF processing are *Staphylococcus aureus, Pseudomonas aeruginosa*, and *Yersinia enterocolitica*. The mortality of *S. aureus* suspended in a phosphate buffer solution (pH = 7.0), when treated with 30 pulses of 20 kV/cm, more than three log reductions were achieved.[19] *S. aureus* was inactivated by about 4–5 log cycles with a field strength of 16 kV/cm and 50 pulses.[33] For *P. aeruginosa*, a PEF treatment of 30 pulses at 20 kV/cm was given to achieve about 3.5 log reductions in a phosphate buffer solution.[19] *Y. enterocolitica* inoculated in a NaCl solution (pH =7.2), after a treatment of 250 pulses at 75 kV and 1 Hz of frequency, about 7 log reductions were achieved.[21]

TABLE 13.2   Inactivation of Microorganisms by Pulse Electric Field Treatment.

Microor-ganisms	Suspension media	Maximum Log reduction achieved	Temp. of processing (°C)	Process conditions	Reference
E. coli *O157:H7*	Apple juice	2.6	29	31 kV cm^{-1} pulse intensity, 4 µs pulse width for 202 µs pulse time	[13]
E coli	Simulated milk ultrafiltrate (SMUF)	4	7	36 kV cm^{-1} intensity, 2 µs pulse width for 100 µs pulse duration	[34]

TABLE 13.2   *(Continued)*

Microor-ganisms	Suspension media	Maximum Log reduc-tion achieved	Temp. of processing (°C)	Process conditions	Reference
E. coli O157:H7	Skim milk	1.3–2.0	30	24 kV cm^{-1}, 2.8 µs pulse width for 141 µs pulse duration	[13]
E. coli	Watermelon juice	4	<40	35 kV cm^{-1} pulse intensity, 4 µs pulse width for 2000 µs pulse duration	[28]
E. coli ATCC 8739	Orange juice-milk beverage	3.8	<55	15–40 kV cm^{-1} pulse intensity, 2.5 µs pulse width for 0–700 µs pulse duration	[40]
Saccha-romyces cerevisiae	15% Apple juice	3.3	Not reported	25 kV cm^{-1} pulse intensity, 8.3 number of pulses for 2 µs pulse duration	[8]
Bacillus cereus	0.15% NaCl	1.3	Not reported	20 kV cm^{-1} pulse intensity, 10.4 number of pulses for 2 µs	[8]
Staphy-lococcus aureus	simulated milk ultrafiltrate (SMUF)	3–4	Not reported	16 kV cm^{-1} pulse intensity and 60 pulses with 200–300 µs pulse width	[34]
L. mono-cytogenes ATCC	Watermelon juice	3.7	<40	35 kV cm^{-1} pulse intensity, 4 µs pulse width for 2000 µs pulse duration	[28]

13.3   HIGH-INTENSITY PULSE LIGHT TECHNOLOGY

High-intensity pulsed light (HIPL) is an innovative method of purification and sterilization of food items, decontamination of surfaces of processing and packaging materials by using very high-power and very short-duration pulses of light emitted by inert gas flash lamps. The short-duration and high-power electric pulses are converted into short-duration and high-power pulses of radiation or intense broad spectrum of light included in the spectra of ultraviolet (UV), visible (VL), and infrared (IR) light. PL is produced using technologies that multiply the power many fold. Power is magnified by storing electricity in a capacitor over relatively long times (fractions of a second) and releasing it in a short time (millionths or thousandths of a second). The emitted light flash has a high peak power and consists of wavelengths from 200 to 1100 nm.[9] The technique used to produce flashes originates, besides high peak power, a greater relative production of light with shorter bactericidal wavelengths. The lethal effect of PL on microorganisms is mostly attributed to the photochemical action of the UV part of the spectrum emitted by the flash lamp. Microbial DNA absorbs UV light that induces chemical modifications in its structure, resulting in damage of genetic information, impaired replication and gene transcription, and eventually death of the cell.

TABLE 13.3 Comparison of Continuous and Pulsed UV Light.

Parameters	Continuous light	Pulsed light
Wavelength	254 nm	100–1100 nm
Instantaneous energy	Less	Magnified several 1000 folds
Inactivation mechanism	Photochemical DNA damage Thymine dimer formation	Photochemical changes and by local heating
Natural cooling of lamp	No	Enables lamp to cool between pulses
Mercury	Commonly used as a source	Provides mercury free alternatives
Inactivation efficiency	Normal	Up to four times increased as compared to continuous UV light
Temperature increase during UV treatment	No significant temperature increase	Significant temperature increase due to infrared

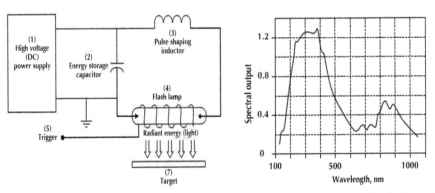

FIGURE 13.4 Schematic diagram of pulsed light equipment (left); spectral distribution of IFP-800 flash lamp (right).[29]

13.3.1   WORKING PRINCIPLE OF PULSED LIGHT TECHNOLOGY

The most important feature of delivering energy in the form of pulses is that apart from the number and duration of pulses, power provided by the pulses is greater than that provided by a continuous light radiation of equivalent total energy, total energy being equal. Another important consequence of the short duration of light pulses is the reduced time availability for thermal conduction inside the material. This results in a very rapid heating of a thin-surface layer up to a temperature much higher than the steady state temperature achieved by a continuous light radiation of equivalent total energy, without significantly increasing the bulk temperature. Table 13.3 illustrates the comparative difference between continuous light and PL.

13.3.1.1   PULSED LIGHT COMPONENTS (FIG. 13.4)

   a. High-voltage power supply.
   b. Storage capacitor.
   c. Flash lamps.
   d. PFN.
   e. Trigger.

13.3.1.2   PULSE LIGHT GENERATION

An electrical power supplier is generally used to convert line low-voltage AC power into high-voltage DC power. Energy storage is normally performed by using a capacitor bank, that is, a number of high-voltage capacitors connected in parallel, which accumulate energy from the electrical power supplier during the charge phase and release it during the discharge phase, thus supplying large amounts of current. The conversion of the continuous low- into the pulsed high-electric power is obtained by means of special switches capable of handling very high power and performing opening/closing cycles of very short duration, by instantaneously passing from a perfect insulating condition to a perfect conducting condition. The action of the switches is regulated by a controller that determines the pulse shape and the electrical operating conditions in order to yield the optimum PL wavelength for a particular application.[31] The high-power pulsed electric energy delivered by the switches is usually converted into high-power light pulses by means of gas-filled flash lamps or other PL sources. The current associated with the high-power electric pulses passes through the gas in the lamp transferring energy to some atoms of the gas which are carried in an "excited-state"; afterward, they tend to go back spontaneously in conditions of lower energy, that is, to the ground state, giving off energy in the form of intense pulses of light. The obtained energy is finally delivered to the target by various systems depending on the different applications.

13.3.1.3   MICROBIAL INACTIVATION MECHANISM

The lethal effects of PL can be attributed to its rich broad spectrum UV content, its short duration, high peak power, and the ability to regulate both the pulse duration and frequency output of flash lamps which play a major role in microbial destruction.[10,45] However, it appears that both the visible and infra-red regions, combined with the high peak power of PL, contribute to killing microorganisms. The effects of HIPL emissions of high or low UV content on the survival of predetermined populations of *L. monocytogenes*, *E. coli*, *Salmonella enteritidis*, *B. Cereus*, and *S. aureus*, *P. aeruginosa* were investigated and inactivation increased with increasing treatment intensity.[42] The improved the inactivation was again observed. Wang et al.[48] determined the germicidal efficiency against *E. coli* as a function of wavelength over the range 230–360 nm using about 6 mJ/cm^2. Furthermore, the authors concluded that the rich UV content from 220 to 290 nm in the UV spectrum provides the major contribution to inactivation, whichever type of UV source is used.

Despite the increasing efforts of scientists to fully elucidate the mechanisms of microbial inactivation by PL, the specific mechanisms by which PL causes cellular inactivation are not yet fully understood. Various mechanisms have been proposed to explain the lethal effect of PL, all of them related to UV part of the spectrum and its photochemical and photothermal effect.[3,49,50] It is possible that both mechanisms coexist, and the relative importance of each one would depend on the target microorganism.

13.3.1.4   PHOTOCHEMICAL MECHANISM

Considerable research has been performed on the mechanism of microbial inactivation by light pulses. The primary cell target of PL is nucleic acids because DNA is the target molecule of these UV wavelengths. The germicidal effect of UV light has been attributed primarily to a photochemical transformation of pyrimidine bases in a DNA of bacteria, viruses, and other pathogens to form dimers. The formation of such bonds prevents DNA unzipping for replication and the organism becomes incapable of reproduction. Without sufficient repair mechanism, such damage results in mutation, impaired replication and gene transcription, and ultimately the death of the organism.[24] Many authors also suggested that the microbial cells are disintegrated due to overheating of its constituents[49] and membrane disruption due to steam production in the cell.[15]

13.3.1.5 PHOTO-THERMAL MECHANISM

With energy exceeding 0.5 J cm^{-2}, the disinfection is achieved through bacterial disruption during their temporary overheating resulting from the absorption of UV light from the flash lamp. Due to this overheating, the water content of the bacteria vaporized generating a small steam flow that induces membrane disruption.[46] Moreover, a ruptured top of treated *Aspergillus niger* spores evidently punctured by an escape of overheated contents of the spore, which is empty after such an explosion during the light pulse.[49]

13.3.2 APPLICATION OF PULSED LIGHT TECHNOLOGY IN FOOD PROCESSING

The PL is a surface treatment. Its action is limited to a 1–2-μm UV penetration in the surface layer. The effectiveness of the PL has been proven on many types of microorganisms, such as bacteria, mold, virus, etc. The result of the light flash is a combination of photothermal and photochemical effects. Pulsed UV light can be used for sanitization, decontamination, and sterilization of smooth, dry surfaces such as aluminum, paper, glass, medical devices, and packaging materials with implementation in clean room pass-through tunnels, and above mail conveyor belts. Additionally, pulsed UV light can be used for decontamination of rough surfaces found on food and other surfaces such as laboratory benches and inside safety hoods. It can also be used for sanitization, decontamination, and sterilization of UV transmissive liquids, such as water, process chemicals, clear liquid pharmaceutical products, buffers, and dilute protein solutions for virus inactivation procedures. In order to increase the light absorption of materials being treated by PL, some absorption enhancing agents can be used.[11] These agents can be sprayed, vaporized or spread in the form of powder on the product surface or applied as a dissolved liquid. Although the absorption-enhancing agents may be easily removed from the product after processing, food items require usage of edible enhancing agents, generally based on approved food colorants such as carotene, lime green, black cherry, as well as natural cooking oils.[11]

There are two major areas for PL treatments: decontamination (killing of microorganisms) and reduction of mycotoxins (substances secreted by molds such as *Aspergillus*, *Penicillium*, *Fusarium*). For decontamination process PL can be treated for bakery products such as bread, pastries and cakes for enhancement of shelf life under refrigerated storage conditions up to 7–10 days. Eggs and raw meats are surface sterilized by pulsed UV and pulsed broad spectrum lights for killing microorganisms. PL can be treated to improve safety and increasing shelf life like lightly preserved fish products like cold smoked fish. Patulin, a mycotoxins generated by *Aspergillus* and *Penicillium*, and some other mycotoxins namely, Trichothecens, Fumonisin, Zearaleon, are also reduced to a significant level by treating the food materials with broad spectrum PL.

Investigators[16] examined celeriac, green bell peppers, iceberg lettuce, radicchio, soybean sprouts, spinach, and white cabbage treated with up to 2700 pulses with a xenon flash lamp with a pulse duration of 30 µs and an intensity of 7 J at a distance of 128 mm from the lamp. Microbial reduction ranged from 0.56 to 2.04 log for the variety of produce examined. Regarding fruits, a reduction of 1.1 and 4.3 log CFU/g of *E. coli O157:H7* with 1.9 and 22.6 J/cm^2, respectively, on blueberries treated with a SteriPulse-XL 3000 lamp at a distance of 80 mm from the lamp.[5] No significant differences were seen at other fluence doses between 1.9 and 22.6 J/cm^2. The reduction of *L. monocytogenes* and *E. coli O157:H7* was reported[30] on the muscle and skin side of raw salmon fillets using a SteriPulse-XL 3000. For *E. coli O157:H7*, a reduction of 0.30 and 1.09 log CFU/g was seen on the muscle and skin side, respectively, when treated at a distance of 80 mm with 180 pulses.

13.3.3 EFFECTS OF PULSE LIGHT ON FOOD QUALITY

Some studies related to the effect of PL treatment on different foods had been conducted and the effect on food nutritional components such as proteins and vitamins, and inactivation of some enzymes present in some food products within a layer of 0.1 mm deep were evaluated. The key enzyme for enzymatic browning in lots of fruits and vegetables is polyphenol oxidase (PPO). It was reported to be inactivated significantly potato slices by treatment of

2–3 flashes of broad spectrum PL at a fluence of 30 kJ/m^2.[11] It was also observed that the samples retained their color after a prolonged storage, while the untreated samples began to brown through the action of PPO in a few minutes. However, the opposite unexposed surfaces of treated samples showed visible browning. In the similar work, the activity of alkaline phosphatase, which catalyzes the hydrolysis of phosphatase esters, was reduced by 60–70% with a single pulse of broad spectrum light at fluence of 10 kJ/m^2 and completely inactivated by 5 pulses at the same fluence.[11]

In addition to microbial destruction, one can also question whether pulsed-light treatment induced conformational changes in food components. Concerning nutritional properties of PL-treated food materials, the nutritional analysis by testing protein, riboflavin, nitrosamine, benzopyrene, and vitamin C[10] on frankfurters exposed up to 300 kJ/m^2 of PL was conducted. Total treatment showed that no differences between treated and untreated samples were observed. But a strong loss of riboflavin was observed in foods because of heat, oxygen, and light. Again, there were no effects on concentrations of riboflavin in beef, chicken, and fish. Results from tests performed at Xenon Corp's lab in Wilmington, MA on Portobello and White mushrooms, show that as few as 2 pulses, applied in under one second (3 pulses/second flash rate), increase vitamin D to over 100% RDA. The system used for this study was a commercially available model RC-847 controller and 16-in. lamp housing, model LH-840.

13.4 HIGH HYDROSTATIC PRESSURE PROCESSING TECHNOLOGY

HPP is a nonthermal processing method, subject foods, which can be liquid or solid, packaged or unpackaged, up to a pressure range of 40,000 and 80,000 pounds/square inch (PSI), usually for five minutes or less. The effects of high pressures on foods were first discovered in 1899 by Bert Hite of the Agriculture Research Station in Morgan town, West Virginia, USA, designed and constructed a high-pressure unit to pasteurize milk and other food products in excess of 680 MPa. The application of HPP is recently been widely recognized and commercialized though the potential was largely ignored through most of the last century. Most of the research studies of the research regarding the use of high pressure for food preservation has concerned inactivation of microorganisms; the pressure stability of food enzymes. HPP inactivates the microorganisms. The advantage of the HPP is the expanded shelf-life and improvement of food safety due to the inactivation in the microbial population. The loss of viability of microorganisms through HPP is probably the result of a combination of injuries in the cell. HPP is equally effective on molds, bacteria, viruses, and other parasites and also has achieved some success in treating bacterial spores, which are extremely resistant to many types of thermal processing treatments. In addition, the process helps food retain its quality of freshness and maintains the nutrient content retaining, for example, heat-sensitive vitamins. HPP of foods is of interest to food processing industries, because it permits microbial inactivation at moderate temperatures with minimum degradation in quality. The energy required in HPP is far less than that required in the thermal treatment process.

13.4.1 WORKING PRINCIPLE AND SYSTEM COMPONENTS OF HHPP

There are two basic principles which determine the effect of high pressure on food material:

i. Pascal's isostatic principle: that pressure applied to a sample including biological products is transmitted in a uniform and quasi-instantaneous manner.
ii. Le Chatelier Principle: When a system at equilibrium is subjected to change in concentration, temperature, volume, or pressure, then the system readjusts itself to (partially) counteract the effect of the applied change and a new equilibrium is established.

The energy required for pressurization at 400 MPa is comparable to that required for raising the temperature of the material from 0 to 30°C. Therefore, HPP consumes less energy compared to thermal processing. The effect of high pressure in a food product is dependent on the amount of pressure applied, duration of compression, depressurization rate, temperature of treatment, product pH, water activity, and salt concentration.

A high-pressure system generally consists of a high-pressure vessel, its closure, pressure-generation system, temperature-control device, and material-handling system. The vessel is filled with a pressure-transmitting medium after loading and closing by its closure and then air is removed from the vessel by means of a low-pressure fast-fill-and-drain pump, in combination with an automatic deaeration valve, and high hydrostatic pressure is then generated. High pressure can be generated by direct or indirect compression, or by heating of the pressure medium. Pressurizing a medium with the small diameter end of a piston generates direct compression and the large diameter end of the piston is driven by a low-pressure pump (Fig. 13.5). This method allows very fast compression, but the limitations of the high-pressure dynamic seal between the piston and the vessel internal surface restrict the use of this method to small-diameter laboratory or pilot plant systems.[25] Indirect compression uses a high-pressure intensifier to pump a pressure medium from a reservoir into a closed high-pressure vessel until the desired pressure is reached (Fig. 13.5). Indirect method of compression is generally used in industries for most isostatic processing.

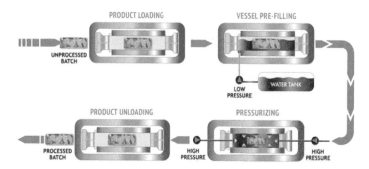

FIGURE 13.5 Process flow of high-pressure processing of food material.

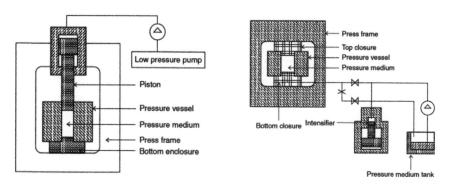

FIGURE 13.6 Generation of high pressure by direct (left) and indirect (right) compression of the pressure-transmitting medium.[25]

In processing operation, a sterile container filled with food material is sealed with Ethylene-vinyl alcohol copolymer and polyvinyl alcohol films and placed in the pressure chamber for pressurizing. Multilayer plastic and some aluminum packages may be used for HPP. The basis for applying high pressure to foods is to compress the water surrounding the food. When the required operating pressure is attained, the pumping rate is reduced. At the end of the specified holding time, the pressure vessel is decompressed in two stages to avoid sudden release of pressurized water (Fig. 13.6). The holding time in the pressure vessel depends on the type of food and process temperature.

13.4.2 APPLICATION OF HIGH-PRESSURE PROCESSING IN FOOD PROCESSING

High hydrostatic pressure does not affect the low energy, covalent bonds, which have low compressibility and do not break within the ranges of pressures normally used in food. Therefore, the structures of protein and fatty acids remain intact. But the secondary and tertiary structures of these complex molecules for maintaining the secondary and tertiary structure of proteins are disrupted, and this event is associated with decreases in volume because the structures are made up of ionic bonds and hydrophobic interactions. However, vitamins, amino acids, flavor molecules, or other low-molecular-weight compounds are hardly affected and as a result, the organoleptic and nutritional properties are not significantly changes. Due to alteration of the functional properties the activity of certain enzymes is inhibited, the nutrient digestibility and bioavailability modified. Thus, the application of HPP relates to enzyme inactivation, changes in the functional capacity, heat transfer, microbial inactivation, etc.

The major application of HPP in food processing is the inactivation of microorganisms and denaturation of enzymes, which allows a substantial increase in shelf-life and improves food safety. The study of pressure effects on living organisms is called barobiology and the effect of HPP on inactivation efficiency of microorganisms and the process parameters are given in the Table 13.4. The effects of pressure on microorganisms in foods are determined by the effect of pressure on water, temperature of treatment, constituents of foods, and the properties and physiological state of microorganisms. Most of the studies carried out in the 1960s and 1970s on the effects of high pressure on intact cells focused on microorganisms at pressures naturally present in the biosphere,[18] where species can withstand pressures of 20–30 MPa. The microorganisms capable of growth above 40 MPa are called barophiles, those that can grow within the range of 0.1 and 50 MPa are called eurybaric, and those that can survive, but cannot grow at pressures of 50–202 MPa are called baroduric.[4] The acidophiles, which can withstand acid pH and that microbial growth, can occur under such conditions, as in the case of some yeasts, molds, and bacteria. However, the application of high pressure on fruits and vegetable juices has been effective on deteriorative and pathogen microorganisms when pressure is applied with sufficient intensity.

TABLE 13.4   Inactivation of Microorganisms by High Hydrostatic Pressure Treatment.

Microorganisms	Suspension media	Inactivation (log CFU)	Pressure (MPa)	Time (min)	Temperature (°C)	Reference
E. coli O157:H7	Poultry meat	3	600	15	20	[32]
E. coli O157:H7	Broth	8.14	345	10	35	[1]
E. coli O157:H7	Mango juice	2	400	15	20–23	[17]
E. coli O157:H7	Fish slurry	6	400	18-30	20–25	[36]
Salmonella enteritidis	Broth	8.22	345	10	35	[1]
Salmonella **spp.**	Ground chicken	5	450	10	4–6	[44]
Salmonella **Senftenberg 775W**	Strained baby food	<2	340	10	23	[26]
Staphylococcus aureus	Poultry meat	3	600	15	20	[32]
Listeria monocytogenes	Fish slurry	7	400	10	20–25	[36]

HPP affects the cell morphology in various ways, given as (1) collapsing of intracellular gas vacuoles, (2) cell enlargement in case of *E. coli* and *Pseudomonas*, (3) movement of mobile microorganisms may inhibited, (4) thinning of cell walls, and (5) reduction in number of ribosomes.

HPP of foods also leads to inactivation of enzymes, influenced by the pH, substrate concentration, subunit structure of enzyme, and temperature of treatment. The mechanism of high-pressure inactivation of enzymes involves series of events such as formation or disruption of interactions, change in the native structure of enzymes,[22] which eventually leads to a change in enzyme activity as its specificity is related to the structure of its active site. The inactivation of enzymes in fruits and purees by using HPP is reviewed in Table 13.5.[7]

TABLE 13.5 Denaturation of Enzymes by High Hydrostatic Pressure Treatment.[7]

Sample (medium)	Enzyme	Target P/T in (MPa/°C)	CUT/DCT> in (min/min)	Activation or inactivation	Maximum PE (log) values (treatment condition)	Reference
Apple juice (pH 3–4; 12 °Brix)	Amylase	100–400/6–40	0.5–3/<0.15	Inactivation	1.79 (400 MPa/30°C/pH 3	[39]
Pineapple puree (pH 3–4; 12 °Brix)	PPO and POD	200–500/30–60	0.8–1.8/<0.15	Inactivation	0.332 (PPO) and 0.319 (POD) (500 MPa/60°C/pH 3	[7]
Strawberry puree (pH 3.52; 6.5 °Brix)	PPO, PME, and PG	200–600/40–80	0.25–0.75/<0.15	Inactivation	0.135 (PPO), 0.223 (PME), and 0.315 PG) (600 MPa/80°C/30% added sugar)	[6]
Litchi juice (pH 4)	PPO and POD	300–600/30	1.1–2.11/<0.15	Inactivation	Activation for both (max. > 130% and 225%, respectively, at 300 MPa)	[20]
Apple juice (pH 3.5, 12 °Brix)	PME	250–400/25	1.5–3/<0.25	Inactivation	1.05 (400 MPa/25 °C/–)	[38]

P: pressure; T: temperature; CUT: pressure come-up time; DCT: decompression time; PE: pulsed effect.

13.5 IONIZING RADIATION

Radiation generally means the transmission of energy from a source into a surrounding medium or destination. Irradiation is the condition of being exposed to radiation. Ionizing radiations are part of the electromagnetic spectrum having relatively short wavelengths and high energy. These radiations generate electrically charged ions by ejecting electron from atom of a molecule in any matter leading to further ionization. Ionization can also break molecules into smaller fragments with unpaired electron. These small fragments with unpaired electrons are called free radicals. The end product of radiation affected molecules of a material are called radiolytic products. Ionization radiation is highly penetrating and effective. Ionizing radiation, which are permitted by international regulatory bodies such as *Codex General Standards for food Irradiation*, are as follows:

a. Gamma rays from radioisotope Cobalt^{60} or Cesium^{137}.
b. X-rays at or below an energy level of 5 MeV.
c. Electrons generated from machine sources at or below energy level of 10 MeV.

13.5.1 GAMMA RAYS

Gamma rays are at the high-energy end of the electromagnetic spectrum which is produced during changes inside the atomic nucleus and which carry more than 1 million eV. Gamma rays are produced from radioisotopes, and hence, they have fixed energies. In practice, Cobalt^{60} is the major isotope source which is manufactured for irradiation. An alternative radioisotope is Cesium^{137}, which is a byproduct of nuclear fuel reprocessing and is used very much less widely than Cobalt^{60}. Gamma rays can penetrate the food product to a great depth exposing the pathogens in deep layers of the irradiated product. The isotope Cesium^{137} and Co^{60} has been used as a gamma ray source in medical applications too.

13.5.2 X-RAYS

X-rays are electromagnetic waves with energies in the hundred thousand electron-volt range and with wavelengths in the order of 10^{-10} m. The X-rays are produced by energy state transitions of inner electrons close to the nucleus. X-rays are machine sources, which are powered by electricity. Hence, they exhibit a continuous spectrum of energies depending on the type and conditions of the machinery. They hold a major advantage over isotopes in that they can be switched on and off and can in no way be linked to the nuclear industry.

X-rays are generated by bombarding heavy metal targets like tungsten and molybdenum with fast moving electrons. These electrons lose energy after colliding with electrons of tungsten or molybdenum or better to mention as target materials. The electrons of the target material get excited and are ejected from the atom utilizing the energy from the incident high-velocity electron. The emitting radiations appear as electromagnetic radiation of a wide and almost continuous range of wavelengths superimposed on a few "characteristic" lines. Since each element has distinctive energy levels, these "characteristic" X-rays are specific for the element bombarded. The fact that X-rays consist of a wide variety of wavelengths is the main difference from gamma radiation.

At the present time, the use of X-rays for the treatment of food on an industrial scale is not proposed, but the principle involved is the same as that in the employment of gamma radiation. For food treatment the X-ray machines should be operated at an energy level of 5 MeV, or lower.

13.5.3   UV LIGHT

In UV light processing, radiation is obtained from an electromagnetic spectrum's UV region. Microbial inactivation can be obtained by exposing the food material to 400 J/m^2 in all parts of the product. Critical factors of the system include the transmissivity of the product, geometric configuration of the reactor, power, wavelength, and physical arrangement of the UV sources, product flow profile and radiation path length. The UV radiations also react synergistically with oxidizing agent such as ozone, and so can also be used to treat fruit juices and cider microbiologically, although, ozone can oxidize desirable moieties. The radiation suffers from an inability to penetrate food as all action is on the surface. The UV radiation is frequently employed to irradiate the surfaces of food package materials ostensibly to sterilize or at least sanitize them prior to aseptic extended shelf-life (ESL) packaging. Some of the microbicidal effects are attributable to the generation of ozone, a powerful oxidizing agent. In some ESL operation for dairy products, UV is applied in synergy with chemical sterilants on surfaces of packaging material to increase the microbicidal effect and to reduce the residual chemicals.

13.5.4 APPLICATION OF IONIZING RADIATION IN FOOD PROCESSING

Ionizing radiation is used as a nonthermal food pasteurization process that reduces or eliminates spoilage and pathogenic microorganisms. Application of irradiation process may be applied for disinfecting various agricultural commodities such as legumes, cereal grains, spices, fruits, and vegetables (Table 13.6). Apart from disinfection from insects, irradiation can be used to retain color of meat and control microbial activity in poultry and meat. Of the freshness enhancing nonthermal technologies, many consider irradiation to be the most effective approach to eliminate pathogens and spoilage microorganisms from the food supply.[23]

TABLE 13.6 Application of Radiation in Food Processing.[41]

Dose	Application
Low dose (less than 1 kGy)	Inhibition of sprouting or germination in certain crops (onions, potatoes, etc.)
	Delay of senescence or control of ripening of some tropical foods
	Killing insects in cereal grains, fruit, cocoa beans, and other crops
Medium Dose (1–10 kGy)	Killing food-poisoning bacteria, particularly *Salmonella* and *Campylobacter*, in raw poultry, prawns and shellfish
	Killing parasites such as *Trichinellaspiralis* and *Taeniasaginata* in raw meat
	Extension of product life by reduction of microbial populations (by ca. 10^6) that spoil meat, fish, fruit, and vegetables
	Improvement of technological properties (increased juice yield from fruits, reduced cooking times for dehydrated vegetables, etc.)
	Sterilization of insects and parasites
	Sterilization of packaging materials
High dose (greater than 10 kGy)	Sterilization of food
	Reduction of bacterial contamination of herbs and spices (by ca. 10^6)
	Reduction in the number of viruses (by ca. 10^6)
	Enzyme inactivation

There are three approaches to the use of radiation in food products; low doses, up to 1 kGy, to inhibit sprouting and delay fruit ripening; medium doses, 1–10 kGy, for partial destruction of microbial flora, to reduce the risk of food pathogens and to increase shelf life; and high doses more than 10 kGy, needed for complete destruction of microorganisms to achieve sterility of a particular food product. In addition to the three levels of radiation application, the following trade type names have been given to these general ranges, which relate more to the desired function than to the actual dose.[43]

Radurization refers to treatment of foods with ionizing radiation sufficient to lengthen shelf life by reducing the initial number of spoilage organisms before or immediately after packaging. This amount varies with individual food products because spoilage conditions and storage conditions change with each commodity. Radurization doses are usually considered low dose less than 2 kGy.

Radicidation is the irradiation treatment required to sufficiently reduce the level of nonspore-forming pathogens, including parasites, to an undetectable level, thus reducing the risk of food borne illness to near zero. This level of irradiation is generally considered in the medium range, <5 kGy, and may vary depending on the product and possible suspected pathogens.

Radappertization is the highest level of irradiation processing required to achieve sterility in a food product. This application allows for shelf stability at ambient temperatures much as does "canning" or "aseptic packaging." Doses required for radappertization generally are >10 kGy for most food products.[2]

Irradiation is likely to be generally accepted in the future as useful to the public's health as pasteurization of milk is today. Its benefits exceed its risk from a Risk–Benefit analysis perspective. However, there are still lingering questions about some aspects of the new technology.

Irradiation uses electron beams, which could in turn generate X-rays as they interact with matter, or gamma rays sources to irradiate food products. X-rays and gamma rays are short wave length electromagnetic radiation that is not capable, at the energies used, of transmuting nuclei and forming radioactive isotopes in food. Thus, they do not increase human exposure to radiation. Once the food has been irradiated, the food pathogens are destroyed, and the radiation does not remain in the irradiation food.

13.6 CONCLUSIONS

The effectiveness of nonthermal food processing has been proven in the laboratory and there are a number of examples of scale-up. Compared to other thermal processing technology, nonthermal technologies have emerged as the most promising field for microbial safety, enzyme inactivation, and various extraction purposes with additional benefit of retention of nutritional and sensory qualities. Nonthermal technologies are being emphasized as consumer nowadays demands minimally processed high quality and safe food. To expand the applicability and to increase the efficiency, these technologies can be combined with traditional food preservation methods. The limitation associated with nonthermal technologies regarding cost of the equipment and spore injury rather than complete death, which leads to further development of microbes during storage in ambient temperatures. To overcome the problems associated with the technologies, research is still going on and the hence these are the most promising technologies till date.

KEYWORDS

- **barophiles**
- **electromagnetic rays**
- **electroporation**
- **flash lamp**
- **high-pressure processing**
- **hydrostatic pressure**
- **irradiation**
- **nonthermal technology**
- **pulse effect**
- **pulse electric field**
- **pulse light**
- **radappertization**
- **radicidation**
- **radurization**

REFERENCES

1. Alpas, H.; Kalchayanand, N.; Bozoglu, F.; Ray, B. Interactions of High Hydrostatic Pressure, Pressurization Temperature and pH on Death and Injury of Pressure-resistant and Pressure-sensitive Strains of Foodborne Pathogens. *Int. J. Food Microbiol.* **2000,** *60*(1), 33–42.
2. Anderson, A. W. Irradiation in the Processing of Food. In *Food Microbiology*; Rose, A. H. Ed.; Academic Press: London, 1983; pp 145–169.
3. Anderson, J. G.; Rowan, N. J.; MacGregor, S. J.; Fouracre, R. A.; Farish, O. Inactivation of Food Borne Enteropathogenic Bacteria and Spoilage Fungi Using Pulsed Light. *IEEE Trans. Plasma Sci.* **2000,** *28*, 83–88.

4. Barbosa-Ca´ Novas, G. V.; Pothakamury, U. R.; Palou, E.; Swanson, B. G. *Non-thermal Preservation of Foods*. Marcel Dekker, Inc.: New York, 1998.

5. Bialka, K. L.; Demirci, A. Decontamination of *Escherichia coli* O157:H7 and *Salmonella enterica* on Blueberries using Ozone and Pulsed UV-light. *J. Food Sci.* **2007,** *72,* M391–M396.

6. Chakraborty, S.; Mishra, H. N.; Knorr, D. *Strawberry Enzyme Inactivation by HPP: Models & Contours*. Lambert Academic Publishing: Germany, 2012.

7. Chakraborty, S.; Kaushik, N.; Rao, P. S.; Mishra, H. N. High-Pressure Inactivation of Enzymes: A Review on Its Recent Applications on Fruit Purees and Juices. *Comprehen. Rev. Food Sci. Food Saf.* **2014,** *13*(4), 578–596.

8. Cserhalmi, Z. S.; Vidacs, I.; Beczner, J.; Czukor, B. Inactivation of *Saccharomyces cerevisiae* and *Bacillus cereus* by Pulsed Electric Fields Technology. *Innov. Food Sci. Emerg. Technol.* **2002,** *3*, 41–45.

9. Dunn, J.; Bushnell, A.; Ott, T.; Clark, W. Pulsed White Light Food Processing. *Cereal Foods World* **1997,** *42*, 510–515.

10. Dunn, J.; Clark, W.; Ott, T. Pulsed-light Treatment of Food and Packaging. *Food Technol.* **1995,** *49*(9), 95–98.

11. Dunn, J. E.; Clark, W. R.; Asmus, J. F. *Methods for Preservation of Foodstuffs*. Maxwell Laboratories Inc.: San Diego, CA, 1989. US Patent 4871559.

12. EPRI. *Pulsed Electric Field Processing in the Food Industry: A Status Report on PEF*. Industrial and Agricultural Technologies and Services: Palo Alto, CA, 1998, CR-109742.

13. Evrendilek, G. A.; Zhang, Q. H. Effect of Pulse Polarity and Pulse Decaying Time on Pulsed Electric Fields induced Pasteurized of *E. coli* O157:H7. *J. Food Eng.* **2005,** *6*, 271–276.

14. FDA. Effect of Preservation Technologies and Microbiological Inactivation in Foods. *Evaluation and Definition of Potentially Hazardous Foods*, 2006.

15. Fine, F.; Gervais, P. Efficiency of Pulsed UV Light for Microbial Decontamination of Food Powders. *J. Food Protect.* **2004,** *67*, 787–792.

16. Gomez-Lopez, V. M.; Devileghere, F.; Bonduelle, V.; Debevere, J. Intense Light Pulses Decontamination of Minimally Processed Vegetables and their Shelf-life. *Int. J. Food Microbiol.* **2005,** *103*, 79–89.

17. Hiremat, N. D.; Ramaswamy, H. S. High-pressure Destruction Kinetics of Spoilage and Pathogenic Microorganisms in Mango Juice. *J. Food Process. Preserv.* **2012,** *36*(2), 113–125.

18. Hoover, D. G.; Metrick, C.; Papineau, A. M.; Farkas, D. F.; Knorr, D. Biological Effects of High Hydrostatic Pressure on Food Microorganisms. *Food Technol.* **1989,** *43*(3), 99–107.

19. Hulsheger, H.; Potel, J.; Niemann, E. G. Electric Field Effects on Bacteria and Yeast Cells. *Radiation Environ. Biophys.* **1983,** *22*, 149–162.

20. Kaushik, N.; Kaur, B. P.; Rao, P. S. Application of High Pressure Processing for Shelf Life Extension of Litchi Fruits (*Litchi chinensis* cv. Bombai) during Refrigerated Storage. *Food Sci. Technol. Int.* **2013,** *20*(7), 527–541.

21. Lubicki, P.; Jayaram, S. High Voltage Pulse Application for the Destruction of the Gram-negative Bacterium *Yersinia enterocolitica. Bioelectrochem. Bioenergy* **1997,** *43*, 135–141.

22. Ludikhuyze, L.; Van Loey, A.; Denys, I. S.; Hendrickx, M. E. Effects of High-pressure on Enzymes Related to Food Quality. *Ultrahigh Pressure Treatments of Foods*; Springer: New York, pp 115–166.

23. Mathee, F. N.; Marais, P. G. Preservation of Food by Means of Gamma Rays. *Food Irradiation* **1963**, *4*, A10–A11.

24. Mcdonald, K. F.; Curry, R. D.; Clevenger, T. E.; Unklesbay, K.; Eisenstark, A.; Golden, J.; Morgan, R. D. A Comparison of Pulsed and Continuous Ultraviolet Light Sources for the Decontamination of Surfaces. *Plasma Sci.* **2000**, *28*, 1581–1587.

25. Mertens, B. *New Methods of Food Preservation.* Blackie Academic and Professional: New York, 1995; p 135.

26. Metrick, C.; Hoover, D. G.; Farkas, D. Effects of High Hydrostatic Pressure on Heat-sensitive Strains of *Salmonella. J. Food Sci.* **1989**, *54*, 1547–1564.

27. Mohamed, M. E. A.; Eissa, A. H. A. Pulsed Electric Fields for Food Processing Technology. *Structure and Function of Food Engineering*; License Intech, 2012; pp 275–306.

28. Mosqueda-Melgar, J.; Raybaudi-Massilia, R. M.; Martin-Belloso, O. Influence of Treatment Time and Pulse Frequency on *Salmonella enteritidis, Escherichia coli* and *Listeria monocytogenes* Populations Inoculated in Melon and Watermelon Juices Treated by Pulsed Electric Fields. *Int. J. Food Microbiol.* **2007**, *117*(2), 192–200.

29. Nederita, V. Studii privindimbunatatirea operatiilor si a instalat-iilordin industria alimentara prinutilizareaimpulsurilorultrascurte de lumina de intensitateinalta. Teza de doctorat, UniversitateaDunarea de Jos Galati (Studies concerning food operations and equipment improvement by using ILP treatment, *Ph. D thesis*, Galati Dunarea de Jos University), 1995.

30. Ozer, N. P.; Demirci, A. Inactivation of *Escherichia coli* O157:H7 and *Listeria monocytogenes* Inoculated on Raw Salmon Fillets by Pulsed UV-light Treatment. *Int. J. Food Sci. Technol.* **2006**, *41*, 354–360.

31. Pai, S. T.; Zhang, Q. Energy Storage and Switch. *Introduction to High Power Pulse Technology.* World Scientific Publishing: Singapore, 1995.

32. Patterson, M. F.; Quinn, M.; Simpson, R. K.; Gilmour, A. Sensitivity of Vegetative Pathogens to High Hydrostatic Pressure Treatment in Phosphate Buffered Saline and Foods. *J. Food Protect.* **1995**, *58*, 524–529.

33. Pothakamury, U. R.; Monsalve-Gonzàlez, A.; Barbosa-Canovas, G. V.; Swanson, B. G. Inactivation of *Escherichia coli* and *Staphylococcus aureus* in Model Foods by Pulsed Electric Field Technology. *Food Res. Int.* **1995**, *28*(2), 167–171.

34. Pothakamury, U. R.; Vega, H.; Qinghua, Z.; Barbosa-Canovas, G. V.; Swanson, B. G. Effect of Growth Stage and Processing Temperature on the Inactivation of *E. coli* by Pulsed Electric Fields. *J. Food Protect.* **1996**, 1153–1247.

35. Qin, B. L.; Chang, F.; Barbosa-Cfinovas, G. V.; Swanson, B. G. Non-thermal Inactivation of *Saccharomyces cerevisiae* in Apple Juice using Pulsed Electric Fields. *Lebensm.-Wiss. Technol.* **1995**, *28*, 564–568.

36. Ramaswamy, H. S.; Zaman, S. U.; Smith, J. P. High Pressure Destruction Kinetics of *Escherichia coli* (O157:H7) and *Listeria monocytogenes* (Scott A) in a Fish Slurry. *J. Food Eng.* **2008**, *87*(1), 99–106.

37. Reitler, W. Conductive Heating of Foods, Ph.D. Thesis, TU Munich, Munich, Germany, 1990.

38. Riahi, E.; Ramaswamy, H. S. High-pressure Processing of Apple Juice: Kinetics of Pectin Methyl Esterase Inactivation. *Biotechnol. Progr.* **2003**, *19*(3), 908–914.

39. Riahi, E.; Ramaswamy, H. S. High-pressure Inactivation Kinetics of Amylase in Apple Juice. *J. Food Eng.* **2004**, *64*(2), 151–60.

40. Rivas, A.; Sampedro, F.; Rodrigo, D.; Martinez, A.; Rodrigo, M. Nature of the Inactivation of *Escherichia coli* Suspended in an Orange Juice and Milk Beverage. *Eur. Food Res. Technol.* **2006,** *223*(4), 541–545.

41. Rosenthal, I. Electromagnetic Radiations in Food Science. *Advanced Series in Agricultural Sciences*; Springer-Verlag, 1992; p 19.

42. Rowan, N. J.; MacGregor, S. J.; Anderson, J. G.; Fouracre, R. A.; McIlvaney, L.; Farish, O. Pulsed-Light Inactivation of Food-related Microorganisms. *Appl. Environ. Microbiol.* **1999,** *65*(3), 1312–1315.

43. Satin, M. *Food Irradiation*; Technomic Publishing: Lancaster, 1993; pp 1–182.

44. Sheen, S.; Cassidy, J.; Scullen, B.; Uknalis, J.; Sommers, C. Inactivation of *Salmonella* spp. in Ground Chicken using High Pressure Processing. *Food Contr.* **2015,** *57*, 41–47.

45. Takeshita, K.; Shibato, J.; Sameshima, T.; Fukunaga, S.; Isobe, S.; Arihara, K.; Itoh, M. Damage of Yeast Cells Induced by Pulse Light Irradiation. *Int. J. Food Microbiol.* **2003,** *85*, 151–158.

46. Takeshita, K.; Yamanaka, H.; Sameshima, T.; Fukunaga, S.; Isobe, S.; Arihara, K.; Itoh, M. Sterilization effect of Pulsed Light on Micro-organisms. *BokinBobai* **2002,** *30*, 277–284.

47. Vega-Mercado, H.; Gongora-Nieto, M. M.; Barbosa-Canovas, G. V.; Swanson, B. G. Non-thermal Preservation of Liquid Foods using Pulsed Electric Fields. In *Handbook of Food Preservation*; Rahman, M. S., Ed.; Marcel Dekker, Inc.: New York, 1999.

48. Wang, T.; MacGregor, S. J.; Anderson, J. G.; Woolsey, G. A. Pulsed Ultra-violet Inactivation Spectrum of *Escherichia coli*. *Water Res.* **2005,** *39*, 2921–2925.

49. Wekhof, A.; Trompeter, F. J.; Franken, O. Pulsed UV Disintegration (PUVD): A New Sterilization Mechanism for Packaging and Broad Medical-hospital Applications. In The First International Conference on Ultraviolet Technologies, 2001, 14–16 June 2001, Washington, USA.

50. Wuytack, E. Y.; Phuong, L. D.; Artsen, A.; Reyn, K. M.; Marquenie, D. Comparison of Sublethal Injury Induced in *Salmonella entericasarovar Typhimurium* by Heat and by Different Non-thermal Treatments. *J. Food Protect.* **2003,** *66*, 31–37.

51. Zhang, Q. H.; Qiu, X.; Sharma, S. K. Recent Developments in Pulsed Electric Processing. In *New Technologies Yearbook*; Chandrana, D. I., Ed.; National Food Processors Association: Washington, DC 1997; pp 31–42.

GLOSSARY

High-pressure processing is a method of preserving and sterilizing food, in which a product is processed under very high pressure, leading to the inactivation of certain microorganisms and enzymes in the food.

Ionizing radiation is radiation that carries enough energy to free electrons from atoms or molecules, thereby ionizing them.

Pulse electric field is a nonthermal preservation methods, in which short electrical pulses to the disruption of cells are used.

Pulse light is a broad spectrum light source, with a general spectral range of 200–1100 nm.

CHAPTER 14

BIOSENSORS IN FOOD ENGINEERING

MINH-PHUONG NGOC BUI, SNOBER AHMED, and
ABDENNOUR ABBAS*

*Biosensors and Bionanotechnology Laboratory, Department of
Bioproducts and Biosystems Engineering, University of Minnesota,
2004 Folwell Hall, Falcon Heights, MN 55108, USA.
E-mail: mbui@umn.edu, ahmed580@umn.edu,
aabbas@umn.edu, http://www.abbaslab.com*

*Corresponding author.

CONTENTS

ABSTRACT

There are many advantages of using biosensors as an analytical tool in food processing which enable easy automation, simple and portable equipment, and fast analysis. Biosensors can be applied in quality control laboratories or in industrial facilities at any stage of food processing. The chapter highlighted the basic principle of different format of biosensors with classification of various biosensor generation and applications in food processing. Also, commercial biosensors are included along with top tier manufacturers involved in the development of biosensors for the food industry. Furthermore, the chapter discussed some major fields where biosensors can play critical role in the future of food safety and food defense.

14.1   INTRODUCTION

The increasing global demand of food has put pressure on farmers and processors to adopt new technological tools for enabling rapid control and efficient determination of inherent hazards and providing information on the safety and quality of food products. According to the World Health Organization (WHO), each year millions of people around the world are affected by foodborne diseases from caused by bacteria, viruses, biotoxins, and chemical substances that enter the food during processing, shipping, or commercialization. Biosensors are becoming an important tool in the agricultural and food sections to control production processes and to ensure food quality and safety. The global biosensor market is currently worth over 10 billion dollars annually and is a blossom of interdisciplinary research. Traditional analytical methods such as high-performance liquid chromatography (HPLC) and gas chromatography (GC) are usually off-line testing methods. This means that food samples have to be taken out from mainstream process to be analyzed in laboratory setting. This not only jeopardizes the safety and sterility of the food product but also increases the time needed for analysis. In addition, those types of testing require trained personals and expensive instrumentation. In this context, biosensors are promising alternative to conventional analytical tools as they offer advantages in size, cost, specificity, fast response, precision, sensitivity, and ease of handling. Biosensors can be integrated along the food supply chain, from the producer to the consumer. The integration of biosensors in the food chain is necessary for the development of modern industrial and manufacturing processes. They can provide real-time monitoring and manipulation of process variables (pH, temperature, oxygen, pressure, volatile substance), quantify the different compounds in food (concentration of sugar, amino acid, alcohols) as they become limiting factors in final products.[3] Biosensors can also be used for the implementation of hazardous analysis, biosafety food packaging, and pathogens detection in animals and plants. The detection of genetically modified organism (GMO) is another useful application for biosensors since many countries have established laws regulating GMO products and their derivatives.[7,15] With the advancement in new technologies, especially nanotechnology, biosensors have been qualitatively improved in term of selectivity, sensitivity, cost-effectiveness, and stability. These properties will be discussed later.

This chapter discusses components, properties, types of transducer, and fabrication of biosensors in food processing. The chapter also includes biosensor applications, and quality control of biosensors in food engineering, and food defense.

14.2   BIOSENSOR ARCHITECTURE AND FABRICATION

A biosensor is an analytical device that operates by converting a biological response into a detectable and measurable signal. Biosensors are often described as a 3-component system consisting of a bioreceptor, a transducer, and a signal processing unit. When the analyte interacts with the bioreceptor, the transducer converts biological interactions into measurable signals, which are then quantified with a signal processing unit. The signal can be obtained by optical, mechanical, or electrochemical methods according to the transducer type (Fig. 14.1). The performance of biosensors is evaluated by measuring the analytical parameters including:

- *Specificity and selectivity*: The biosensor should contain receptors that are specific to the target analyte and show minimum or no cross reactivity with molecules having analogous structure.
- *Sensitivity*: It is the ratio of change in sensor output to the corresponding change in the stimulus.
- *Operational range (linearity)*: The linear response range of the system should cover the concentration range over which the target analyte is to be measured.
- *Limit of detection:* It is the lowest possible sample concentration that can be detected by the biosensor with a signal three times higher than the noise or background signal.

- *Reproducibility*: The biosensor should be able to reproduce the signal response with the same analyte.
- *Response time and recovery time:* It is the time necessary for having 95% of the response. The response time of the biosensor should be short enough to monitor the analyte both in stationary or flow system. The recovery time should be small for reusability of the biosensor system.
- *Stability:* The biosensor should be stable in its environmental or storage conditions. Active components of the biosensors should remain active for a long time for practical usage and transportation.
- *Drift:* It refers to the variation in the output of a sensor independent of any change in the stimulus.

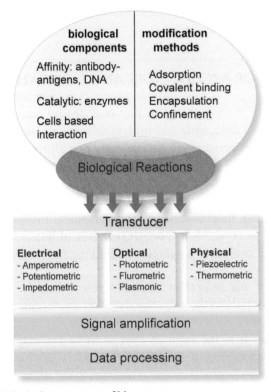

FIGURE 14.1　Principal components of biosensors.

Biosensors can be fabricated on a variety of materials including metal, glass, polymer, and cellulose papers. The bioreceptor that provides specificity usually consists of antibodies, enzyme, oligonucleotides, cells, specific chemical groups, or bacteriophage receptors. It provides unique and targeted interaction with the analyte in samples. For many years, enzyme-based biosensors have dominated the market with products such as glucose sensors and chemical sensors for toxins and pesticides. Many efforts to improve the sensitivity and stability of biosensors have been attempted using functional polymers and nanomaterials.[17,19,37] The second important part of a biosensor is a transducer. It can be categorized into three main groups depending on how it transfers biological interaction to measurable signals.

14.2.1 OPTICAL TRANSDUCERS

Optical sensors perform analysis by employing visible or ultraviolet light. They are usually composed of a light source and a detector. Based on the technique, the detector monitors transmittance, scattering, absorbance, diffraction, and/or reflectance of light after interaction with the sample. Visual color formation or color change due to molecule interaction is the easiest and inexpensive detection system. Colorimetric transducers can be used for the detection by the naked eye and do not require specific equipment. Instrumentation might be applied in some cases for quantitative analysis such as in case of ELISA. The color change can be detected in the human visible range from 400 nm to 700 nm. Color can be produced by different phenomena including chemical reactions of substrates with conjugated enzyme, fluorescence dye tags or plasmonic colorimetry of noble metal nanoparticles.[23,30] Plasmonic sensors can be developed either in propagative mode using continuous metal wires or thin films on a transparent substrate or in localized mode known as localized surface plasmon resonance (LSPR), which requires the use of noble metal nanoparticles[2]. The assembly or aggregation of gold nanoparticles results in the coupling of plasmonic fields surrounding each particle, and the generation of enhanced electrical field and improved sensitivity in the inter-particle space, called plasmonic hot spots. At the macroscale level, this phenomenon translates into a change in light absorbance and thus a change in color of the nanoparticle solution. The use of plasmonic colorimetry in biosensing mainly involves the detection of a target that directly or indirectly causes nanoparticle aggregation. This strategy has been largely used in lateral flow assays.[32] Optical biosensors have grown exponentially over the last decade with important improvement in the analytical parameters, cost effectiveness, and portability.

14.2.2 ELECTROCHEMICAL TRANSDUCERS

Electrochemical biosensors use an electrochemical transducer where a signal is generated during a biochemical reaction and is monitored using electrochemical workstation in the potentiometric, amperometric, or impedance modes. The electrodes are usually modified with a bioreceptor, where electro-active species on the electrode surface are produced or removed in response to the interaction between the bioreceptor and the analyte.[33] More details on the fabrication of electrochemical biosensor can be found in a comprehensive review by Ronkainen et al.[26] These types of biosensors are the most commonly used in the food industry because of their lower limit of detection and high sensitivity. They have already been used to identify various foodborne pathogens such as *Escherichia coli* O157:H7, *Salmonella*, *Listeria monocytogenes*, and viruses.[13,28,29,31,38]

14.2.3 PIEZOELECTRIC TRANSDUCERS

The piezoelectric transducers rely on the account of piezoelectric crystal which vibrates at distinct frequency when an electrical signal or it's a mass is applied. The binding of chemical on the functionalized surface of the crystal results in an increase in mass, which in turn changes the oscillation frequency of the crystal. This change can be measured electrically and utilized in the determination of the analyte concentration. Quartz crystal microbalance is one of the most successful commercial biosensor instruments. Although the technique is simple, fast, and provide economic readout, its low specificity lags it behind other biosensing technologies.[10]

14.2.4 CALORIMETRIC TRANSDUCERS

Calorimetric biosensors yield a measurable signal in terms of temperature change when the analyte comes in contact with the reactive surface of the sensor. Since enzymes involved in catalyzed reaction are exothermic, the heat generation (or dissipation) is usually used to measure the concentration of analyte. There are two kind of devices that are commonly used in calorimetric detection of metabolites: thermistor-based devices which measure the change in the resistance with changing temperature, and thermopile-based devices which are used to measure the changes.[8] Calorimetric sensors are advantageous in a way that they are not optically interfered by the color or the turbidity of sample. However, they are expensive since larger amounts of enzymes are required to carry out the reaction.

14.3   INTEGRATION OF BIOSENSORS IN FOOD PROCESSING

Food process refers to series of actions carried in sequence on food from raw materials to final products with different methods for maintaining food at a desired level of properties for their maximum benefits. During the past years, food processing has been overwhelmed by the engineering methods being employed to improve each step. It has become a rapidly growing field to meet up the demands of consumers. Monitoring safety and nutritional quality of food requires the development of rapid, sensitive, and reliable detection techniques. Biosensors play a vital role in food industries both in quantification and qualification of food components during manufacturing process. The biosensors help food manufacturers to access the quality of food and guide them to follow specific downstream process. It is important to maintain quality at every food processing stage. Manufacturers try to keep deviation between different batches and within batches at minimum to keep the trust of consumer in the product including carbohydrate, protein, fat, and vitamin contents of food in general. For specific industries such as brewing industries, alcoholic content of drink is very important to produce a large range of products. To assure quality control, biosensors have played a remarkable role since most of the sensors are based on enzyme–substrate reactions. Different enzymes have been used as bioreceptors in biosensor. These enzymes catalyze specific substrate reactions, which usually results in reduction or oxidation. As a result, electrochemical biosensors have been mostly used in monitoring food quality. A range of oxidative enzymes such as glucose oxidase, lactate oxidase, glutamate oxidase, tryptophan monoxygenase, lysine oxidase, and xanthine oxidase have been used to determine the concentration of glucose, lactate, glutamate, tryptophan, lysine, and hypoxanthine in food processing.[27] In the same way, alcohol and glycerol dehydrogenases have also been used to evaluate ethanol and glycerol contents.[16] Essential fatty acids such as linoleic acid and arachidonic are quantified by using lipases.

In addition, the presence of pesticides and other toxic compounds also need to be monitored. Pesticides such as chloripyrifos, dichlorvos 2,4-dichlorophenoxyacetic acid (2,4-D) and atrazine have shown to decrease the activity of enzyme called acetylcholinesterase. This decrease in activity has been used to monitor the presence of carbaryl and dichlorvos by Andreou and Clonis with a limit of detection of 5.2 ppb and 118 ppb, respectively.[5]

Another important use in food processing is in animal agriculture. When processing animal-derived food products, it is necessary to evaluate the presence of veterinary drugs in them. An immunoassay based optical biosensor was developed for the detection of chloramphenicol glucuronide in poultry, milk and prawn with a limit of detection of 5 ng/kg and 40 ng/kg for both milk and prawn.[14] The biogenic amines such as catecholamines (dopamine, norepinephrine, epinephrine) and tyramine are useful indicators of meat products spoilage especially fish and fish products. Biosensors for tyramine and xanthine are usually associated with microbial decarboxylation activity in fish meat and have already been used to check fish spoilage.[6,12] Wen et al. devised a microbial biosensor for the detection of ethanol in food products using *Methylobacterium organophilium*[36]. A comprehensive list of biosensors used for quality control of fresh produce, juices and other raw food materials can be found in literature.[33] Table 14.1 summarizes selected companies involved in the development of biosensors for food analysis.

TABLE 14.1   Commercial Biosensors for Food Analysis.

Company	Biosensor products
3M	3M™ Microbial Luminescence System (MLS)
Beckton Dickinson Inc.	BD Affirm™ APIII, BD ProbeTec™ GC, CT
Biomerieux	VIDAS
DuPont	BAX® System, Danisco®, Lateral Flow System
Hologic	Accuprobe®, APTIMA® CT, PACE2 CT, GASDirect®
Neogen	Reveal®, GeneQuene®, ANSR™
PathSensors	Zephyr BioFlash-E®, BioFlash-AF®
Roche	COBAS AMPLICOR MTB

14.4   BIOSENSORS IN FOOD SAFETY AND QUALITY CONTROL

Due to the complexity of raw materials and their processing products, the role of biosensors is becoming increasingly important in the food and agricultural production and control of food processes in a cost-effective manner. According to food safety market analysis in 2010, global revenues for biosensor will exceed $14 billion in 2016, mostly from applications in security and biodefense, environmental monitoring, home diagnostics and food process industry (Fig. 14.2).[1] Consumer demands have dramatically changed in the production and commercialization of foods, leading producers to seek for innovative products and technologies that improve food quality, food safety, and better eating habits. Compared to conventional analytical tools that are tedious, time-consuming, and require trained personal, biosensors offer a promising and effective alternative approach for food quality and food safety analysis as well as controlling production process.

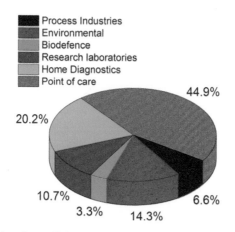

FIGURE 14.2   Market share of biosensors (percent revenues) in 2016.

Food quality and food safety including nutritional value, appearance, taste, texture, and other characteristics are of high importance to consumers. Therefore, methods for assessing food composition, abnormal flavors, and harmful microbial growth are essential for the food industry. There are an estimated 48 million cases of foodborne illness resulting in about 3000 deaths reported every year, with a total cost of about 78 billion dollars for treatment annually.[1] In particular, foodborne bacterias (such as *Campylobacter jejuni, Literia monocytogenes, Salmonella enterica, E. coli* O157:H7, and other *Vibrio* spp.) are leading causes of foodborne diseases. According to a market report published by BCC Research LLC, the global food safety testing market reached $10.5 billion in 2014 and should reach about $13.6 billion in 2019 with a 5-year compound annual growth of 5.3% through 2019. *Campylobacter* is the most common cause of intestinal and diarrheal disease in the United States, with an estimated 2.4 million cases per year.[18] *C. jejuni* and *Campylobacter coli* together are responsible for ~95% of these cases due to the consumption of unpasteurized milk and milk products and undercooked poultry.[22] Figure 14.3 summaries the spectrum of food contamination from different sources.

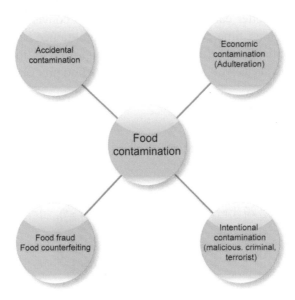

FIGURE 14.3   The spectrum of food contamination.

Mycotoxin and microbial contamination usually occurs in trace amounts ranging from nanograms to micrograms per gram of food samples. Detection and identification of these contaminations require a highly sensitive technique. Most of traditional methods for mycotoxin analysis based on chromatographic techniques including GC and HPLC, which are time-consuming and require costly instrumentation. Furthermore, traditional methods based on plate counting, isolation and selective growth media require long time, high cost, and trained personal. Enzyme-linked immunosorbent assays (ELISA) offer better alternative methods for pathogen detection in term of qualitative and quantitative analysis. However, the use of enzyme conjugation, and antibody increases the testing cost and the possibility of unreliable result in a rapid test. Aptamer-based biosensors have been recognized as promising alternative elements for antibodies in the bioassay area. Such innovation in the development of faster and more effective detection method to control biological hazards in food samples would offer valuable advantages in term of controlling and managing food safety products. Recently,

Ma et al. developed an electrochemical biosensor using aptamers specific for *Salmonella* with a limit of detection as low as 3 cfu/mL.[20] Multiplex optical detection based on surface-enhanced Raman spectroscopy on silver and gold core–shell nanoparticles was employed to detect *S. typhi*, *S. aureus*, and *E. coli*, with a limit of detection ranging from 10^2 to 10^3 cfu/mL.[11]

The major companies that are developing and producing biosensors for food analysis include 3M™, Neogen, PDS Biophage Pharma and Stratophase Ltd. (UK). 3M has developed a Microbial Luminescence System which detects the presence of microbes in ultra-heat treated and extended shelf life dairy products in less than 1 h. Stratophase Ltd. also developed a product named Ranger™ Probe which offers quick, in-line bioprocessing, and controlling of fermentation in food processes.[4]

14.5   BIOSENSORS IN FOOD PROTECTION AND DEFENSE

The outbreak of *Listeria monocytogenes* in cantaloupe in 2011 and cyclospora in 2013, multiple illness outbreak caused by *Vibrio parahaemolyticus* from oyster harvested from East Coast in 2013, and the anthrax-tainted letters after 9/11 have raised concerns about the threats of bioterror attacks and natural disasters to the safety and security of food supply in the United States. Food allergy is also considered to be an emergent public health problem. An estimated 1–3% of adults and 4–6% of children suffer from allergy due to the consumption of hidden allergens in food processing.[9] Cow's milk, eggs, fish, peanut butter, soybean, and wheat are the most common allergenic food and their very small amounts represent a health risk to allergic people.[25]

Early detection of related foodborne pathogens is crucial to enable efficient prevention and is vital to the development of an appropriate and timely response to disease outbreaks and bioterrorist attacks. Current biosensor techniques mostly use ELISA and antigen–antibody interaction or DNA-based approaches for identify foodborne pathogens.[24,35] PathSensor Inc. has been developed biosensor kit with rapid, sensitive, and specific identification of biological threat agents (bacteria, viruses, and toxins) from environmental surface samples and food products. Wang et al. developed a label-free real-time nanopore sensing method for anthrax toxin detection at subnanomolar concentration within 1 min.[34] A comprehensive summary of biosensors application for food content analysis including glucose, carbohydrates, alcohols, phenols, organic acids, amino acids, biogenic compounds, contaminants, and additives compounds has been given by Mello et al.[21]

14.6 CONCLUSIONS

Over the last decade, the biosensor market has been mainly dominated by the medical arena. Applications to food production and food processing represent only a very limited portion, but they are expected to significantly grow over the next few years due to the increasing demand for process and quality control. The integration of modern sensing technologies in the food industry is contingent upon the development of low cost and reliable analytical devices that can be easily used or integrated in food processing units. The adoption of biosensors by the food industry will have a significant impact on the food supply chain, including food production and preservation, food quality and safety, improved livestock monitoring, effective control of foodborne diseases, and precision farming.

KEYWORDS

- biodetection
- biosensor
- colorimetric sensors
- electrochemical sensors
- ELISA
- food contamination
- food defense
- food pathogens
- food processing
- food safety
- nanosensors
- nanotechnology
- PCR
- quality control

REFERENCES

1. http://www.cdc.gov/foodborneburden/2011-foodborne-estimates.html.

2. Abbas, A. K. R.; Tian, L; Singamaneni, S. Molecular Linker-mediated Self-assembly of Gold Nanoparticles: Understanding and Controlling the Dynamics. *Langmuir* **2012**, *29*, 56–64.

3. Adley, C. Past, Present and Future of Sensors in Food Production. *Foods* **2014**, *3*, 491.

4. Adley, C. C. Past, Present and Future of Sensors in Food Production. *Foods* **2014**, *3*, 491–510.

5. Andreou, V. G.; Clonis, Y. D. A Portable Fiber-optic Pesticide Biosensor based on Immobilized Cholinesterase and sol–gel Entrapped Bromcresol Purple for in-field Use. *Biosens. Bioelectron.* **2002**, *17*, 61–69.

6. Apetrei, I. M.; Apetrei, C. The Biocomposite Screen-printed Biosensor based on Immobilization of Tyrosinase onto the Carboxyl Functionalised Carbon Nanotube for Assaying Tyramine in Fish Products. *J. Food Eng.* **2015**, *149*, 1–8.

7. Arugula, M. A.; Zhang, Y.; Simonian, A. L. Biosensors as 21st Century Technology for Detecting Genetically Modified Organisms in Food and Feed. *Anal. Chem.*, **2014**, *86*, 119–129.

8. Bataillard, P.; Steffgen, E.; Haemmerli, S.; Manz, A.; Widmer, H. M. An Integrated Silicon Thermopile as Biosensor for the Thermal Monitoring of Glucose, Urea and Penicillin. *Biosens. Bioelectron.* **1993**, *8*, 89–98.

9. Bock, S. A.; Muñoz-Furlong, A.; Sampson, H. A. Fatalities due to Anaphylactic Reactions to Foods. *J. Allergy Clin. Immunol.*, **2001**, *107*, 191–193.

10. Bunde, R. L.; Jarvi, E. J.; Rosentreter, J. J. Piezoelectric Quartz Crystal Biosensors. *Talanta* **1998**, *46*, 1223–1236.

11. Craig, A. P.; Franca, A. S.; Irudayaraj, J. Surface-Enhanced Raman Spectroscopy Applied to Food Safety. *Annu. Rev. Food Sci. Technol.* **2013**, *4*, 369–380.

12. Devi, R.; Yadav, S.; Nehra, R.; Yadav, S.; Pundir, C. S. Electrochemical Biosensor based on Gold Coated Iron Nanoparticles/Chitosan Composite Bound Xanthine Oxidase for Detection of Xanthine in Fish Meat. *J. Food Eng.* **2013**, *115*, 207–214.

13. Dong, S.; Zhao, R.; Zhu, J.; Lu, X.; Li, Y.; Qiu, S.; Jia, L.; Jiao, X.; Song, S.; Fan, C.; Hao, R.; SongH. Electrochemical DNA Biosensor Based on a Tetrahedral Nanostructure Probe for the Detection of Avian Influenza A (H7N9) Virus. *ACS Appl. Mater. Interfaces.* **2015**, *7*, 8834–8842.

14. Ferguson, J.; Baxter, A.; Young, P.; Kennedy, G.; Elliott, C.; Weigel, S.; Gatermann, R.; Ashwin, H.; Stead, S.; Sharman, M. Detection of Chloramphenicol and Chloramphenicol Glucuronide Residues in Poultry Muscle, Honey, Prawn and Milk using a Surface Plasmon Resonance Biosensor and Qflex® kit Chloramphenicol. *Anal. Chim. Acta* **2005**, *529*, 109–113.

15. Huang, L.; Zheng, L.; Chen, Y.; Xue, F.; Cheng, L.; Adeloju, S. B.; Chen, W. A Novel GMO Biosensor for Rapid Ultrasensitive and Simultaneous Detection of Multiple DNA Components in GMO Products. *Biosens. Bioelectron.* **2015**, *66*, 431–437.

16. Ismail-Hakk, B.; Mehmet, M. Amperometric Biosensors in Food Processing, Safety, and Quality Control, In *Biosensors in Food Processing, Safety, and Quality Control*, Ismail-Hakk, B., Mehmet, M., Eds.; CRC Press: Boca Raton, FL, 2010; pp 1–51.

17. Kim, H. U.; Kim, H.; Ahn, C.; Kulkarni, A.; Jeon, M.; Yeom, G. Y.; Lee, M. H.; Kim, T. In situ Synthesis of MoS_2 on a Polymer Based Gold Electrode Platform and Its Application in Electrochemical Biosensing. *RSC Adv.* **2015**, *5*, 10134–10138.

18. Kothary, M. H.; Babu, U. S. Infective Dose of Foodborne Pathogens in Volunteers: A Review. *J. Food Safety* **2001**, *21*, 49–68.

19. Li, L.; Wang, Y.; Pan, L.; Shi, Y.; Cheng, W.; Shi, Y.; Yu, G. A Nanostructured Conductive Hydrogels-based Biosensor Platform for Human Metabolite Detection. *Nano Lett.* **2015**, *15*, 1146–1151.

20. Ma, X.; Jiang, Y.; Jia, F.; Yu, Y.; Chen, J.; Wang, Z. An Aptamer-based Electrochemical Biosensor for the Detection of *Salmonella*. *J. Microbiol. Methods* **2014**, *98*, 94–98.

21. Mello, L. D.; Kubota, L. T. Review of the use of Biosensors as Analytical Tools in the Food and Drink Industries. *Food Chem.* **2002**, *77*, 237–256.

22. Mortari, A.; Lorenzelli, L. Recent Sensing Technologies for Pathogen Detection in Milk: A Review. *Biosens. Bioelectron.* **2014**, *60*, 8–21.

23. Narsaiah, K.; Jha, S.; Bhardwaj, R.; Sharma, R.; Kumar, R. Optical Biosensors for Food Quality and Safety Assurance: A Review. *J Food Sci. Technol.* **2012**, *49*, 383–406.

24. Patel, P. D. Biosensors for Measurement of Analytes Implicated in Food Safety: A Review. *TrAC, Trends Anal. Chem.* **2002**, *21*, 96–115.

25. Powell, D. A.; Jacob, C. J.; Chapman, B. J. Enhancing Food Safety Culture to Reduce Rates of Foodborne Illness. *Food Control* **2011**, *22*, 817–822.

26. Ronkainen, N. J.; Halsall, H. B.; Heineman, W. R. Electrochemical Biosensors. *Chem. Soc. Rev.* **2010**, *39*, 1747–1763.

27. Scampicchio, M.; Ballabio, D.; Arecchi, A.; Cosio, M. S.; Mannio, S. Amperometric Electronic Tongue for Food Analysis. *Microchim. Acta* **2010**, *163*, 11–21.

28. Shakoori, Z.; Salimian, S.; Kharrazi, S.; Adabi, M.; Saber, R. Electrochemical DNA Biosensor Based on Gold Nanorods for Detecting Hepatitis B Virus. *Anal. Bioanal. Chem.* **2015**, *407*, 455–461.

29. Sharma, H.; Mutharasan, R. Review of Biosensors for Foodborne Pathogens and Toxins. *Sens. Actuators, B* **2013**, *183*, 535–549.

30. Silletti, S.; Rodio, G.; Pezzotti, G.; Turemis, M.; Dragone, R.; Frazzoli, C.; Giardi, M. T. An Optical Biosensor based on a Multiarray of Enzymes for Monitoring a Large Set of Chemical Classes in Milk. *Sens. Actuators, B* **2015**, *215*, 607–617.

31. Singh, A.; Poshtiban, S.; Evoy, S. Recent Advances in Bacteriophage Based Biosensors for Food-borne Pathogen Detection. *Sensors* **2013**, *13*, 1763.

32. Sun, J.; Xianyu, Y.; Jiang, X. Point-of-care Biochemical Assays using Gold Nanoparticle-implemented Microfluidics. *Chem. Soc. Rev.* **2014**, *43*, 6239–6253.

33. Terry, L. A.; White, S. F.; Tigwell, L. J. The Application of Biosensors to Fresh Produce and the Wider Food Industry. *J. Agric. Food Chem.* **2005**, *53*, 1309–1316.

34. Wang, L.; Han, Y.; Zhou, S.; Wang, G.; Guan, X. Nanopore Biosensor for Label-free and Real-time Detection of Anthrax Lethal Factor. *ACS Appl. Mater. Interfaces* **2014**, *6*, 7334–7339.

35. Wang, W.; Han, J.; Wu, Y.; Yuan, F.; Chen, Y.; Ge, Y. Simultaneous Detection of Eight Food Allergens Using Optical Thin-film Biosensor Chips. *J. Agric. Food Chem.* **2011**, *59*, 6889–6894.

36. Wen, G.; Li, Z.; Choi, M. M. F. Detection of Ethanol in Food: A New Biosensor Based on Bacteria. *J. Food Eng.* **2013**, *118*, 56–61.

37. Yadav, S. K.; Agrawal, B.; Chandra, P.; Goyal, R. N. In Vitro Chloramphenicol Detection in a *Haemophilus influenza* Model using an Aptamer-polymer based Electrochemical Biosensor. *Biosens. Bioelectron.* **2014**, *55*, 337–342.

38. Yu, Y.; Chen, Z.; Jian, W.; Sun, D.; Zhang, B.; Li, X.; Yao, M. Ultrasensitive Electrochemical Detection of Avian Influenza A (H7N9) virus DNA based on Isothermal Exponential Amplification Coupled with Hybridization Chain Reaction of DNAzyme Nanowires. *Biosens. Bioelectron.* **2015**, *64*, 566–571.

CHAPTER 15

MILK PASTEURIZATION BY MICROWAVE

ASAAD R. S. AL-HILPHY*, AMMAR B. R. AL-TEMIMI, and ALAA A. AL-SERAIH

Food Science Department, Agriculture College, Basrah University, Basrah, Iraq. E-mail: aalhilphy@yahoo.co.uk, asaad197013@gmail. com, ammaragr@siu.edu, alseraihalaa@yahoo.com

*Corresponding author.

CONTENTS

ABSTRACT

In spite of all valuables advantages for pasteurization processing, the shelf life of pasteurized milk is still very short due to growth of pathogenic bacteria. Hence, there is an economical loss, in addition to rejection of oxidized milk by most consumers. Because of the importance of milk for human nutrition and possibility of storage for long time, this chapter includes research results on the use of microwave to produce microwaved milk. The goals of this study were using microwave equipment to produce pasteurized cow milk and study its chemical and microbiological contents; the ability of microwaved milk to prevent oxidation compared to traditional pasteurized method; and inhibition aspects of microwave against pathogenic bacteria in pasteurized milk through different periods of time.

It was concluded that pasteurization by microwave was faster than by traditional method. The microbial content of microwaved milk was less compared to a traditional method. The TBA and FFA values were less in microwaved milk compared to a traditional method. Alkaline phosphatase enzyme was not present in microwaved milk compared to a traditional method. It can be concluded that milk pasteurization by microwave is beneficial in terms of milk quality.

15.1   INTRODUCTION

Since a long time, the human knew milk as a natural whole product, which contains all nutritional elements. The flavor of milk is very good, when the storage conditions are controlled. However, oxidized flavor is still one of major problems, which develops in a negative way under bad storage conditions and leads to produce volatile compounds with bad odors because of fat oxidation.[5] Ryan et. al.[6] indicated that there are lot of microorganisms in milk, which can negatively affect the quality of milk, health, and safety of consumers.[1] In 1860, Louis Pasteur found a method called later "pasteurization processing," which can kill most of the bacteria that cause spoilage.[6]

In spite of all valuables advantages for pasteurization processing, the shelf life of pasteurized milk is still very short due to growth of pathogenic bacteria. Hence, there is an economical loss, in addition to rejection of oxidized milk by most consumers.[6]

Microwave processing has been found to be one of the nontraditional heating systems, which applies electromagnetic waves in range of 300 MHz.

International Telecommunication Union determined the vibration range for heating by microwave into 2450 MHz. Water, proteins, and carbohydrates are considered polar compounds, which can cluster within electric field for microwave equipment.

The vibration of molecules for billions time in a second leads to generation of heat, which is also produced by electric charged ions within food. Heating foods by microwave is totally dependent on natural characterization of water, because hydrogen bonds between molecules thereby make the heating by microwave very easily due to exchange of energy with photons. The food spoilage is minimum by microwave because exposing heating and radiation time is very short compared to traditional methods.[4]

Because of the importance of milk for human nutrition and possibility of storage for long time, this chapter includes research results on the use of microwave to produce microwaved milk. The goals of this study are listed:

1. Using microwave equipment to produce pasteurized cow milk and study its chemical and microbiological contents.
2. The ability of microwaved milk to prevent oxidation compared to traditional pasteurized method.
3. Inhibition aspects of microwave against pathogenic bacteria in pasteurized milk through different periods of time.

15.2   CHEMICAL COMPOSITION

Table 15.1 shows that fat percentage was higher in microwaved milk compared with traditional pasteurization, which caused loss of part of fat, especially if the mixing milk is stopped, causes bununing fat. Microwaved milk becomes isothermal during pasteurization. Therefore, there is no need of mixing. There is high pH value of microwaved milk and traditional method, due to loss of CO_2 during heating that reduces acidity.

TABLE 15.1   Chemical Composition of Cow Milk by Al-Hilphy.[2]

Pasteurization type	Fat	Protein	Moisture content	Lactose	Ash	Acidity	pH
Microwave	3.8	3.6	85.22	5.4	0.55	0.14	6.7
Traditional	3.6	3.7	85.5	5.2	0.63	0.15	6.7
Control sample	3.7	3.6	86.83	5.5	0.58	0.15	6.6

15.2.1 TBA

The thiobarbituric acid (TBA) was reduced after pasteurization by microwave and traditional method, thereafter, TBA of microwaved milk and
traditional pasteurization was increased with increasing storage time at
temperatures of 5 (Table 15.2) and 25°C (Table 15.3), which is attributable
to abundance of voids in the milk granulars. Fat content was significantly
increased in cow milk, which gave high fat percentage at storage time of
6 days and was 1.6, 4.2, and 2.5 mg monoaldehyde/g for control sample,
microwaved, and traditional pasteurization at 5°C, respectively, but at 25°C
the fat content was 12.2, 8.5, 4.6 mg monoaldehyde/g), respectively.

15.2.2 FREE FATTY ACIDS

Table 15.4 and 15.5 showed that free fatty acids (FFA) of milk in the
control sample stored at temperatures of 5 and 25°C was higher compared
with pasteurized milk stored at the same conditions. FFA of microwaved
milk was significantly increased compared with traditional pasteurization,
because penetration of microwaves into all milk molecules.

TABLE 15.2   Effect of pasteurization on TBA of cow milk stored at 5°C.[2]

Pasteurization type	Storage time (days)			
	0	**2**	**4**	**6**
Microwave	1.5 ± 0.28	2.3 ± 0.42	3 ± 0.70	4.2 ± 0.28
Traditional	0.6 ± 0.14	1.3 ± 0.42	1.8 ± 0.28	2.5 ± 0.28
Control sample	0.3 ± 0.14	0.8 ± 0.28	1.2 ± 0.28	1.6 ± 0.28

TABLE 15.3   Effect of Pasteurization on TBA of Cows Milk Stored at 25°C.[2]

Pasteurization Type	Storage time (days)			
	0	**2**	**4**	**6**
Microwave	4.5 ± 0.91	6.3 ± 0.28	9 ± 0.28	12.2 ± 0.28
Traditional	2.6 ± 0.14	3.3 ± 0.28	5.8 ± 0.28	8.5 ± 0.70
Control sample	0.8 ± 0.28	1.8 ± 0.28	2.4 ± 0.28	4.6 ± 0.28

TABLE 15.4 Effect of Pasteurization on FFA of Cows Milk Stored at 5°C.[2]

Pasteurization type	Storage time (days)			
	0	2	4	6
Microwave	2 ± 0.14	2.7 ± 0.28	3.8 ± 0.28	4.9 ± 0.42
Traditional	1.5 ± 0.28	2 ± 0.28	2.6 ± 0.28	3.3 ± 0.28
Control sample	0.7 ± 0.14	1.3 ± 0.14	1.8 ± 0.28	2.3 ± 0.28

TABLE 15.5 Effect of Pasteurization on FFA of Cows Milk Stored at 25°C.[2]

Pasteurization type	Storage time (days)			
	0	2	4	6
Microwave	4 ± 1.41	7.7 ± 0.28	9.8 ± 0.28	11.9 ± 0.14
Traditional	2.5 ± 0.42	4 ± 1.41	6.6 ± 0.28	9.3 ± 0.28
Control sample	1.2 ± 0.28	2.3 ± 0.28	3.8 ± 0.28	5.3 ± 0.28

15.2.3 MICROBIAL TESTS

Microbial content in pasteurized milk and traditional pasteurization was significantly increased with increasing storage time (Tables 15.6 and 15.7). The highest microbial content was after 90 days of storage and was reached 121×10^3, 77×10^3, and 56.5×10^3 CFU/ml at 5°C and 135.5×10^3, 81×10^3, and 72.5×10^3 CFU/ml at 25°C for control sample, traditional pasteurization and microwaved milk, respectively. A.O.A.C.[3] mentioned that the high heat treatment of milk has essential role in reducing total count of microbes as a result of killing. Also, the differences among pasteurization methods in the total count of microbes was significantly increased with increasing storage time. Therefore, microwaved milk has total count of microbes lesser than traditional pasteurization and control sample. The slow growth of microbe cells in the microwaved milk is due to inactivation effect of microwaves on the microbes, and this improved the milk quality.

TABLE 15.6 Effect of Pasteurization on Microbial Count of Cows Milk Stored at 25°C.[2]

Pasteurization type	Total count ×103			
	Storage time (days)			
	0	2	4	6
Microwave	80.5 ± 0.70	92.5 ± 0.70	112 ± 2.82	135.5 ± 0.70
Traditional	56.5 ± 0.70	61 ± 1.41	72.5 ± 0.70	81 ± 1.41
Control sample	45 ± 1.41	54 ± 1.41	65.5 ± 2.12	72.5 ± 0.70

TABLE 15.7 Effect of Pasteurization *Psychrotrophic Bacteria* of Cows Milk stored at 5°C.[2]

Pasteurization Type	Total count ×103			
	Storage time (days)			
	0	2	4	6
Microwave	9 ± 1.4	11.5 ± 0.70	13 ± 1.4	16.5 ± 0.70
Traditional	6 ± 1.4	6.5 ± 0.70	8.5 ± 0.70	11.5 ± 0.70
Control sample	2 ± 1.4	3.5 ± 0.70	5 ± 1.4	6.5 ± 0.70

TABLE 15.8 Effect of Pasteurization *Psychrotrophic Bacteria* of Cows Milk Stored at 25°C.[2]

Pasteurization type	Total count ×103			
	Storage time (days)			
	0	2	4	6
Microwave	14 ± 1.40	17.5 ± 0.70	20 ± 2.82	23.5 ± 2.12
Traditional	8 ± 2.82	10.5 ± 0.70	13.5 ± 2.12	16.5 ± 2.12
Control sample	4 ± 1.42	6.5 ± 2.12	8 ± 2.82	10.5 ± 2.12

15.2.4 PSYCHROTROPHIC BACTERIA

The results on the total count of *psychrotrophic bacteria* of milk (Tables 15.7 and 15.8) indicated that the differences in the total count of bacteria between pasteurization methods were significant. On the contrary, increasing total count of bacteria followed the order: control sample of milk > traditional pasteurization > microwaved milk, at storage temperatures of 5 and 25°C. The results on the total count of *psychrotrophic bacteria* of milk are in agreement with those mentioned by *Australia standard specifications* that total count of *psychrotrophic bacteria* was reached to 10^2 CFU/ml in pasteurized milk.[7] Therefore, Australia New Zealand Food Authority confirms results of milk heat treatment using microwave and traditional pasteurization for reducing total count of *psychrotrophic bacteria* after 6 days of storage.[7]

15.2.5 TOTAL COUNT OF ESCHERICHIA COLI AND DETECTION OF SALMONELLA

The tests carried on pasteurized milk by microwave and traditional pasteurization stored for 2, 4, and 6 days at temperatures of 5 and 25°C for detection of *E. coli* and *Salmonella* showed that tests were negative. While total count of *E. coli* and *Salmonella* were 40 and 20 CFU/ml, respectively, in the control sample. These values are higher than the standard values of 0 and 10 for *E. coli* and *Salmonella*, respectively.

15.2.6   DETERMINATION OF ALKALINE PHOSPHETASE

Figure 15.1 shows absence of alkaline phosphetase activity in microwaved milk while presence of this activity in traditional method and control sample. The values were 15 and 30 IU/L in the traditional method and control sample, respectively, under same conditions of temperature and storage time.

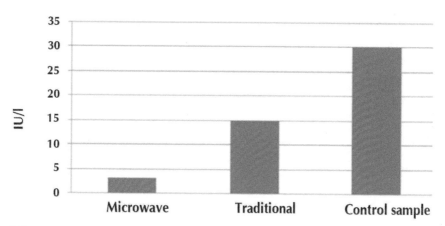

FIGURE 15.1   Activation of phosphetase in pasteurized milk.[2]

15.3   CONCLUSIONS

1. Pasteurization by microwave was faster than by traditional method.
2. The microbial content of microwaved milk was less compared to a traditional method.
3. The TBA and FFA values were less in microwaved milk compared to a traditional method.
4. Alkaline phosphatase enzyme was not present in microwaved milk compared to a traditional method.

It can be concluded that milk pasteurization by microwave is beneficial in terms of milk quality.

KEYWORDS

- A.O.A.C.
- alkaline phosphetase
- autoxidation
- cow milk
- *dairy chemistry*
- fats
- FFA
- microbial content
- microwave
- pasteurization
- pH
- storage time
- TBA
- UHT

REFERENCES

1. Akhtar, S.; Zahoor, T.; Hashmi, A. Physio-chemical Changes in UHT Treated and Whole Milk Powder During Storage at Ambient Temperature. *J. Res. Sci.* **2003,** *14*(1), 97–101.
2. Al-Hilphy, A. R. S.; Al-temimi, A. B. R.; Al-Seraih, A. A. Milk Pasteurization by Microwave and Study its Chemical and Microbiological Characteristics During Different Storage Times. *Basrah Res. J.* **2010,** *36*(3), 66–76.
3. Association Official Analytical Chemists (A.O.A.C.). *Official Method of Analysis.* 13th Ed. A.O.A.C.: Washington, DC, 1980.
4. *Hassan, B. H. Heat Transfer Coefficients for Particles in Liquid in Axially Rotating Cans, Ph.D. Thesis, University of California: Davis, CA, 1984.*
5. Parks, O. W. Autoxidation. In *Fundamentals of Dairy Chemistry*; Wedd, B. H, Johnson, A. H., Alford, J. A., Eds.; The AVI Publishing Company, Inc., 1978; Chapter 11, p 240.
6. Ryan, Y.; Barbano, D.; Galton, M.; Rudan, A.; Boor, K. Effects of Somatic Cell Count on Quality and Shelf-life of Pasteurized Fluid Milk. *J. Dairy Sci.* **2000,** *83*, 264–274.
7. The Australia New Zealand Food Authority (ANZFA). *The Regulation of Microbial Hazards in Food: Discussion Paper.* Review of Microbiological Standards Full Assessment Report, 2001.

APPENDIX 15.1

Examples of microwaved foods.

APPENDIX A

A. MAJOR SOURCES OF POLLUTION

1. Asbestos manufacturing units
2. Bullion refining (waste water discharge)
3. Calcium carbide
4. Caustic soda
5. Carbon black
6. Cement plants
7. Ceramic industry
8. Coffee industry
9. Coke ovens
10. Composite woolen mills
11. Copper, lead, and zinc smelting
12. Cotton textile industries
13. Dairy
14. Dye and dye intermediate industry
15. Electroplating industries
16. Fermentation industry
17. Fertilizer industry
18. Flour mills
19. Food and fruit processing industry
20. Foundries
21. Glass industry
22. Hotel industry
23. Inorganic chemical industry (waste water discharge)
24. Integrated iron and steel plants
25. Iron and steel (integrated)
26. Large pulp and paper
27. Leather tanneries
28. Lime kiln
29. Man-made fiber industry
30. Natural rubber processing industry
31. News print/rayon grade plants
32. Nitric acid (oxides of nitrogen)
33. Paint industry waste water discharge
34. Pesticide manufacturing and formulation industry
35. Re-heating furnaces
36. Slaughter house
37. Small boilers
38. Small pulp and paper industry
39. Soft coke industry
40. Starch industry
41. Stone crushing unit
42. Sugar industry
43. Sulphuric acid plant
44. Synthetic rubber
45. Tanneries
46. Tannery (after primary treatment)
47. Thermal power plants

B.   CATEGORY GREEN A

List of industries in approved industrial areas which may be directly considered for issue of no objection certificate without referring to *Ministry of Environment and Forests.* (In case of doubts reference will be made to: Ministry of Environment and Forests, Government of India.)

1. All nonobnoxious and nonhazardous industries employing up to 100 persons. The obnoxious and hazardous industries are those using inflammable, explosive, corrosive, or toxic substances.
2. All industries, which do not discharge industrial effluents of a polluting nature and which do not undertake any of the following process:

a. Alcohol distillation, spillage, evaporation

b. Bleaching

c. Cooking of fibers and digesting

d. Degreasing

e. Designing of fabric

f. Dyeing

g. Electroplating

h. Galvanizing

i. Juicing of sugar cane, extraction of sugar, filtration, centrifugation, distillation

j. Phosphating

k. Pickling, tanning

l. Polishing

m. Processing of fish

n. Pulp making, pulp processing, paper making, cocking of coal washing of blast furnace flue gases

o. Pulping and fermenting of coffee beam

p. Separated milk, buttermilk, and whey

q. Slaughtering of animals, rendering of bones, washing of meat

r. Solvent extraction

s. Stopping and processing of grain

t. Stripping of oxides

u. Trimming, puling, juicing, and blanching of fruits and vegetables

v. Unhairing, Soaking, deliming, and bating of hides

w. Washing of equipment and regular floor washing, considerable cooling water

x. Washing of fabric

y. Washing of latex

z. Washing of used sand by hydraulic discharge

3. All the industries, which do not use fuel in their manufacturing process or in any subsidiary process and which do not emit fugitive emissions of a diffused nature. The following industries appear to fall in non-hazardous, non-obnoxious and non-polluting category, subject to fulfillment of above three conditions:

a. Apparel making
b. Assembly of air coolers
c. Atta-chakkies
d. Automobile servicing and repair stations
e. Bamboo and cane products
f. Block making for printing
g. Candles
h. Card board and paper products
i. Carpentry
j. Carpet weaving
k. Chilling
l. Cold storage (small scale)
m. Cotton and woolen hosiery
n. Cotton spinning and weaving
o. Dal mills
p. Domestic electrical appliances
q. Electronics equipment (assembly)
r. Fountain pens
s. Furniture (wooden and steel)
t. Gold and silver smithy
u. Gold and silver thread and sari work
v. Groundnut decorticating (dry)
w. Handloom weaving
x. Ice boxes
y. Ice-cream
z. Insulation and other coated papers

aa. Jobbing and machining
bb. Manufacture of mirror
cc. Manufacture of steel units and suit cases
dd. Mineralized water
ee. Musical instruments manufacturing
ff. Oil-ginning/expelling
gg. Optical frames
hh. Paper pins and U-clips
ii. Polythene, plastic, and PVC goods
jj. Radio assembling
kk. Railway sleepers (only concrete)
ll. Restaurants
mm. Rice millers
nn. Rope (cotton and plastic)
oo. Scientific and mathematical instruments
pp. Shoe lace manufacturing
qq. Small nonmotorized vehicles assembly
rr. Sports goods
ss. Surgical gauges and bandages
tt. Tailoring and garment making
uu. Toys
vv. Wires, pipes-extruded shapes metals

Source: This appendix is summarized from "Central Pollution Control Board, Ministry of Environment and Forests, Govt. of India, Delhi, 2010."

INDEX

Printed and bound by CPI Group (UK) Ltd, Croydon, CR0 4YY

23/10/2024

01777696-0015